Obtaining and Characterization of New Materials, Volume IV

Obtaining and Characterization of New Materials, Volume IV

Editor

Andrei Victor Sandu

Basel • Beijing • Wuhan • Barcelona • Belgrade • Novi Sad • Cluj • Manchester

Editor
Andrei Victor Sandu
Faculty of Materials Science and Engineering
Gheorghe Asachi Technical University of Iasi
Iasi
Romania

Editorial Office
MDPI AG
Grosspeteranlage 5
4052 Basel, Switzerland

This is a reprint of articles from the Special Issue published online in the open access journal *Materials* (ISSN 1996-1944) (available at: https://www.mdpi.com/journal/materials/special_issues/7EH16X88E3).

For citation purposes, cite each article independently as indicated on the article page online and as indicated below:

Lastname, A.A.; Lastname, B.B. Article Title. *Journal Name* **Year**, *Volume Number*, Page Range.

ISBN 978-3-7258-2072-6 (Hbk)
ISBN 978-3-7258-2071-9 (PDF)
doi.org/10.3390/books978-3-7258-2071-9

© 2024 by the authors. Articles in this book are Open Access and distributed under the Creative Commons Attribution (CC BY) license. The book as a whole is distributed by MDPI under the terms and conditions of the Creative Commons Attribution-NonCommercial-NoDerivs (CC BY-NC-ND) license.

Contents

About the Editor vii

Preface ix

Hang Lin, Rui Li and Su Li
Fabrication of Lead–Zinc Tailings Sintered Brick and Its Effect Factors Based on an Orthogonal Experiment
Reprinted from: *Materials* **2024**, *17*, 2352, doi:10.3390/ma17102352 1

Anna V. Ruseikina, Maxim V. Grigoriev, Ralf J. C. Locke, Vladimir A. Chernyshev, Alexander A. Garmonov and Thomas Schleid
Synthesis, Crystal Structure, and Optical and Magnetic Properties of the New Quaternary Erbium Telluride EuErCuTe$_3$: Experiment and Calculation
Reprinted from: *Materials* **2024**, *17*, 2284, doi:10.3390/ma17102284 19

Tai-Nan Lin, Pin-Hsun Liao, Cheng-Chin Wang, Hung-Bin Lee and Leu-Wen Tsay
Corrosion Resistance of Fe-Based Amorphous Films Prepared by the Radio Frequency Magnetron Sputter Method
Reprinted from: *Materials* **2024**, *17*, 2071, doi:10.3390/ma17092071 36

Alexander Delp, Shuang Wu, Jonathan Freund, Ronja Scholz, Miriam Löbbecke, Thomas Tröster, et al.
Characterization of Interfacial Corrosion Behavior of Hybrid Laminate EN AW-6082 ∪ CFRP
Reprinted from: *Materials* **2024**, *17*, 1907, doi:10.3390/ma17081907 49

Csaba Bús, Marianna Kocsis, Áron Ágoston, Ákos Kukovecz, Zoltán Kónya and Pál Sipos
Application of Alcohols to Inhibit the Formation of Ca(II) Dodecyl Sulfate Precipitate in Aqueous Solutions
Reprinted from: *Materials* **2024**, *17*, 1806, doi:10.3390/ma17081806 62

Yuanwen Pang, Hong Li, Yue Hua, Xiuling Zhang and Lanbo Di
Rapid Synthesis of Noble Metal Colloids by Plasma–Liquid Interactions
Reprinted from: *Materials* **2024**, *17*, 987, doi:10.3390/ma17050987 74

Darya Ilieva, Lyudmila Angelova, Temenuzhka Radoykova, Andriana Surleva, Georgi Chernev, Petrica Vizureanu, et al.
Characterization of Bulgarian Copper Mine Tailing as a Precursor for Obtaining Geopolymers
Reprinted from: *Materials* **2024**, *17*, 542, doi:10.3390/ma17030542 86

Elena D. Fakhrutdinova, Anastasia V. Volokitina, Sergei A. Kulinich, Daria A. Goncharova, Tamara S. Kharlamova and Valery A. Svetlichnyi
Plasmonic Nanocomposites of ZnO-Ag Produced by Laser Ablation and Their Photocatalytic Destruction of Rhodamine, Tetracycline and Phenol
Reprinted from: *Materials* **2024**, *17*, 527, doi:10.3390/ma17020527 107

Adrian-Victor Lăzărescu, Andreea Hegyi, Alexandra Csapai and Florin Popa
The Influence of Different Aggregates on the Physico-Mechanical Performance of Alkali-Activated Geopolymer Composites Produced Using Romanian Fly Ash
Reprinted from: *Materials* **2024**, *17*, 485, doi:10.3390/ma17020485 127

Miguel Ángel López-Álvarez, Pedro Ortega-Gudiño, Jorge Manuel Silva-Jara, Jazmín Guadalupe Silva-Galindo, Arturo Barrera-Rodríguez, José Eduardo Casillas-García, et al.
DyMnO$_3$: Synthesis, Characterization and Evaluation of Its Photocatalytic Activity in the Visible Spectrum
Reprinted from: *Materials* 2023, *16*, 7666, doi:10.3390/ma16247666 148

Chi-Hung Tsai, Wen-Tien Tsai and Li-An Kuo
Effect of Post-Washing on Textural Characteristics of Carbon Materials Derived from Pineapple Peel Biomass
Reprinted from: *Materials* 2023, *16*, 7529, doi:10.3390/ma16247529 165

Shahin Jalali, Catarina da Silva Pereira Borges, Ricardo João Camilo Carbas, Eduardo André de Sousa Marques, João Carlos Moura Bordado and Lucas Filipe Martins da Silva
Characterization of Densified Pine Wood and a Zero-Thickness Bio-Based Adhesive for Eco-Friendly Structural Applications
Reprinted from: *Materials* 2023, *16*, 7147, doi:10.3390/ma16227147 176

Nur Bahijah Mustapa, Romisuhani Ahmad, Mohd Mustafa Al Bakri Abdullah, Wan Mastura Wan Ibrahim, Andrei Victor Sandu, Ovidiu Nemes, et al.
Effect of the Sintering Mechanism on the Crystallization Kinetics of Geopolymer-Based Ceramics
Reprinted from: *Materials* 2023, *16*, 5853, doi:10.3390/ma16175853 195

Cristian Petcu, Andreea Hegyi, Vlad Stoian, Claudiu Sorin Dragomir, Adrian Alexandru Ciobanu, Adrian-Victor Lăzărescu and Carmen Florean
Research on Thermal Insulation Performance and Impact on Indoor Air Quality of Cellulose-Based Thermal Insulation Materials
Reprinted from: *Materials* 2023, *16*, 5458, doi:10.3390/ma16155458 208

Anastasiia V. Shabalina, Ekaterina Y. Gotovtseva, Yulia A. Belik, Sergey M. Kuzmin, Tamara S. Kharlamova, Sergei A. Kulinich, et al.
Electrochemical Study of Semiconductor Properties for Bismuth Silicate-Based Photocatalysts Obtained via Hydro-/Solvothermal Approach
Reprinted from: *Materials* 2022, *15*, 4099, doi:10.3390/ma15124099 234

Darya Y. Savenko, Mikhail A. Salaev, Valerii V. Dutov, Sergei A. Kulinich and Olga V. Vodyankina
Modifier Effect in Silica-Supported FePO$_4$ and Fe-Mo-O Catalysts for Propylene Glycol Oxidation
Reprinted from: *Materials* 2022, *15*, 1906, doi:10.3390/ma15051906 252

About the Editor

Andrei Victor Sandu

Dr. Eng. Andrei Victor Sandu is an Associate Professor at the Faculty of Materials Science and Engineering, Technical University "Gheorghe Asachi" of Iași. He obtained his PhD in Materials Engineering in 2012 with summa cum laudae. He has published over 450 scientific articles, with over 400 indexed by SCOPUS and more than 350 indexed by ISI Web of Science. His H-index is 32. He is the co-author of 45 patents and 9 other patent applications (Romania, R. Moldova and Malaysia), and he has published 13 books, 5 of them in the USA. He is a Publishing Editor for the *International Journal of Conservation Science* (indexed in Web of Science and Scopus) and the *European Journal of Materials Science and Engineering*; he is a reviewer for more than 20 Web of Science-indexed journals. He is a Visiting Professor at Universiti Malaysia Perlis and also the President of the Romanian Inventors Forum. Based on his expertise, he is also a Senior Researcher for the National Institute for Research and Development for Environmental Protection INCDPM and Representative for Romania at IFIA (International Federation of Inventors' Associations) and WIIPA (World Invention Intellectual Property Associations).

Preface

After the success of the first three volumes of this Special Issue, entitled "Obtaining and Characterization of New Materials", we have decided to create the fourth volume in order to collect and publish a series of state-of-the-art research in the field of new materials and their understanding.

The fourth volume of this Special Issue, like the first three, covers a wide range of topics, such as obtaining and characterizing new materials, from the nano- to macro-scale, involving new alloys, ceramics, composites, biomaterials, and polymers, as well as procedures and technologies for enhancing their structure, properties, and functions. In order to be able to select the future use of the new materials, we first must understand their structure to understand their characteristics, involving modern techniques such as microscopy (SEM, TEM, AFM, STM, etc.), spectroscopy (EDX, XRD, XRF, FTIR, XPS, etc.), mechanical tests (tensile, hardness, elastic modulus, toughness, etc.), and their behavior (corrosion, thermal—DSC, STA, DMA, magnetic properties, biocompatibility—in vitro and in vivo), among many others.

Andrei Victor Sandu
Editor

Article

Fabrication of Lead–Zinc Tailings Sintered Brick and Its Effect Factors Based on an Orthogonal Experiment

Hang Lin, Rui Li and Su Li *

School of Resources and Safety Engineering, Central South University, Changsha 410083, China; hanglin@csu.edu.cn (H.L.); 215511001@csu.edu.cn (R.L.)
* Correspondence: lisu1996@csu.edu.cn

Abstract: The existence of lead-zinc tailings threatens the social and ecological environment. The recycling of lead–zinc tailings is important for the all-round green transformation of economic society. In this study, the possibility of fabricating sintered ordinary bricks with lead–zinc tailings was studied based on orthogonal experimentation, and the phase composition and micromorphology of sintered products were analyzed by X-ray diffraction (XRD) and scanning electron microscope (SEM). With lead–zinc tailings as the main material, and clay and fly ash as additives, the effect of clay content, forming pressure, sintering temperature, and holding time on physical properties of sintered bricks was analyzed. The results show that clay content and sintering temperature have a major effect on compressive strength, while sintering temperature and holding time play an important role in water absorption. During sintering, mica, chlorite, and other components in lead–zinc tailings are decomposed to form albite, hematite, maghemite, and anhydrite, which play a role in the strength of bricks. The optimal process parameters were found to be a ratio of lead–zinc tailings:clay:fly ash = 6:3:1, forming pressure of 20 MPa, firing temperature of 1080 °C, and holding time of 60 min. The corresponding compressive strength and water absorption were 34.94 MPa and 16.02%, which meets the Chinese sintered ordinary bricks standard (GB/T 5101-2017).

Keywords: lead–zinc tailings; sintered ordinary brick; orthogonal experiment; compressive strength

Citation: Lin, H.; Li, R.; Li, S. Fabrication of Lead–Zinc Tailings Sintered Brick and Its Effect Factors Based on an Orthogonal Experiment. *Materials* **2024**, *17*, 2352. https://doi.org/10.3390/ma17102352

Academic Editor: Dimitrios Papoulis

Received: 9 April 2024
Revised: 11 May 2024
Accepted: 12 May 2024
Published: 15 May 2024

Copyright: © 2024 by the authors. Licensee MDPI, Basel, Switzerland. This article is an open access article distributed under the terms and conditions of the Creative Commons Attribution (CC BY) license (https://creativecommons.org/licenses/by/4.0/).

1. Introduction

China's lead–zinc ore production accounts for more than 40% of the world's total production. Lead–zinc tailings are the residual parts of lead–zinc ore after grinding and flotation, which are very representative in China's solid ore waste dumps [1]. Tailings storage is the most direct disposal method at present. The annual production of tailings in China is about 1.5 billion tons, and there are more than 7000 tailings ponds in China. However, after being disturbed, tailings reservoirs may cause geological disasters such as debris flows, landslides, and ground collapse, resulting in heavy casualties and property losses [2,3]. Furthermore, long-term accumulation will lead to the leaching of harmful components containing heavy metal ions such as Pb^{2+}, Zn^{2+}, and Cd^{2+}, threatening the environmental safety of water and soil near the mining area [4]. Therefore, there is an urgent need to carry out research on the reutilization of lead–zinc tailings, which is of great significance to improve resource utilization efficiency, improve environmental quality, and promote the comprehensive green transformation of economic and social development.

At present, the comprehensive utilization of lead–zinc tailings mainly has two aspects. First, lead–zinc tailings are selected as secondary resources to recover the valuable components and improve the recovery rate. The second is the direct utilization of lead–zinc tailings, mainly including the production of building materials and backfill materials. The grades of lead, zinc, sulfur, and fluorite in lead–zinc tailings are relatively high, and these components have a high recovery value. At present, the secondary recovery of lead–zinc tailings is achieved by a variety of processes, including chemical leaching, microbial leaching, magnetizing roasting–magnetic separation, and flotation technology [5–9]. However, the existence of difficult-to-recover metals in tailings will result in secondary tailings, which will also threaten human society. Therefore, the direct utilization of tailings without reselection value should be considered.

Using tailings as the main raw material to produce building materials can not only consume a large quantity of tailings, but also bring good economic benefits to mining enterprises. This direction has already attracted the extensive attention of scholars [10,11]. Guo et al. [12] studied the preparation of ceramic bricks from tungsten tailings. Kim et al. [13] and Wei et al. [14] studied the possibility of fabricating bricks from gold tailings. Luo et al. [15] used iron tailings, sludge, and other materials to prepare sintered bricks, systematically analyzed the influence of many factors on the properties of the sintered bricks, and proposed the optimal process parameters. In addition, other scholars have conducted extensive research on the use of tailings to prepare ceramics [16,17], filling materials [18,19], cement material [20–23], etc.

Due to the low grade of lead–zinc ore in China, the output of lead–zinc tailings is usually more than ten times that of lead–zinc concentrate. The disposal of lead–zinc tailings has become a key problem restricting the development of the lead–zinc industry. Si, Al, and other elements contained in lead–zinc tailings are essential components of building materials production [24,25]. Therefore, if lead–zinc tailings can be used as a substitute for building materials, this can solve the problem of tailings storage and maximize the effective utilization of resources [26,27]. In recent years, many scholars have carried out research on the use of lead–zinc tailings as raw materials to fabricate building materials. Liu et al. [28–30] fabricated foam ceramics from lead–zinc tailings, red mud, and fly ash, and studied the influence of various process parameters on ceramic performance. By studying the geopolymers with lead–zinc tailings, Zhao et al. [31] found that the curing rates of Zn^{2+}, Pb^{2+}, and Cd^{2+} were all higher than 97.80%, and the leaching concentrations only fluctuated within the limited environmentally acceptable range. In addition, Wang et al. [32] analyzed the fixation behavior of heavy metal ions in the sintering process of lead–zinc tailings brick and found that a high temperature (over 1050 °C) can play a positive role in the fixation of heavy metal ions. Li et al. [33] and Zhang et al. [34] studied the leaching behavior of heavy metal ions when lead–zinc tailings were used as raw materials to prepare building materials. Wang et al. [35] studied the use of lead–zinc tailings to prepare ultra-high-performance concrete and found that the addition of lead–zinc tailings can significantly reduce the early auto-shrinkage of concrete and is conducive to the development of its microstructure. By studying the effect of temperature on the performance of foam ceramics mixed with lead–zinc tailings, Liu et al. [28] found that foam ceramics with sintering temperature at 970 °C had the best performance, with higher porosity (76.2%), higher mechanical strength (5.3 MPa), and lower thermal conductivity (0.21 W/(m K)). In summary, the current use of lead–zinc tailings as raw materials to prepare ceramics, cementing materials, and fillers has been widely reported. However, there are few reports on the fabrication of sintered bricks with lead–zinc tailings.

In this study, lead–zinc tailings were used as primary raw materials, and clay and fly ash as auxiliary materials, to fabricate sintered bricks that reached the highest strength grade in the Chinese sintered ordinary bricks standard (GB/T 5101-2017 [36]). The approximate range of the raw material ratio was determined by the single-factor test. The effects of clay content, forming pressure, sintering temperature, and holding time on the properties of sintered bricks were analyzed based on orthogonal experiments. The brick sintering process was studied with XRD, SEM, and thermogravimetry/differential scanning calorimetry technology (TG-DSC). The research results are expected to realize the secondary utilization of lead–zinc tailings and reduce the environmental and safety threats caused by the accumulation of lead–zinc tailings.

2. Materials and Experiments

The lead–zinc tailings used in this study are tailings waste residues after reconcentration of tailings, which are from the Sanguikou lead–zinc tailings of Ulat Houqi Zijin

Mining Co, Inner Mongolia, China (Figure 1). Clay and fly ash, the auxiliary materials, were purchased from the market.

Figure 1. Lead–zinc tailings powder used in this study.

2.1. Physicochemical Properties Test

The phase composition of lead–zinc tailings was determined by XRD (Bruker D8 ADVANCE, Bruker, Mannheim, Germany, angular accuracy is 0.0001°), and the chemical compositions of three raw materials (lead–zinc tailings, clay, and fly ash) were examined by an X-ray fluorescence spectroscopy analyzer (XRF, Bruker S2 PUMA, Bruker, Germany). The thermodynamic characteristics of lead–zinc tailings were analyzed through TG-DSC (Netzsch/STA 449 F5 Jupiter, Netzsch, Germany, balance resolution is 0.1 μg and temperature resolution is 0.001 K). The surface morphology and pore structure of lead–zinc tailings particles at different sintering temperatures were scanned by SEM (MIRA3 LMH eds: One Max 20, Tescan, Czech Republic). A laser particle size analyzer (LPSA, Mastersizer 2000 with Hydro2000M, Malvern, UK) was adopted to analyze the particle size distribution of lead–zinc tailings. According to the provisions of the liquid–plastic limit combined test method in the Standard for geotechnical testing method (GB/T50123-2019 [37]), the plasticity index (the difference between the liquid limited water content and the plastic limited water content) of lead–zinc tailings was measured by a liquid–plastic limit combined device. The particle size should be less than 0.5 mm, and the plasticity index was calculated from the average of three parallel tests. The tailings bodies fabricated with different levels of forming moisture were fully dried and a uniaxial compression test was carried out. The drying performance of lead–zinc tailings with different levels of forming moisture was studied by analyzing the shape and strength changes of the tailings bodies, and then the appropriate forming moisture was determined (the corresponding forming moisture of tailing bodies with a smaller shape change and higher strength).

The uniaxial loading experiments of sintered bricks were carried out on a HUALONG WHY-300/10 test system (Hualong, Shanghai, China, relative error of force is less than 0.5%). The maximum load is 300 kN, and the measurement range is 2–100% fullscale, which meets the experimental requirements. According to the Test Method for Wall Bricks (GB/T2542-2012 [38]), the loading method adopts force control and the loading rate is set at 0.8 kN/s. Based on the Chinese sintered ordinary bricks standard (GB/T 5101-2017 [36]), the strength requirements of sintered bricks are divided into five grades named M10, M15,

M20, M25, and M30, and the corresponding uniaxial compressive strengths are 10, 15, 20, 25, and 30 MPa, respectively. The phase composition and microstructure of the sintered bricks with lead–zinc tailings prepared by optimal process parameters were analyzed by XRD and SEM.

2.2. Specimen Preparation

Figure 2 shows the process flow and sintering scheme of sintered bricks from lead–zinc tailings. Before preparation began, all raw materials were dried to remove moisture and impurities. As can be seen in Figure 2a, the dried tailings, clay, and fly ash were first screened through a 65-mesh sieve, and then evenly mixed based on the set ratio. According to the set forming moisture, a certain amount of water was added to the mixture to be stirred, and the evenly mixed mixture was put into a sealed bag and aged at room temperature (25 °C) for 24 h. The aged mixture was placed into a Φ50 × 50 mm mold, and compacted to different molding pressures at the loading rate of 0.1 kN/s and maintained for 100 s after forming. The molded bricks were placed at room temperature (25 °C) for 24 h, then placed in a thermostatic drying oven (Supor 101 s, Supor, China, maximum temperature is 300° C and temperature resolution is 0.1 °C) and dried at 105 °C for 12 h. After drying, the bricks were put into a box-type resistance furnace (Yiheng SX2-10-12NP, Yiheng, Shanghai, China, maximum temperature is 1200 °C and temperature resolution is 1 °C) for sintering. After rising to the specified sintering temperature in a certain sintering process, it was kept warm for a certain time, and finally cooled naturally to room temperature. Figure 2b shows the sintering scheme of sintered bricks from lead–zinc tailings, which is divided into four stages: low temperature dehydration stage, stable heating stage, high temperature sintering stage, and natural cooling stage.

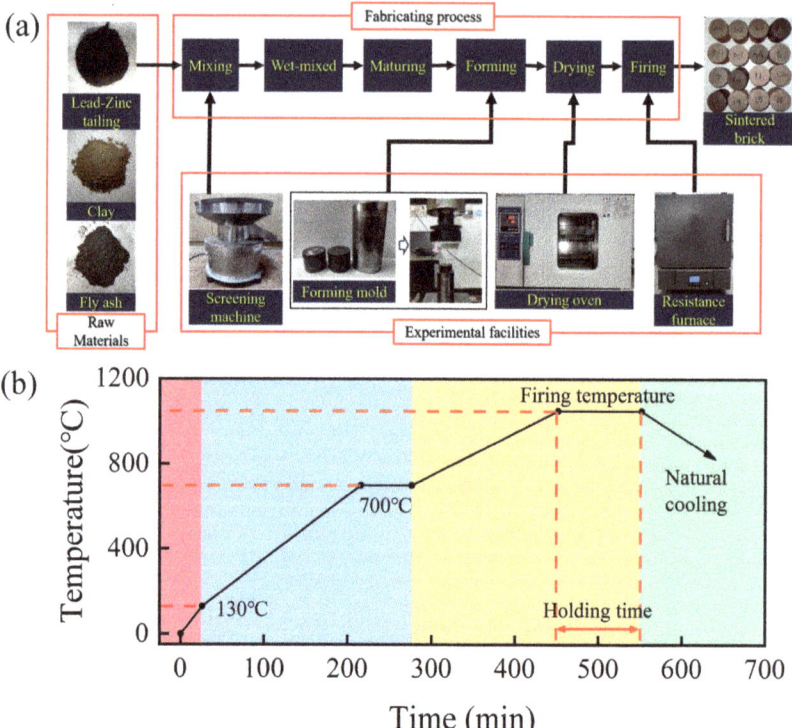

Figure 2. Fabrication of sintered ordinary brick from lead–zinc tailings: (**a**) fabrication process, (**b**) sintering scheme.

2.3. Experimental Scheme

The loading experiment was divided into two parts: a single-factor experiment and an orthogonal experiment. For the single-factor experiment, the content of different raw materials is shown in Table 1. The proportion of fly ash was fixed at 10% and the clay content was 0%, 10%, 20%, 30%, and 40%, in order. The forming parameters were as follows: forming moisture was 12.5% and forming pressure was 15 MPa. The sintering parameters were: sintering temperature was 1050 °C and holding time was 60 min. In addition, the results were compared with those of specimens with pure lead–zinc tailings. The ID was given using a-b-c, where a, b, and c, respectively, represented the proportion of tailings, clay, and fly ash.

Table 1. Mass ratio of different raw materials (unit: %).

Specimen ID	Lead–Zinc Tailings	Clay	Fly Ash
10-0-0	100	0	0
9-0-1	90	0	10
8-1-1	80	10	10
7-2-1	70	20	10
6-3-1	60	30	10
5-4-1	50	40	10

The orthogonal experiment selected four important parameters in the fabrication process, which were clay content, forming pressure, sintering temperature, and holding time. For each process parameter, four different levels were set, which are shown in Table 2. According to Wang et al. [32], when the sintering temperature is higher than 1050 °C, the leaching rate of Pb and Zn tends to 0. Therefore, the sintering temperature in this experiment was equal to or greater than 1050 °C. According to different process parameters and corresponding levels, the L16 (4^4) orthogonal test table was used to fabricate sintered brick specimens from lead–zinc tailings. The name of each specimen represents its corresponding process parameters. For example, the corresponding process parameters of specimen $A_3B_1C_3D_4$ are that the clay content is 30%, the forming pressure is 15 MPa, the sintering temperature is 1110 °C, and the holding time is 120 min.

Table 2. Process parameters and corresponding values of orthogonal experiment.

Level	A: Clay Content (%)	B: Forming Pressure (MPa)	C: Sintering Temperature (°C)	D: Holding Time (min)
1	20	15	1050	30
2	25	20	1080	60
3	30	25	1110	90
4	35	30	1140	120

3. Result Analysis

3.1. Raw Materials Characteristics

From Table 3, the lead–zinc tailings used in this study belong to the SiO_2–Al_2O_3–metal oxide system, which is similar to the clay used in traditional sintered bricks. Clay and fly ash were added as auxiliary materials to make sintered bricks with good performance. The chemical compositions of clay and fly ash selected in this study are also shown in Table 3. The clay is brown-yellow and the fly ash is gray. Both of them are mainly powdery particles with a size less than 2 mm, which meet the requirements for fabricating sintered bricks and can be directly mixed with lead–zinc tailings.

In Figure 3, the XRD pattern of lead–zinc tailings shows that the main phase compositions are quartz, mica, chlorite, and calcite, in addition to dolomite and pyrite, which are consistent with the main chemical composition obtained by XRF analysis. Among them, the diffraction peaks of quartz and mica are sharp and clear, indicating that they have higher content and better crystallinity. Quartz, mica, chlorite, and calcite can be used as raw materials for sintered building materials, which proves the feasibility of applying lead–zinc tailings to the fabrication of sintered ordinary bricks.

Table 3. Main chemical compositions of raw materials (unit: %).

Materials	SiO$_2$	Al$_2$O$_3$	Fe$_2$O$_3$	MgO	CaO	K$_2$O	P$_2$O$_5$	Na$_2$O	MnO$_2$	TiO$_2$	ZnO	Others
Tailings	48.17	10.79	14.15	4.14	4.20	3.01	/	0.456	0.73	0.312	0.493	13.55
Clay	61.37	14.32	4.74	2.36	12.40	2.59	0.21	1.03	/	/	/	0.98
Fly ash	48.80	26.26	4.87	1.84	4.95	2.00	0.15	1.67	/	/	/	9.46

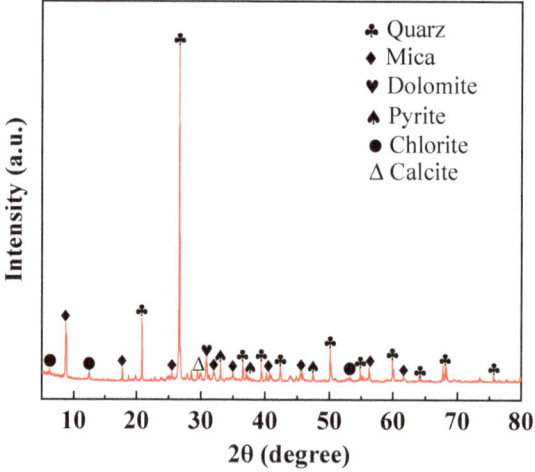

Figure 3. XRD pattern of lead–zinc tailings.

The particle size distribution of lead–zinc tailings is shown in Figure 4. It can be seen that the particle size of lead–zinc tailings is relatively fine, and the maximum particle size is less than 1 mm. The cumulative proportion of particles with size less than 75 μm reaches 61%, the average particle size is 89.5 μm, and the median particle size is 43.2 μm. The particle size composition of lead–zinc tailings used in this study meets the granularity requirement for preparing sintered ordinary bricks.

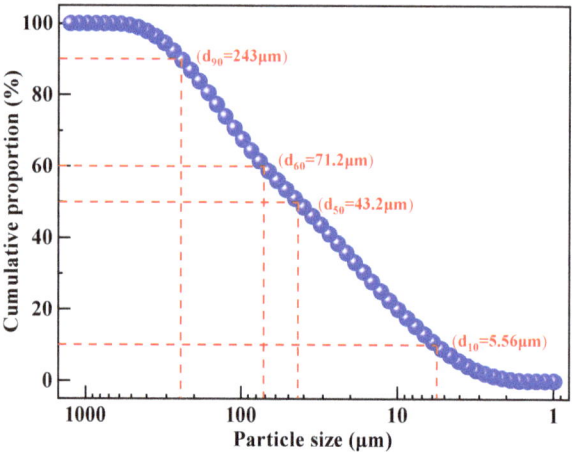

Figure 4. Particle size distribution of lead–zinc tailings.

The plasticity index of lead–zinc tailings is shown in Figure 5. It can be seen that the average plasticity index is 10.4, and the index ranges from 7 to 15, which indicates medium plasticity. The plasticity index of tailings meets the requirements of fabricating sintered ordinary brick, and a better molding effect can be achieved when assisted by clay.

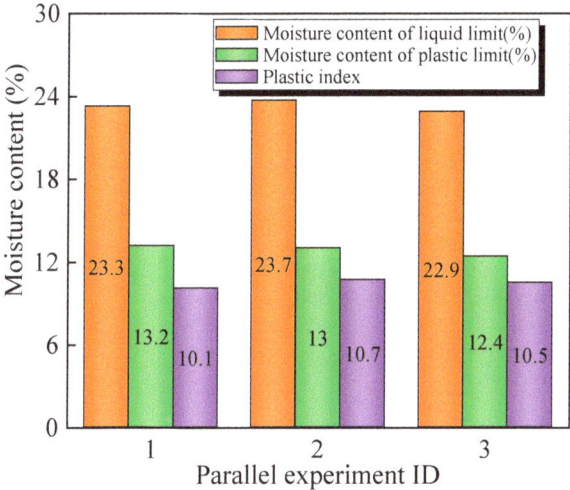

Figure 5. Plasticity index of lead–zinc tailings.

Figure 6 shows the morphology of the tailings body before and after drying, and the strength variation of the tailings body with different levels of forming moisture. It can be seen that the appearance of the tailings body has no obvious change, and the volume shrinkage is basically less than 1%. Forming moisture has a certain influence on the strength of the lead–zinc tailings body. As shown in Figure 6b, with forming moisture ranging from 5% to 17.5%, the uniaxial compressive strength presents a trend of first increasing and then decreasing, and is at a relatively high level in the range of 12.5% to 15%. Generally, the strength variation in the tailings body is relatively stable. When the forming moisture ranges from 12.5% to 15%, the drying performance is relatively good.

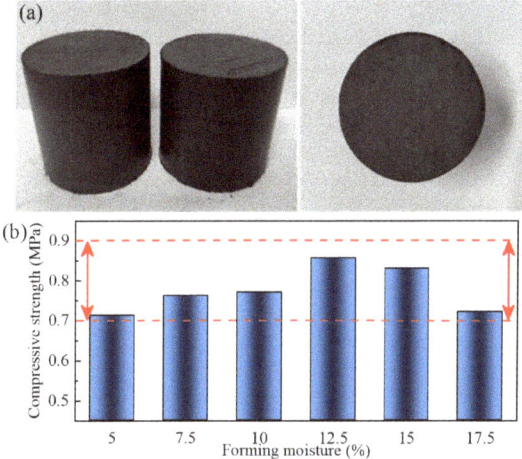

Figure 6. Drying performance of lead–zinc tailings: (**a**) shape comparison; (**b**) relationship between uniaxial compressive strength and forming moisture.

3.2. Analysis of Single-Factor Experiment

Figure 7 shows the influence of clay content on the main performance indices (compressive strength, water absorption, bulk density, and mass loss rate) of sintered ordinary bricks from lead–zinc tailings. In Figure 7a, the addition of fly ash can improve the performance of sintered ordinary bricks, but the improvement is weak. The addition of clay obviously improves the performance of sintered ordinary brick. When the clay content reaches 10%, the compressive strength of the sintered brick is 10.5 MPa and the water absorption rate is 18.7%, which meets the requirements of the MU10 strength level and water absorption in the Chinese sintered ordinary bricks standard (GB/T 5101-2017). With a further increase in clay content, the strength of sintered brick gradually increases, and the water absorption gradually decreases. When the clay content is 30%, the compressive strength of the sintered brick is 22.4 MPa, which reaches the strength level of MU20 in the Chinese sintered ordinary bricks standard (GB/T 5101-2017). Figure 7b shows the influence of raw material ratio on bulk density and mass loss rate of sintered ordinary brick. It can be seen that the addition of fly ash slightly reduces the bulk density, but the subsequent bulk density continues to increase with the increase in clay content. In addition, the mass loss rate of sintered brick decreases continuously, and is less than 15% under six raw material ratios.

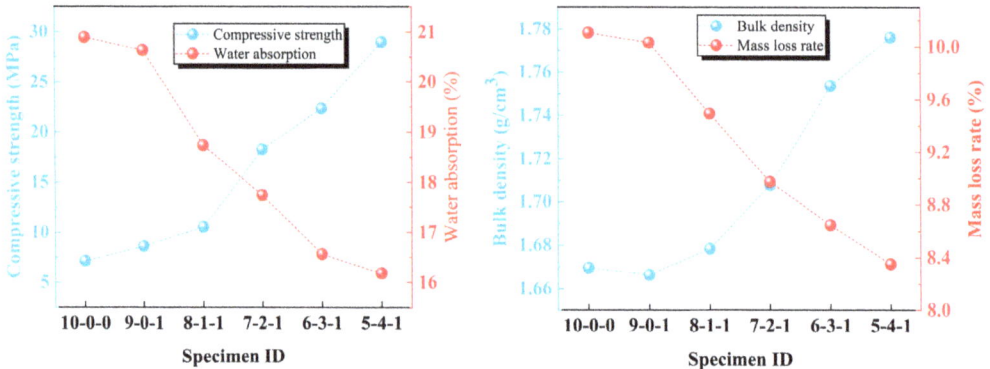

Figure 7. Influence of clay content on performance indices of sintered ordinary bricks from lead–zinc tailings: (**a**) compressive strength and water absorption; (**b**) bulk density and mass loss rate.

3.3. Analysis of Orthogonal Experiment

The performance indices of 16 specimens fabricated according to the orthogonal experiment table are shown in Table 4. It can be seen that the compressive strength of specimen $A_4B_2C_3D_1$ is the highest, which is 34.43 MPa. When used as a load-bearing brick, it meets the requirements of the MU30 level in the Chinese sintered ordinary bricks standard (GB/T5101-2017). When used as road bricks, it also meets the strength requirements of the MX category in the Chinese fired paving unit standard (GB/T26001-2010 [39]). Considering the requirements of environmental protection and construction in mining areas, the sintered bricks with lead–zinc tailings can replace the pavement bricks around the mining area. For all specimens, except for the slightly larger water absorption of $A_1B_1C_1D_1$, $A_1B_2C_2D_2$, and $A_2B_1C_2D_3$, the water absorption values of other specimens are all less than 17%, and thus meet the requirement. It can be found that the change in process parameters has little influence on mass loss rate and bulk density. Mass loss rate is mainly concentrated in the range of 9%~11%, and the bulk density is between 1.67 g/cm^3 and 1.81 g/cm^3.

Table 4. Performance indices of sintered ordinary bricks from lead–zinc tailings.

Specimen ID	Compressive Strength/MPa	Water Absorption/%	Mass Loss Rate/%	Bulk Density/(g/cm^3)
$A_1B_1C_1D_1$	9.40	18.38	7.64	1.69
$A_1B_2C_2D_2$	25.37	17.16	9.60	1.73
$A_1B_3C_3D_3$	22.91	14.92	10.75	1.74
$A_1B_4C_4D_4$	11.16	4.26	10.40	1.68
$A_2B_1C_2D_3$	23.50	18.48	10.09	1.67
$A_2B_2C_1D_4$	25.65	16.97	10.03	1.71
$A_2B_3C_4D_1$	16.33	9.55	10.84	1.76
$A_2B_4C_3D_2$	22.6	14.57	10.64	1.73
$A_3B_1C_3D_4$	32.17	12.39	10.97	1.79
$A_3B_2C_4D_3$	26.15	4.62	11.15	1.76
$A_3B_3C_1D_2$	32.01	16.67	9.20	1.73
$A_3B_4C_2D_1$	30.66	15.99	9.20	1.76
$A_4B_1C_4D_2$	31.89	6.08	10.84	1.81
$A_4B_2C_3D_1$	34.43	16.55	10.10	1.71
$A_4B_3C_2D_4$	33.96	15.20	10.18	1.71
$A_4B_4C_1D_3$	31.95	16.53	9.33	1.74

3.3.1. Mean Value Analysis

Figure 8 shows the trend of the mean value of performance indices with the variation in each process parameter. In Figure 8a, the strength of sintered ordinary brick is significantly affected by clay content, and the average compressive strength increases linearly with clay content. When clay content is 35%, the uniaxial compressive strength reaches 33 MPa. The average compressive strength of sintered ordinary bricks increases and then decreases with the variation in the three other parameters. The influence of these three process parameters on compressive strength is relatively low, and the maximum points are 1080 °C, 20 MPa, and 60 min, respectively. When the sintering temperature is greater than 1080 °C, the main reason that the strength will gradually decrease is that the increasing sintering temperature leads to the increase in liquid phase content, which results in the degradation of the bricks' skeleton structure.

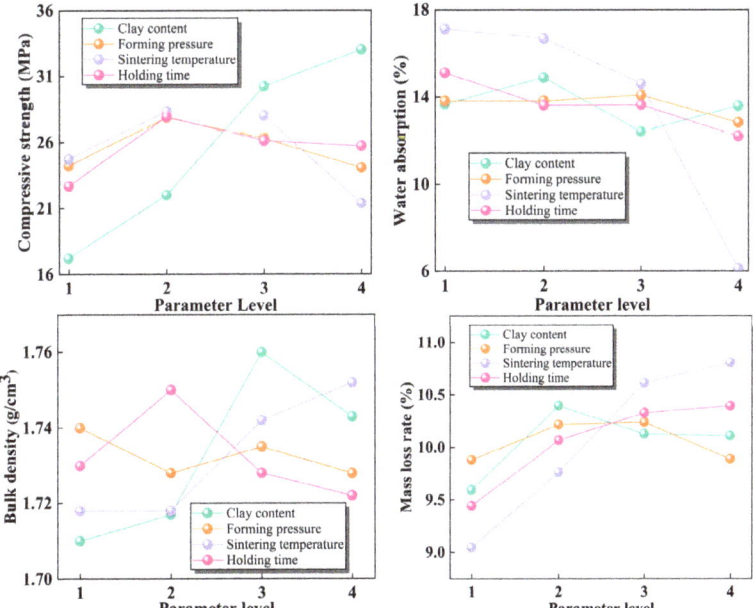

Figure 8. Influence of four process parameters on the average value of performance indices: (a) compressive strength; (b) water absorption; (c) bulk density; (d) mass loss rate.

In Figure 8b, the mean water absorption of sintered bricks decreased significantly with the variation in sintering temperature, and the decreasing trend is more obvious with the increasing sintering temperature. When the sintering temperature increases from 1050 °C to 1140 °C, the mean water absorption decreases from 17.1% to 6.1%. With the increase in temperature, lead–zinc tailings continue to melt, and liquid phase in sintered brick accumulates, which continuously fills the pores and reduces the porosity of sintered bricks, and finally leads to the weakening of water absorption. The influence of clay content, forming pressure, and holding time on water absorption of sintered bricks is relatively weak, and average water absorption decreases slightly with the increase in the three factor levels.

The variation trends of mean bulk density with the four factor levels are shown in Figure 8c. It is not difficult to see that the bulk density is more affected by sintering temperature and clay content. With the increase in sintering temperature and clay content, the bulk density shows an increasing trend. With the increase in sintering temperature, the amount of liquid phase increases, which filles the pores in sintered bricks and results in the increase in bulk density. When the sintering temperature reaches 1040 °C, the bulk density of sintered brick reaches 1.75 g/cm^3. Compared with lead–zinc tailings, clay contains a higher proportion of silicon oxides, which melt during the sintering process. The increase in clay content will produce more molten substances, resulting in smaller pores and higher bulk density of sintered bricks. Bulk density was less affected by forming pressure and holding time, which shows a slowly decreasing trend with the increase in these two process parameters.

Figure 8d shows the trend of mass loss rate with the variation in four different factors. During the high-temperature sintering process, raw materials undergo complex physical and chemical reactions. The decomposition of calcite and dolomite produces gas, and water evaporation occurs during the decomposition of mica. These behaviors will result in mass loss of sintered bricks. With the increase in sintering temperature and holding time, the decomposition action will be more complete, and the mass loss rate of sintered brick will increase. When the sintering temperature is 1140 °C, the mass loss rate of sintered brick with lead–zinc tailings reaches 10.8%. With the increase in sintering temperature and holding time, the reaction between chemical components of raw materials will gradually reach completion. This explains why the trend of mass loss rate gradually flattens with the increase in sintering temperature and holding time. In addition, the forming pressure has almost no effect on mass loss rate, which was always maintained near 10% with the increase in forming pressure.

3.3.2. Range Analysis

Ranges of the influences of the four process parameters on the performance index of sintered bricks are shown in Figure 9. It can be seen that the order of the influence of the four parameters on compressive strength of sintered brick is clay content > sintering temperature > holding time > forming pressure. The order of the influence of the four parameters on water absorption is sintering temperature > holding time > clay content > forming pressure. Careful selection of clay content and sintering temperature can optimize sintered bricks with better performance. The order of the influence of the four parameters on mass loss rate is sintering temperature > holding time > clay content > forming pressure. The sensitivity order of the four parameters on bulk density is clay content > sintering temperature > holding time > forming pressure. However, there is little difference between the range of these four parameters on mass loss rate and bulk density, indicating that the difference in the influence of these four parameters on mass loss rate and bulk density is not obvious. Overall, the sensitivity of the four performance indices to the variation in process parameters is in the order of compressive strength > water absorption > mass loss > volume density.

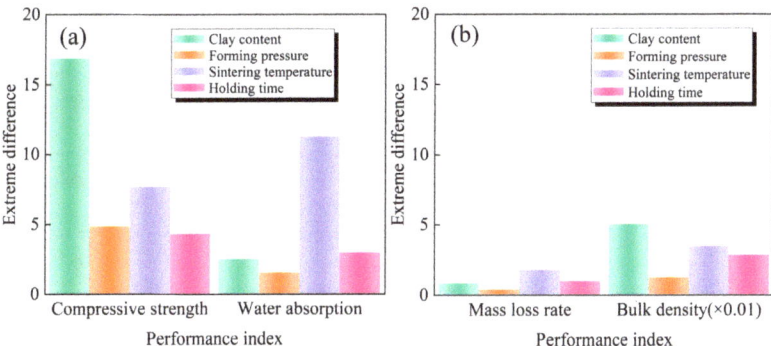

Figure 9. Range of process influences on performance indexes: (**a**) compressive strength and water absorption; (**b**) mass loss rate and bulk density.

3.3.3. Variance Analysis

The variance analysis of the four process parameters on performance indices of sintered bricks with lead–zinc tailings are shown in Tables 5 and 6. The significance level $\alpha = 0.05$ is used in this study, that is, the confidence probability is 95%. When the F-ratio of a parameter is greater than the threshold, it indicates that this parameter has a significant impact on the performance index.

Table 5. Variance of parameters' influences on compressive strength and water absorption.

Performance Indices	Parameters	Sum of Square	Degree of Freedom	F-Ratio	Threshold of F-Ratio ($\alpha = 0.05$)	Significance
Compressive strength	Clay content	641.670	3	13.158		Significant
	Forming pressure	39.604	3	0.812		None
	Sintering temperature	128.322	3	2.631	9.280	None
	Holding time	57.112	3	1.171		None
	Error	48.77	3			
Water absorption	Clay content	12.269	3	7.360		None
	Forming pressure	3.653	3	2.191		None
	Sintering temperature	316.063	3	189.600	9.280	Significant
	Holding time	16.970	3	10.180		Significant
	Error	1.67	3			

Table 6. Variance of parameters' influences on mass loss rate and bulk density.

Performance Indices	Parameters	Sum of Square	Degree of Freedom	F-Ratio	Threshold of F-Ratio ($\alpha = 0.05$)	Significance
Mass loss rate	Clay content	1.349	3	4.557		None
	Forming pressure	0.470	3	1.588		None
	Sintering temperature	7.890	3	26.655	9.280	Significant
	Holding time	2.254	3	7.615		None
	Error	0.30	3			
Bulk density	Clay content	0.006	3	0.667		None
	Forming pressure	0.000	3	0.000		None
	Sintering temperature	0.004	3	0.444	9.280	None
	Holding time	0.002	3	0.222		None
	Error	0.01	3			

Taking the compressive strength as an example, the sensitivity order of the four process parameters on compressive strength i: clay content (13.158) > sintering temperature (2.631) > holding time (1.171) > molding pressure (0.812). It can be concluded that this order is consistent with that of the range analysis (Figure 9a). According to the significance results in Tables 5 and 6, it can be seen that the clay content has a significant effect on the compressive strength of sintered ordinary bricks, sintering temperature and holding time have a significant effect on water absorption, and sintering temperature also has a significant effect on mass loss rate. The influence of all process parameters on bulk density of sintered ordinary bricks is not significant. Considering that the compressive strength of sintered ordinary brick is the main index affecting its performance, the clay content was selected to be 30%. The average compressive strength of sintered ordinary bricks with this clay content can meet the requirements of the highest strength grade, MU30, in the Chinese sintered ordinary bricks standard (GB/T 5101-2017), and also ensure large utilization of lead–zinc tailings. The optimal sintering temperature, holding time, and forming pressure are determined according to the parameter levels corresponding to the maximum value of each average strength curve. In summary, the optimal process parameters should be clay content of 30%, forming pressure of 20 MPa, sintering temperature of 1080 °C, and holding time of 60 min.

3.4. Verification and Mechanism Analysis

3.4.1. Verification of Optimal Process Parameters

According to the optimal process parameters of sintered ordinary bricks with lead–zinc tailings determined in Section 3.3.3, corresponding specimens were prepared and their main performance indices are shown in Table 7. Compared with the results of the orthogonal experiment, the compressive strength of sintered brick with optimal process parameters is the highest, which fully meets the requirements of the highest strength grade, MU30, in the Chinese sintered ordinary bricks standard (GB/T5101-2017). According to Wei et al. [14], the optimal strength of sintered brick containing gold tailings is 22.45 MPa. It is clear that the strength of sintered brick fabricated in this study is higher. In terms of raw material proportions, 35% clay is used in the research of Wei et al. [14], and the clay content in this study is 30%. Fly ash and lead–zinc tailings can be regarded as solid waste. Therefore, the strength and utilization rate of the tailings sintered bricks fabricated in this study are competitive. For unfired bricks with lead–zinc tailings [25], the compressive strength reaches 10 MPa (MU10 grade) when the proportion of tailings is 45%. It can be seen that the mechanical properties of sintered bricks with lead–zinc tailings fabricated in this study are superior. In addition, the measured water absorption, mass loss rate, and bulk density of sintered ordinary bricks under this parameter combination also meet the requirements.

Table 7. Performance indices of sintered brick with optimal process parameters.

Performance Indices	Value
Compressive strength	34.94 MPa
Water absorption	16.02%
Mass loss rate	9.85%
Bulk density	1.75 g/cm^3

3.4.2. Sintering Mechanism of Lead–Zinc Tailings

The thermogravimetry/differential scanning calorimetry (TG-DSC) curves of lead–zinc tailings are shown in Figure 10. It can be seen from the TG curve that lead–zinc tailings show a trend of weight loss during the sintering process, and the cumulative mass loss rate is 14.72%. The mass loss process can be divided into four stages: (1) The temperature rises from room temperature to around 370 °C during the first stage. The TG curve declines slowly in this period, and the cumulative mass loss rate is only 0.24%. The mass loss at this stage is mainly caused by the evaporation of adsorbed water and crystal water. However, the specimen was dried before thermal analysis, and part of the adsorbed and crystalline water was discharged, so the mass loss rate was small. (2) The second stage corresponds to the range in the sintering temperature from 370 °C to about 575 °C. The TG curve begins to decrease obviously, while the DSC curve shows obvious exothermic and endothermic peaks. The exothermic peak at about 490.5 °C may be caused by the combustion of organics contained in lead–zinc tailings, and the endothermic peak at about 554.3 °C is mainly related to the decomposition of pyrite and silicate, and is also affected by the quartz phase change. The quality of quartz does not change during the phase change, but the combustion of organic matter, and the decomposition of pyrite and silicate, will cause mass loss. (3) In the third stage, the sintering temperature rises from 575 °C to 870 °C. The DSC curve shows an obvious peak, and mass loss rate of lead–zinc tailings at this stage is the largest, about 8.99%. It is speculated that the exothermic peak is caused by the further oxidation of pyrite decomposition products, and the endothermic peak is mainly caused by the decomposition of carbonate minerals in lead–zinc tailings. Due to the high content of carbonate minerals in lead–zinc tailings, the mass loss increases significantly at this stage. (4) The sintering temperature above 870 °C corresponds to the fourth stage. In the early part of this stage, the mass loss is relatively slow, and the DSC curve also increases slowly, indicating that the internal changes in lead–zinc tailings are mainly melting, which stay in a stable sintering process. However, when the sintering temperature is higher than 1150 °C, the mass loss rate of tailings is intensified, and there exists an endothermic peak, which may be caused by the decomposition of aluminosilicate minerals.

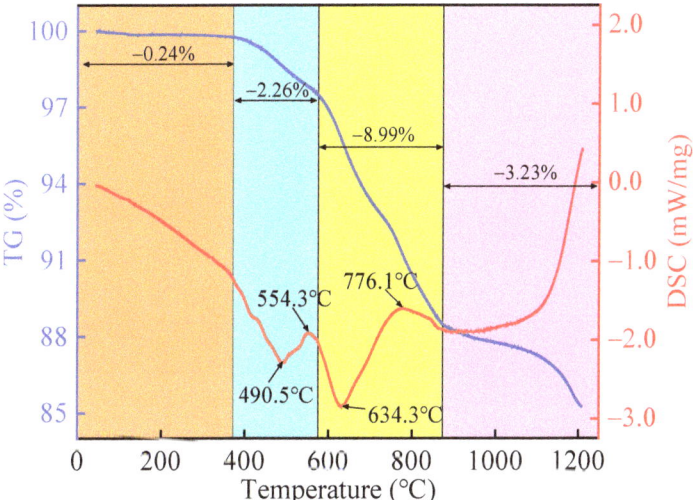

Figure 10. TG-DSC curve of lead–zinc tailings.

3.4.3. XRD and SEM Analysis

XRD and SEM experiments were carried out on sintered ordinary bricks prepared under the optimal combination of process parameters. The phase compositions of sintered ordinary brick are shown in Figure 11. It can be found that the main minerals in sintered ordinary bricks from lead–zinc tailings are quartz, anhydrite, albite, maghemite, and hematite. Comparing pre- and post-sintering XRD patterns (Figures 2 and 11), it can be seen that the diffraction peaks of mica, dolomite, pyrite, chlorite, and calcite in lead–zinc tailings disappeared, while the diffraction peaks of anhydrite, albite, maghemite, and hematite appeared. This indicates that chemical reactions occurred in minerals during the sintering process. According to Table 2, raw materials contain Na_2O, Al_2O_3, and SiO_2, which can be used to form $NaAlSi_3O_8$. Mica ($KAl_2(AlSi_3O_{10})(OH)_2$) is mainly decomposed into SiO_2, Al_2O_3, K_2O, and H_2O. Dolomite is decomposed into CaO, MgO, and CO_2. Pyrite is oxidized to Fe_2O_3 and Fe_3O_4, and then Fe_3O_4 is eventually oxidized to γ-Fe_2O_3 (maghemite) during the cooling process (around 220 °C). Chlorite ($Al_4Si_4O_{10}(OH)_8$) is decomposed into Al_2O_3, SiO_2, and H_2O, and calcite is decomposed into CaO and CO_2. The decomposed CaO reacts with SO_2 to form $CaSO_3$, which is further oxidized to $CaSO_4$. The minerals in the sintered bricks are responsible for the mechanical properties. The mica in lead–zinc tailings is decomposed into SiO_2 and Al_2O_3 during the sintering process, which participate in the formation of quartz and albite to improve the uniaxial compressive strength of bricks. The relevant chemical equations are expressed as:

$$CaMg(CO_3)_2(Dolomite) \rightarrow CaO + MgO + CO_2 \uparrow \tag{1}$$

$$KAl_2(AlSi_3O_{10})(OH)_2(Mica) \rightarrow SiO_2 + Al_2O_3 + K_2O + H_2O \uparrow \tag{2}$$

$$FeS_2(Pyrite) + O_2 \rightarrow Fe_2O_3(Hematite)/Fe_3O_4 + SO_2 \uparrow \tag{3}$$

$$Fe_3O_4 + O_2 \rightarrow \gamma\text{-}Fe_2O_3(Maghemite) \tag{4}$$

$$Al_4Si_4O_{10}(OH)_8(Chlorite) \rightarrow Al_2O_3 + SiO_2 + H_2O \uparrow \tag{5}$$

$$CaCO_3(Calcite) \rightarrow CaO + CO_2 \uparrow \tag{6}$$

$$CaO + SO_2 \rightarrow CaSO_3 \tag{7}$$

$$CaSO_3 + O_2 \rightarrow CaSO_4(Anhydrite) \tag{8}$$

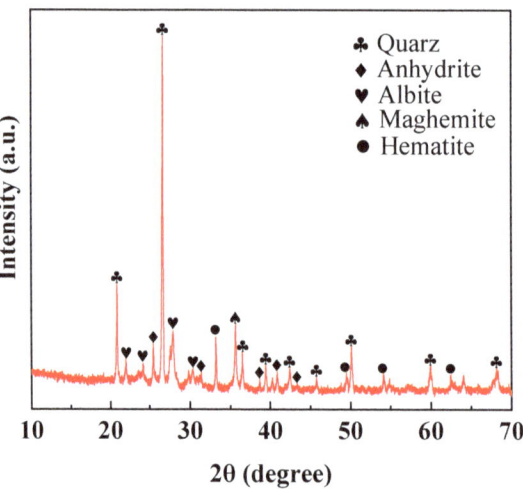

Figure 11. XRD pattern of sintered brick with optimal process parameters.

Figure 12 shows SEM images of raw tailings, sintered tailings at different temperatures, and sintered bricks fabricated according to the optimal process parameters (all magnified by 5000 times). By comparison, raw tailings are characterized by a single particle with flaky debris attached to its surface. From Figure 12a to Figure 12b, at a sintering temperature of 900 °C, the components of lead–zinc tailings are decomposed, the boundary of particles is no longer clear, the surface is rough and presents dense tentacle-like particles, and the micro-pores are very dense. Furthermore, there also exist large cloud-like particles. From Figure 12b to Figure 12c, it can be found that short tentacle-like particles almost disappear, which are replaced by cloud-like particles. Part of the lead–zinc tailings begins to turn into molten liquid, but the amount is small, and micro-pores are still dense. From Figure 12c to Figure 12d, sintered tailings at 1050 °C no longer show a uniform and dense porous structure, mainly because of the increasing molten liquid part, and the generated adhesive material further fills the pores. In addition, the cloud-like particles transform into smaller flocculent particles, and the crystalline boundary appears but is not smooth enough. Lath-like particles also appear, which is due to the high temperature calcination; the particles are fully developed through each other, the crystallization rate is accelerated, and eventually lead to the formation of prismatic crystal phase. From Figure 12d to Figure 12e, when the sintering temperature rises to 1100 °C, it can be seen that the number of lath-like crystals increases greatly, the surface is smoother, and there is less flocculent debris. In addition, micro-cracks can be observed in the figure, which are possibly shrinkage cracks created under high-temperature sintering. The SEM image of sintered brick fabricated according to the optimal process parameters is shown in Figure 12f. It can be seen that the surface of the aggregate is smooth and has a relatively obvious boundary, and the pores are very few, which is mainly attributed to the addition of clay and fly ash to fill the pores in the sintering process. In addition, micro-cracks also appear in the SEM image of the sintered brick, which may not be shrinkage cracks, but may also be caused by the force generated in the loading experiment or grinding. In general, in the range of 900 °C to 1100 °C, with the increase in sintering temperature, the reaction and crystallization of lead–zinc tailings particles will be more adequate, there will be fewer micro-pores, and the particle boundary will be smoother and clearer. Based on the SEM images, it can be determined that the optimal sintering temperature should be above 1050 °C, which also proves the correctness of setting the optimal sintering temperature to 1080 °C.

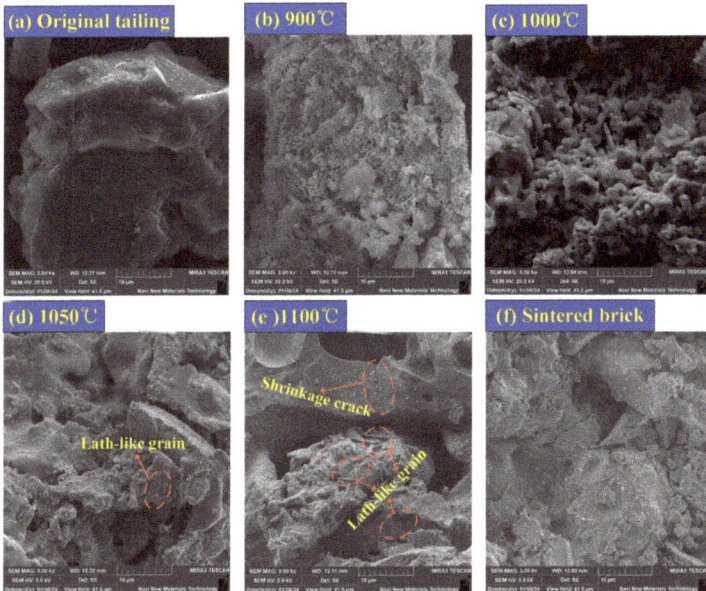

Figure 12. SEM images of lead–zinc tailings and sintered brick: (**a**) raw lead–zinc tailings; (**b**–**e**) lead–zinc tailings at sintering temperature of 900 °C, 1000 °C, 1050 °C, and 1100 °C, respectively; (**f**) sintered brick with optimal process parameters.

4. Conclusions

In this study, lead–zinc tailings were used as the main material, and clay and fly ash as the auxiliary materials, to prepare sintered bricks. Firstly, the effect of clay content on the properties of sintered brick was studied via the single-factor test. On this basis, taking clay content, molding pressure, sintering temperature, and holding time as the influencing factors, an orthogonal experiment was carried out to obtain the optimal process parameters. The main conclusions are as follows:

1. The single-factor experiment shows that with the increase in clay content, uniaxial compressive strength and bulk density increase continuously, while water absorption and mass loss rate decrease gradually. When the clay content is near 30%, the performance of sintered ordinary brick is better.
2. Based on the orthogonal experiment, the mean value analysis shows that the compressive strength increases with the increase in clay content, but increases first and then decreases with the increase in other parameters. Water absorption decreases with the increase in the four parameters, and is significantly affected by sintering temperature. The range analysis shows that the clay content and sintering temperature are the main factors affecting the performance of sintered brick. The variance analysis shows that clay content is a significant factor affecting the compressive strength, while sintering temperature and holding time are significant factors affecting water absorption.
3. According to the XRD pattern, during the sintering process, mica, pyrite, chlorite, calcite, and dolomite are decomposed to form hematite, maghemite, anhydrite, and albitite. Among them, mica with poor hardness is decomposed to SiO_2 and Al_2O_3, which participate in the formation of quartz and albite with higher hardness. SEM images show that when the sintering temperature reaches more than 1050 °C, the crystallization degree of sintered lead–zinc tailings is higher, the surface is smoother and denser, and the porosity is reduced.
4. The optimal process parameters were obtained. That is, the raw material ratio is lead–zinc tailings:clay:fly ash = 6:3:1, molding pressure is 20 MPa, sintering temperature is 1080 °C, and holding time is 60 min. Under this condition, the compressive strength of sintered bricks is 34.94 MPa, which meets the requirements of the highest strength grade of MU30 in "Chinese Sintered Ordinary Bricks" (GB/T5101-2017). The sintered bricks with lead–zinc tailings can be used as pavement bricks around mining areas.

This study only investigated the compressive strength and other physical properties of sintered bricks with lead–zinc tailings at room temperature. In fact, the stress conditions and service environment of bricks are complex, and this manuscript does not consider the fatigue characteristics of bricks (used as road bricks) and durability. Mines are often located in high-altitude areas with harsh weather conditions, so the freeze–thaw resistance and optimization of sintered bricks with tailings can be developed in the future. Furthermore, considering environmental protection and energy conservation, the life-cycle management of bricks with lead–zinc tailings in an environmental context, and determining how to reduce energy consumption while fabricating bricks that fulfill the strength requirement, are of practical significance.

Author Contributions: Methodology, R.L. and S.L.; Software, H.L. and R.L.; Validation, S.L.; Formal analysis, R.L.; Investigation, H.L. and S.L.; Writing—original draft, H.L., R.L. and S.L.; Writing—review & editing, H.L., R.L. and S.L. All authors have read and agreed to the published version of the manuscript.

Funding: This paper gets its funding from Projects (42277175) supported by National Natural Science Foundation of China; Project (NRMSSHR-2022-Z08) supported by Key Laboratory of Natural Resources Monitoring and Supervision in Southern Hilly Region, Ministry of Natural Resources; Project (2023JJ30657) supported by Hunan Provincial Natural Science Foundation of China; Guizhou Provincial Major Scientific and Technological Program (2023-425); Hunan provincial key research and development Program (2022SK2082). The authors wish to acknowledge these sources of support.

Institutional Review Board Statement: Not applicable.

Informed Consent Statement: Not applicable.

Data Availability Statement: The raw data supporting the conclusions of this article will be made available by the authors on request.

Conflicts of Interest: The authors declare no conflicts of interest.

References

1. Li, R.; Yin, Z.; Lin, H. Research Status and Prospects for the Utilization of Lead–Zinc Tailings as Building Materials. *Buildings* **2023**, *13*, 150. [CrossRef]
2. Che, D.; Liang, A.; Li, X.; Ma, B. Remote Sensing Assessment of Safety Risk of Iron Tailings Pond Based on Runoff Coefficient. *Sensors* **2018**, *18*, 4373. [CrossRef] [PubMed]
3. Ruan, S.; Han, S.; Lu, C.; Gu, Q. Proactive control model for safety prediction in tailing dam management: Applying graph depth learning optimization. *Process Saf. Environ. Prot.* **2023**, *172*, 329–340. [CrossRef]
4. Almeida, H.A.; Silva, J.G.; Custódio, I.G.; Karam, D.; Garcia, Q.S. Productivity and food safety of grain crops and forage species grown in iron ore tailings. *J. Food Compos. Anal.* **2022**, *105*, 104198. [CrossRef]
5. Bian, Z.; Zhang, H.; Ye, J.; Ning, Z. Flotation behavior of oleate and dodecylamine as mixed collector for recovery of lithium and rubidium from low-grade spodumene tailings: Experiment, characterization and DFT calculation. *Appl. Surf. Sci.* **2023**, *638*, 158117. [CrossRef]
6. Chen, W.; Yin, S.; Chen, X.; Wang, L.; Zhang, M. Study on comprehensive utilization of tailings by using bioleaching and microbial-cementation. *Case Stud. Constr. Mater.* **2023**, *18*, e02190. [CrossRef]
7. Lei, C.; Yan, B.; Chen, T.; Xiao, X.-M. Recovery of metals from the roasted lead-zinc tailings by magnetizing roasting followed by magnetic separation. *J. Clean. Prod.* **2017**, *158*, 73–80. [CrossRef]
8. Qiu, G.; Ning, X.; Shen, J.; Wang, Y.; Zhang, D.; Deng, J. Recovery of iron from iron tailings by suspension magnetization roasting with biomass-derived pyrolytic gas. *Waste Manag.* **2023**, *156*, 255–263. [CrossRef]
9. Wu, L.; Zhang, J.; Huang, Z.; Zhang, Y.; Xie, F.; Zhang, S.; Fan, H. Extraction of lithium as lithium phosphate from bauxite mine tailings via mixed acid leaching and chemical precipitation. *Ore Geol. Rev.* **2023**, *160*, 105621. [CrossRef]
10. Kang, X.; Gan, Y.; Chen, R.; Zhang, C. Sustainable eco-friendly bricks from slate tailings through geopolymerization: Synthesis and characterization analysis. *Constr. Build. Mater.* **2021**, *278*, 122337. [CrossRef]
11. Yang, C.; Cui, C.; Qin, J.; Cui, X. Characteristics of the fired bricks with low-silicon iron tailings. *Constr. Build. Mater.* **2014**, *70*, 36–42. [CrossRef]
12. Guo, Y.; Wang, C.; Li, S.; He, Y.; Liu, H. Preparation of permeable ceramic bricks with tungsten tailings by two-stage calcination technology. *Constr. Build. Mater.* **2024**, *411*, 134382. [CrossRef]
13. Kim, Y.; Lee, Y.; Kim, M.; Park, H. Preparation of high porosity bricks by utilizing red mud and mine tailing. *J. Clean. Prod.* **2019**, *207*, 490–497. [CrossRef]
14. Wei, Z.; Zhao, J.; Wang, W.; Yang, Y.; Zhuang, S.; Lu, T.; Hou, Z. Utilizing gold mine tailings to produce sintered bricks. *Constr. Build. Mater.* **2021**, *282*, 122655. [CrossRef]
15. Luo, L.; Li, K.; Fu, W.; Liu, C.; Yang, S. Preparation, characteristics and mechanisms of the composite sinteredbricks produced from shale, sewage sludge, coal gangue powder and iron ore tailings. *Constr. Build. Mater.* **2020**, *232*, 117250. [CrossRef]
16. Hui, T.; Sun, H.; Peng, T.; Chen, Y. Preparation and characterization of ceramic foams mainly containing extracted titanium residues and silica tailings. *J. Environ. Chem. Eng.* **2022**, *10*, 108963. [CrossRef]
17. Zhu, Y.; Guo, B.; Zuo, W.; Jiang, K.; Chen, H.; Ku, J. Effect of sintering temperature on structure and properties of porous ceramics from tungsten ore tailings. *Mater. Chem. Phys.* **2022**, *287*, 126315. [CrossRef]
18. Behera, S.K.; Ghosh, C.N.; Mishra, D.P.; Singh, P.; Mishra, K.; Buragohain, J.; Mandal, P.K. Strength development and microstructural investigation of lead-zinc mill tailings based paste backfill with fly ash as alternative binder. *Cem. Concr. Compos.* **2020**, *109*, 103553. [CrossRef]
19. Behera, S.K.; Mishra, D.P.; Singh, P.; Mishra, K.; Mandal, S.K.; Ghosh, C.N.; Kumar, R.; Mandal, P.K. Utilization of mill tailings, fly ash and slag as mine paste backfill material: Review and future perspective. *Constr. Build. Mater.* **2021**, *309*, 125120. [CrossRef]
20. Onuaguluchi, O.; Eren, Ö. Recycling of copper tailings as an additive in cement mortars. *Constr. Build. Mater.* **2012**, *37*, 723–727. [CrossRef]
21. Wang, W.; Zhao, Y.; Liu, H.; Song, S. Fabrication and mechanism of cement-based waterproof material using silicate tailings from reverse flotation. *Powder Technol.* **2017**, *315*, 422–429. [CrossRef]
22. Yin, Z.; Li, R.; Lin, H.; Chen, Y.; Wang, Y.; Zhao, Y. Analysis of Influencing Factors of Cementitious Material Properties of Lead–Zinc Tailings Based on Orthogonal Tests. *Materials* **2022**, *16*, 361. [CrossRef] [PubMed]
23. Zhang, N.; Tang, B.; Liu, X. Cementitious activity of iron ore tailing and its utilization in cementitious materials, bricks and concrete. *Constr. Build. Mater.* **2021**, *288*, 123022. [CrossRef]
24. Li, C.; Zhang, P.; Li, D. Study on low-cost preparation of glass–ceramic from municipal solid waste incineration (MSWI) fly ash and lead–zinc tailings. *Constr. Build. Mater.* **2022**, *356*, 129231. [CrossRef]
25. Wang, P.; Li, J.; Hu, Y.; Cheng, H. Solidification and stabilization of Pb–Zn mine tailing with municipal solid waste incineration fly ash and ground granulated blast-furnace slag for unfired brick fabrication. *Environ. Pollut.* **2023**, *321*, 121135. [CrossRef]
26. Deng, P.; Zheng, Z. Mechanical properties of one-part geopolymer masonry mortar using alkali-fused lead–zinc tailings. *Constr. Build. Mater.* **2023**, *369*, 130522. [CrossRef]
27. Luo, Z.; Guo, J.; Liu, X.; Mu, Y.; Zhang, M.; Zhang, M.; Tian, C.; Ou, J.; Mi, J. Preparation of ceramsite from lead-zinc tailings and coal gangue: Physical properties and solidification of heavy metals. *Constr. Build. Mater.* **2023**, *368*, 130426. [CrossRef]

28. Liu, T.; Li, X.; Guan, L.; Liu, P.; Wu, T.; Li, Z.; Lu, A. Low-cost and environment-friendly ceramic foams made from lead–zinc mine tailings and red mud: Foaming mechanism, physical, mechanical and chemical properties. *Ceram. Int.* **2016**, *42*, 1733–1739. [CrossRef]
29. Liu, T.; Tang, Y.; Han, L.; Song, J.; Luo, Z.; Lu, A. Recycling of harmful waste lead-zinc mine tailings and fly ash for preparation of inorganic porous ceramics. *Ceram. Int.* **2017**, *43*, 4910–4918. [CrossRef]
30. Liu, T.; Tang, Y.; Li, Z.; Wu, T.; Lu, A. Red mud and fly ash incorporation for lightweight foamed ceramics using lead-zinc mine tailings as foaming agent. *Mater. Lett.* **2016**, *183*, 362–364. [CrossRef]
31. Zhao, S.; Lu, W.; Li, D.; Xia, M. Study on acid resistance and high temperature resistance of composite geopolymer-stabilized lead–zinc tailing. *Constr. Build. Mater.* **2023**, *407*, 133554. [CrossRef]
32. Wang, G.; Ning, X.-A.; Lu, X.; Lai, X.; Cai, H.; Liu, Y.; Zhang, T. Effect of sintering temperature on mineral composition and heavy metals mobility in tailings bricks. *Waste Manag.* **2019**, *93*, 112–121. [CrossRef] [PubMed]
33. Li, C.; Wen, Q.; Hong, M.; Liang, Z.; Zhuang, Z.; Yu, Y. Heavy metals leaching in bricks made from lead and zinc mine tailings with varied chemical components. *Constr. Build. Mater.* **2017**, *134*, 443–451. [CrossRef]
34. Zhang, X.; Li, L.; Hassan, Q.U.; Pan, D.; Zhu, G. Preparation and characterization of glass ceramics synthesized from lead slag and lead-zinc tailings. *Ceram. Int.* **2023**, *49*, 16164–16173. [CrossRef]
35. Wang, X.P.; Yu, R.; Shui, Z.H.; Zhao, Z.M.; Song, Q.L.; Yang, B.; Fan, D.Q. Development of a novel cleaner construction product: Ultra-high performance concrete incorporating lead-zinc tailings. *J. Clean. Prod.* **2018**, *196*, 172–182. [CrossRef]
36. GB/T 5101-2017; Fired Common Bricks. National Standardization Administration of China: Beijing, China, 2017.
37. GB/T 50123-2019; Standard for Geotechnical Testing Method. National Standardization Administration of China: Beijing, China, 2019.
38. GB/T 2542-2012; Test Method for Wall Bricks. National Standardization Administration of China: Beijing, China, 2012.
39. GB/T 26001-2010; Fired Paving Units. National Standardization Administration of China: Beijing, China, 2010.

Disclaimer/Publisher's Note: The statements, opinions and data contained in all publications are solely those of the individual author(s) and contributor(s) and not of MDPI and/or the editor(s). MDPI and/or the editor(s) disclaim responsibility for any injury to people or property resulting from any ideas, methods, instructions or products referred to in the content.

Article

Synthesis, Crystal Structure, and Optical and Magnetic Properties of the New Quaternary Erbium Telluride EuErCuTe$_3$: Experiment and Calculation

Anna V. Ruseikina [1,*], Maxim V. Grigoriev [1,2], Ralf J. C. Locke [2], Vladimir A. Chernyshev [3], Alexander A. Garmonov [4] and Thomas Schleid [2,*]

[1] Laboratory of Theory and Optimization of Chemical and Technological Processes, University of Tyumen, Tyumen 625003, Russia; maxgrigmvv@ya.ru
[2] Institute for Inorganic Chemistry, University of Stuttgart, D-70569 Stuttgart, Germany
[3] Institute of Natural Sciences and Mathematics, Ural Federal University Named after the First President of Russia B.N. Yeltsin, Ekaterinburg 620002, Russia; vchern@inbox.ru
[4] Institute of Physics and Technology, University of Tyumen, Tyumen 625003, Russia; gamma125@mail.ru
* Correspondence: adeschina@mail.ru (A.V.R.); thomas.schleid@iac.uni-stuttgart.de (T.S.)

Abstract: This paper reports for the first time on a new layered magnetic heterometallic erbium telluride EuErCuTe$_3$. Single crystals of the compound were obtained from the elements at 1120 K using CsI as a flux. The crystal structure of EuErCuTe$_3$ was solved in the space group *Cmcm* ($a = 4.3086(3)$ Å, $b = 14.3093(9)$ Å, and $c = 11.1957(7)$ Å) with the KZrCuS$_3$ structure type. In the orthorhombic structure of erbium telluride, distorted octahedra ([ErTe$_6$]$^{9-}$) form two-dimensional layers ($^2_\infty\left\{\left[\text{Er}(\text{Te1})^e_{2/2}(\text{Te2})^k_{4/2}\right]^-\right\}$), while distorted tetrahedra ([CuTe$_4$]$^{7-}$) form one-dimensionally connected substructures ($^1_\infty\left\{\left[\text{Cu}(\text{Te1})^e_{2/2}(\text{Te2})^t_{2/1}\right]^{5-}\right\}$) along the [100] direction. The distorted octahedra and tetrahedra form parallel two-dimensional layers ($^2_\infty\left\{[\text{CuErTe}_3]^{2-}\right\}$) between which Eu^{2+} ions are located in a trigonal-prismatic coordination environment ([EuTe$_6$]$^{10-}$). The trigonal prisms are connected by faces, forming chains ($^1_\infty\left\{\left[\text{Eu}(\text{Te1})_{2/2}(\text{Te2})_{4/2}\right]^{2-}\right\}$) along the [100] direction. Regularities in the variations in structural parameters were established in the series of erbium chalcogenides (EuErCuCh_3 with Ch = S, Se, and Te) and tellurides (EuLnCuTe$_3$ with Ln = Gd, Er, and Lu). Ab-initio calculations of the crystal structure, phonon spectrum, and elastic properties of the compound EuErCuTe$_3$ were performed. The types and wavenumbers of fundamental modes were determined, and the involvement of ions in the IR and Raman modes was assessed. The experimental Raman spectra were interpreted. The telluride EuErCuTe$_3$ at temperatures below 4.2 K was ferrimagnetic, as were the sulfide and selenide derivatives (EuErCuCh_3 with Ch = S and Se). Its experimental magnetic characteristics were close to the calculated ones. The decrease in the magnetic phase transition temperature in the series of the erbium chalcogenides was discovered.

Keywords: quaternary erbium telluride; synthesis; crystal structure; magnetic measurements; DFT calculations

Citation: Ruseikina, A.V.; Grigoriev, M.V.; Locke, R.J.C.; Chernyshev, V.A.; Garmonov, A.A.; Schleid, T. Synthesis, Crystal Structure, and Optical and Magnetic Properties of the New Quaternary Erbium Telluride EuErCuTe$_3$: Experiment and Calculation. *Materials* **2024**, *17*, 2284. https://doi.org/10.3390/ma17102284

Academic Editor: Changho Lee

Received: 4 March 2024
Revised: 22 April 2024
Accepted: 26 April 2024
Published: 11 May 2024

Copyright: © 2024 by the authors. Licensee MDPI, Basel, Switzerland. This article is an open access article distributed under the terms and conditions of the Creative Commons Attribution (CC BY) license (https://creativecommons.org/licenses/by/4.0/).

1. Introduction

Heterometallic chalcogenides based on erbium have been of constant interest to researchers due to their structural possibilities, including their channel, tunnel, and layered structures, as well as their potential valuable optoelectronic, magnetic, semiconductor, and thermoelectric properties [1–13]. Doping erbium chalcogenides improves the optical characteristics of the material in terms of conductivity, reduces the bandgap energy, and enhances electrical conductivity and light absorption. Erbium-doped tellurides have found potential applications in solar cell and optical devices [14]. New heterometallic erbium chalcogenides are formed in ternary systems, such as M_2Ch–EuCh–Er$_2Ch_3$ (M = d-element and Ch = chalcogen) [15–18], and they can combine the properties of corresponding binary phases. Magnetic semiconductors based on europium chalcogenides are attractive due to their wide range of magnetic properties, as they can be ferro-, meta-, or antiferromagnetic [19]. They exhibit high saturation magnetization, a strong magneto-optical effect, magnetoresistance, and spin filtration effects, all arising from the unusual combination of electronic, magnetic, and optical properties in the europium chalcogenides [19–21]. Europium chalcogenides have the potential to create new magnetoelectronic devices, including magnetic random-access memory and magnetic tunneling transistors [19,20]. Bulk EuTe is an antiferromagnetic Heisenberg II type with a Néel temperature of T_N = 9.6 K [21–24] and a wide bandgap semiconductor with a bandgap width of 2.5 eV [24]. The optical and magnetic properties of europium tellurides can be controlled by changing their sizes [21,24,25]. In isolated monolayers, both ferrimagnetic and antiferromagnetic phase transitions can occur [23], and nanoparticles exhibit pronounced superantiferromagnetic transitions between 2 K and 20 K [21].

The compounds EuErCuCh_3 (Ch = S, Se, and Te) contain two magnetically active cations, Eu^{2+} and Er^{3+}, which contribute to the magnetic ordering of the chalcogenides. The magnetic moment of the Er^{3+} ion (9.59 μ$_B$) is larger than that of the Eu^{2+} ion (7.94 μ$_B$), suggesting that the EuErCuCh_3 compounds exhibit ferrimagnetic properties [16,18]. Indeed, the EuErCuCh_3 compounds (Ch = S and Se) undergo a transition from ferrimagnetic to paramagnetic states at 4.8 K [18] and 4.7 K [16], respectively. The synthesis and properties of EuErCuTe$_3$ have not been reported yet. However, based on the magnetic ions involved, a ferrimagnetic transition can be expected in this compound. The orthorhombic EuErCuCh_3 compounds (Ch = S and Se) crystallize in the space group $Pnma$ (a structural type Eu$_2$CuS$_3$) and the space group $Cmcm$ (a structural type KZrCuS$_3$), respectively. Thus, with an increase in the chalcogen radius from r_i(S^{2-}) = 1.84 Å to r_i(Se^{2-}) = 1.98 Å [26], a change in the space group to a higher symmetry can occur. According to L.A. Koscielski's review [17], compounds of this type crystallize only in the $Pnma$ and $Cmcm$ space groups, and so it can be assumed that despite a further increase in the chalcogen radius to r_i(Te^{2-}) = 2.21 Å [26], the EuErCuTe$_3$ compound will crystallize in the $Cmcm$ space group. Quantum-mechanical calculations have been previously performed for the EuErCuTe$_3$ compound, assuming it crystallizes in the KZrCuSe$_3$ structural type [27,28].

Recently, the chalcogenidation of a multi-component oxide mixture obtained by the thermolysis of co-crystallized nitrates [3,29–31] or commercial oxides [31,32] has been actively used for the synthesis of quaternary chalcogenides, providing a high yield of the target phase. However, the toxicity and instability of tellurium hydride limit the applicability of this method for obtaining EuErCuTe$_3$. In our opinion, the most effective and safe method for obtaining this compound would be the halide flux method using elements, which allows for the growth of single crystals in a sealed ampoule, with a relatively low synthesis temperature, and the absence of impurities would enable the study of the physical properties of the samples.

The aim of this study was to synthesize single crystals of the quaternary telluride EuErCuTe$_3$, determine its crystal structure, and investigate its magnetic and optical properties. In addition, computational studies were conducted to shed light on the optical properties of EuErCuTe$_3$.

2. Experimental

2.1. Materials

Eu (99.99%), Er (99.9%), CsI (99.9%), and Te (99.9%) were purchased from ChemPur (Karlsruhe, Germany). Cu (99.999%) was obtained from Aldrich (Milwaukee, WI, USA).

2.2. Synthesis

Single crystals of EuErCuTe$_3$ were synthesized by mixing stoichiometric amounts of copper, tellurium, and lanthanide elements (1 Cu: 1 Eu: 1 Er: 3 Te) in the presence of an excess of CsI halide flux in a glovebox under an argon atmosphere. The weighing of the samples into quartz ampoules was carried out in an inert atmosphere inside the glovebox. The glovebox was used to prevent the interaction of the elemental substances with the oxygen and carbon dioxide in the air and water vapors, which would lead to the formation of thermodynamically stable oxides, carbonates, and hydroxycarbonates at room temperature [33,34]. To prevent the formation of silicate oxides during the synthesis process due to the interactions of the starting components with quartz, a pyrolytic thin layer of amorphous carbon was applied to the inner wall of the quartz ampoules prior to beginning. The quartz ampoules were evacuated to a pressure of 2×10^{-3} mbar, sealed, and heated in a muffle furnace from room temperature to 1120 K. The heating was carried out from room temperature to 1120 K for 30 h and kept at this temperature for 96 h, then it was cooled to 570 K for 140 h, and finally, it was cooled to room temperature for 3 h. The reaction product was purified from the residual flux using demineralized water. The product consisted of black needle-like crystals of $EuErCuTe_3$ with sizes of up to 500 μm (Figure 1). The obtained crystals were suitable for single crystal X-ray diffraction analysis and taking measurements of their magnetic and optical properties. Unfortunately, high-quality powder diffraction patterns could not be obtained, as copper compounds are strong absorbers of molybdenum radiation.

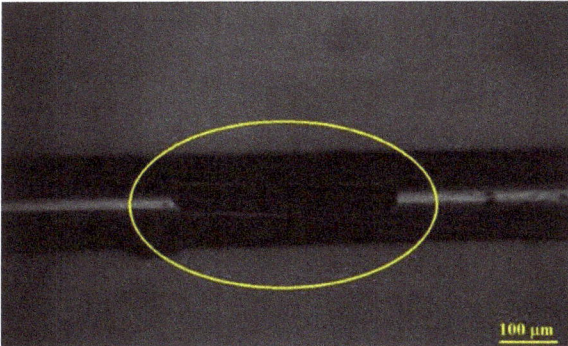

Figure 1. Photograph of an $EuErCuTe_3$ (in yellow circle) crystal placed in a capillary for the X-ray diffraction analysis (the single crystal image was made using a Horiba XploRA Raman spectrometer (HORIBA Scientific, Kyoto, Japan)).

2.3. X-ray Diffraction Analysis

The intensities from a single crystal of the $EuErCuTe_3$ of $0.05 \times 0.05 \times 0.45$ mm^3 dimensions were collected at 293(2) K using a SMART APEX II single-crystal diffractometer (Bruker AXS, Billerica, MA, USA) equipped with a CCD-detector, graphite monochromator, and Mo-K_α radiation source. The parameters of the elementary cell were determined and refined for a set of 11880 reflections. The parameters of the elementary cell corresponded to the orthorhombic crystal system. The space group *Cmcm* was determined from the statistical analysis of all the intensities. Absorption corrections were applied using the SADABS (2008) program. The crystal structure was solved by direct methods using the SHELXS (2013) program and refined in an anisotropic approximation using the SHELXL (2013) program [35]. Structural investigations for the presence of missing symmetry elements were conducted using the PLATON (2009) program [36]. The crystallographic data were deposited in the Cambridge Crystallographic Data Centre. The data can be downloaded from www.ccdc.cam.ac.uk/data_request/cif (accessed on 25 February 2024).

2.4. Electron-Beam Microprobe Analysis

The SEM (scanning electron microscopy) image of the $EuErCuTe_3$ was acquired using an electron-beam X-ray microprobe (SX-100, Cameca, Gennevilliers, France). The EDX (energy dispersive X-ray spectroscopy) spectra for several examples roughly confirmed the 1:1:1:3 stoichiometry of all the investigated EuErCuTe$_3$ compounds.

2.5. Magnetic Measurements

Magnetic measurements were performed with a Quantum Design Magnetic Property Measurement System (MPMS3), San Diego, CA, USA. The SQUID-magnetometer was used to measure the temperature dependence of the EuErCuTe$_3$ sample's magnetic moment under a magnetic field of 40 kA·m^{-1}. These measurements took place in the temperature range 2 to 300 K using the zero-field cooling (ZFC) and field cooling (FC) modes. The isothermal magnetization of up to 4 MA·m^{-1} was measured at 2 K and 300 K.

2.6. Spectroscopy of the Raman Scattering

Raman spectra of the single crystal sample of EuErCuTe$_3$ were acquired using a Horiba XploRa spectrometer (HORIBA Scientific, Kyoto, Japan). The excitation light at a wavelength of 532 nm was used. The acquisition conditions were as follows: filter—10, hole—300, slit—100, and resolution—2400.

2.7. DFT Calculations

Calculations were performed at the theoretical level DFT (density functional theory)/B3LYP. In this approach, we took into account the nonlocality of the exchange interaction, which was necessary for describing the compounds with ionic and covalent bonds. The calculations were carried out in the program CRYSTAL17 [37]. For the rare-earth ions, quasi-relativistic pseudopotentials with attached basis sets were used. We used the pseudopotentials ECP53MWB and ECP57MWB with attached basis sets of the TZVP type for the outer shells ($5s^25p^6$) [38]. The all-electron basis set "Cu_86-4111(41D)G_doll_2000" was used for the copper [37]. For the tellurium, an all-electron basis was also used [39]. Any diffuse functions with exponents smaller than 0.1 were deleted from the basis sets. A self-consistent field was calculated with an accuracy of 10^{-9} a.u. The Monkhorst-Pack mesh used was $8 \times 8 \times 8$.

The use of the pseudopotential for the inner shells of a rare-earth ion reduced the cost of the computer resources and allowed us to calculate the structure and the dynamics of the crystal lattice adequately [15].

The elastic constants and phonon spectrum were calculated for the previously optimized crystal structure.

3. Results

3.1. Crystal Structures of the EuErCuTe$_3$

According to the X-ray crystallographic analysis of the single crystals, the compound EuErCuTe$_3$ crystallized in the orthorhombic space group *Cmcm* with a KZrCuS$_3$ structural type. The crystallographic data, data collection details, atomic coordinates, thermal displacement parameters, bond lengths, and valence angles are presented in Tables 1–4. A similar structural type was observed in the erbium quarter selenide compound EuErCuSe$_3$ [16,40]. The lattice constants obtained from the DFT calculations a = 4.3401 Å, b = 14.2459 Å, and c = 11.2309 Å were in good agreement with those determined experimentally (Table 1).

Table 1. Main parameters of the processing and refinement of the EuErCuTe$_3$ sample.

	EuErCuTe$_3$
Molecular weight	765.56
Space group	Cmcm
Structure type	KZrCuS$_3$
Z	4
a (Å)	4.3086(3)
b (Å)	14.3093(9)
c (Å)	11.1957(7)
V (Å^3)	690.25(8)
ρ_{cal} (g/cm^3)	7.367
μ (mm^{-1})	36.369
Reflections measured	6748
Reflections independent	477
Reflections with $F_o > 4\sigma(F_o)$	411
$2\theta_{max}$ (°)	27.48
h, k, l limits	$-5 \leq h \leq 5, -18 \leq k \leq 18, -14 \leq l \leq 14$
R_{int}	0.070
Refinement results	
Number of refinement parameters	24
R_1 with $F_o > 4\sigma(F_o)$	0.026
wR_2	0.056
Goof	1.084
$\Delta\rho_{max}$(e/Å^3)	1.438
$\Delta\rho_{min}$(e/Å^3)	-1.386
Extinction coefficient, ε	0.0026(2)
CSD-number	2261647

Table 2. Fractional atomic coordinates and isotropic or equivalent isotropic displacement parameters of the EuErCuTe$_3$ sample.

Atom	x/a	y/b	z/c	U_{eq} (Å^2)
Eu	0	0.75485(6)	$1/4$	0.0288(3)
Er	0	0	0	0.0229(2)
Cu	0	0.47077(14)	$1/4$	0.0287(4)
Te1	0	0.08068(6)	$1/4$	0.0204(3)
Te2	0	0.35822(5)	0.06307(6)	0.0218(2)

Table 3. Atomic displacement parameters (Å^2) of the EuErCuTe$_3$ sample.

	U_{11}	U_{22}	U_{33}	U_{12}	U_{13}	U_{23}
Eu	0.0153(4)	0.0231(5)	0.0480(5)	0	0	0
Er	0.0155(3)	0.0205(4)	0.0328(4)	0	0	$-0.0041(3)$
Cu	0.0261(10)	0.0263(10)	0.0337(10)	0	0	0
Te1	0.0166(5)	0.0177(5)	0.0269(5)	0	0	0
Te2	0.0149(3)	0.0185(4)	0.0320(4)	0	0	$-0.0028(3)$

Table 4. Bond lengths (d/Å) and bond angles ($\angle$/°) in the crystal structures of the EuErCuTe$_3$.

Bond lengths

Eu–Te1 [i]	3.294(1)	Er–Te1	3.0277(4)	Cu–Te2	2.641(1)
Eu–Te1 [ii]	3.294(1)	Er–Te1 [v]	3.0277(4)	Cu–Te2 [x]	2.641(1)
Eu–Te2 [iii]	3.3479(6)	Er–Te2 [vi]	3.0423(5)	Cu–Te1 [ii]	2.667(1)
Eu–Te2 [i]	3.3479(6)	Er–Te2 [vii]	3.0423(5)	Cu–Te1 [i]	2.667(1)
Eu–Te2 [ii]	3.3479(6)	Er–Te2 [viii]	3.0423(5)		

23

Table 4. *Cont.*

Bond angles					
Te1 ^{i}–Eu–Te1 ^{ii}	81.68(3)	Te1–Er–Te1 ^{v}	180.0	Te2–Cu–Te2 ^{x}	104.84(7)
Te1 ^{i}–Eu–Te2 ^{iii}	139.03(2)	Te1–Er–Te2 ^{vi}	87.72(2)	Te2–Cu–Te1 ^{ii}	111.08(1)
Te1 ^{ii}–Eu–Te2 ^{iii}	85.03(1)	Te1 ^{v}–Er–Te2 ^{vi}	92.28(2)	Te2 ^{x}–Cu–Te1 ^{ii}	111.08(1)
Te1 ^{i}–Eu–Te2 ^{i}	85.03(1)	Te1–Er–Te2 ^{vii}	92.28(2)	Te2–Cu–Te1 ^{i}	111.08(1)
Te1 ^{ii}–Eu–Te2 ^{i}	139.03(2)	Te1 ^{v}–Er–Te2 ^{vii}	87.72(2)	Te2 ^{x}–Cu–Te1 ^{i}	111.08(1)
Te2 ^{iii}–Eu–Te2 ^{i}	127.56(3)	Te2 ^{vi}–Er–Te2 ^{vii}	180.00(2)	Te1 ^{ii}–Cu–Te1 ^{i}	107.74(8)
Te1 ^{i}–Eu–Te2 ^{ii}	139.03(2)	Te1–Er–Te2 ^{viii}	92.28(2)		
Te1 ^{ii}–Eu–Te2 ^{ii}	85.03(1)	Te1 ^{v}–Er–Te2 ^{viii}	87.72(2)		
Te2 ^{iii}–Eu–Te2 ^{ii}	77.38(2)	Te2 ^{vi}–Er–Te2 ^{viii}	89.84(2)		
Te2 ^{i}–Eu–Te2 ^{ii}	80.10(2)	Te2 ^{vii}–Er–Te2 ^{viii}	90.16(2)		
Te1 ^{i}–Eu–Te2 ^{iv}	85.03(1)	Te1–Er–Te2 ^{ix}	87.72(2)		
Te1 ^{ii}–Eu–Te2 ^{iv}	139.03(2)	Te1 ^{v}–Er–Te2 ^{ix}	92.28(2)		
Te2 ^{iii}–Eu–Te2 ^{iv}	80.10(2)	Te2 ^{vi}–Er–Te2 ^{ix}	90.16(2)		
Te2 ^{i}–Eu–Te2 ^{iv}	77.38(2)	Te2 ^{vii}–Er–Te2 ^{ix}	89.84(2)		
Te2 ^{ii}–Eu–Te2 ^{iv}	127.56(3)	Te2 ^{viii}–Er–Te2 ^{ix}	180.00(2)		

Symmetry codes: (i) $x - 1/2, y + 1/2, z$; (ii) $x + 1/2, y + 1/2, z$; (iii) $x + 1/2, y + 1/2, -z + 1/2$; (iv) $x - 1/2, y + 1/2, -z + 1/2$; (v) $-x, -y, -z$; (vi) $-x + 1/2, -y + 1/2, -z$; (vii) $x - 1/2, y - 1/2, z$; and (viii) $x + 1/2, y - 1/2, z$; (ix) $-x - 1/2, -y + 1/2, -z$; (x) $x, y, -z + 1/2$.

The crystal structure of EuErCuTe$_3$ has a layered-block structure (Figure 2).

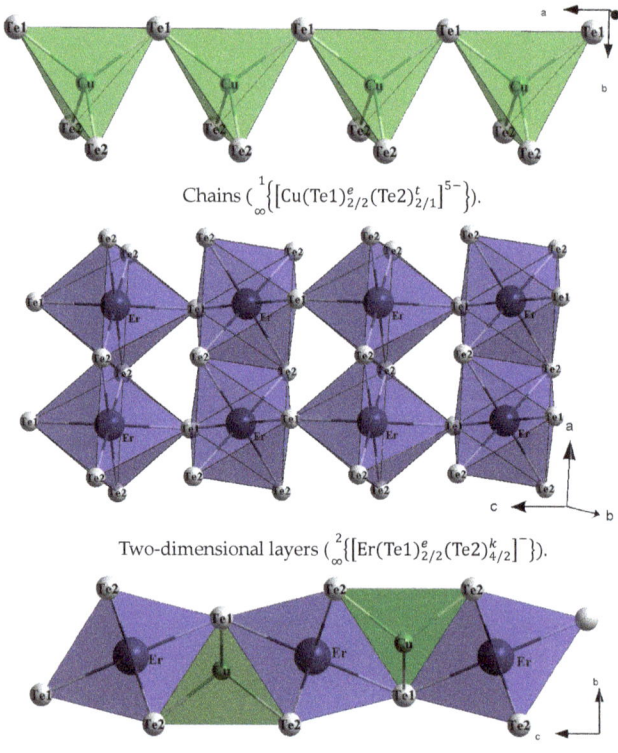

Chains ($^1_\infty\{[Cu(Te1)^e_{2/2}(Te2)^t_{2/1}]^{5-}\}$).

Two-dimensional layers ($^2_\infty\{[Er(Te1)^e_{2/2}(Te2)^k_{4/2}]^-\}$).

Parallel 2D layers ($^2_\infty\{[CuErTe_3]^{2-}\}$).

Figure 2. *Cont.*

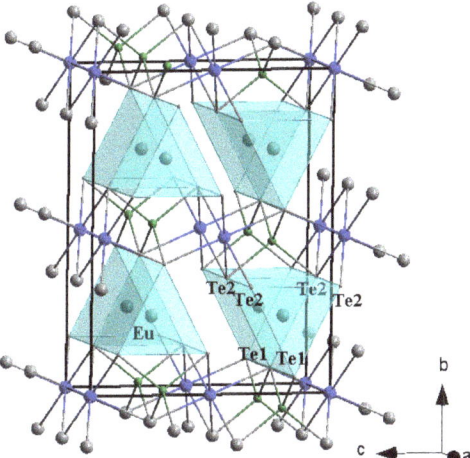

Figure 2. Crystal structure of the EuErCuTe$_3$.

The cations Eu^{2+}, Er^{3+}, and Cu^{+} occupy independent crystallographic positions. In the EuErCuTe$_3$ compound, the structure was formed by distorted tetrahedra ([CuTe$_4$]$^{7-}$), octahedra ([ErTe$_6$]$^{9-}$), and trigonal prisms ([EuTe$_6$]$^{10-}$). The sums of the valence forces for the EuErCuTe$_3$ compound, taking into account coordination, were Eu (1.64), Er (2.99), and Cu (1.42).

Distorted tetrahedra form chains ($\frac{1}{\infty}\left\{\left[\text{Cu}(\text{Te1})_{2/2}^{e}(\text{Te2})_{2/1}^{t}\right]^{5-}\right\}$) along the [100] direction through shared vertex atoms (Te1) (Figure 2). In the tetrahedra, the distances d(Cu–Te) are equal to 2.641(1) Å and 2.667(1) Å (Table 4), indicating a deviation from the theoretical value of d(Cu–Te) = 2.81 Å (calculated based on r_i(Cu$^+$) = 0.6 Å, coordination number (C.N.) = 4; r_i(Te^{2-}) = 2.21 Å) [26]), which is associated with an increase in the covalent component of the chemical bond.

The values of the valence angles ∠(Te–Cu–Te) deviated from the value of the ideal tetrahedral angle (Table 4). The distortions of the tetrahedra in the EuErCuTe$_3$ structure were evaluated using the τ_4 descriptor [41]. The value of τ_4 was 0.978, indicating a distortion in the coordination geometry around the Cu$^+$ by 15% from an ideal tetrahedral structure to a trigonal-pyramidal structure. The degree of distortion in the tetrahedral polyhedra in the telluride EuErCuTe$_3$ was higher than that in the selenide EuErCuSe$_3$ [16] and sulfide EuErCuS$_3$ [15], for which the values of τ_4 were 0.984 and 0.986, and the distortion indexes were 11% and 9%, respectively.

Between the chains of tetrahedra, there were distorted octahedra ([ErTe$_6$]$^{9-}$) with bond lengths of 3.0277(4) Å and 3.0423(5) Å (Table 4) compared to the theoretical value of d(Er–Te) = 3.1 Å (r_i(Er^{3+}) = 0.89 Å, C.N. = 4) [26]. The values of the valence angles ∠(Te–Er–Te) deviated from the value of the ideal octahedral angle (Table 4). The distorted octahedra formed two-dimensional layers ($\frac{2}{\infty}\left\{\left[\text{Er}(\text{Te1})_{2/2}^{e}(\text{Te2})_{4/2}^{k}\right]^{-}\right\}$) through the shared vertex atoms (Te1) along the [001] direction and the shared edges (Te2Te2). The distorted octahedra and tetrahedra formed parallel two-dimensional layers ($\frac{2}{\infty}\left\{[\text{CuErTe}_3]^{2-}\right\}$) in the [101] plane. Between these layers, there were trigonal prisms that were connected by the faces Te2Te2Te1, forming chains ($\frac{1}{\infty}\left\{\left[\text{Eu}(\text{Te1})_{2/2}(\text{Te2})_{4/2}\right]^{2-}\right\}$) along the [100] direction. The theoretical value of d(Eu–Te) is 3.38 Å (r_i(Eu^{2+}) = 0.89 Å, C.N. = 4) [26]. In the EuErCuTe$_3$ structure, six Eu–Te distances were shorter than 3.35 Å, while the seventh and eighth distances were larger than the theoretical value and measured 3.86 Å, which was not accounted for in the coordination polyhedron due to weak interactions.

Thus, the three-dimensional crystal structure of EuErCuTe$_3$ was formed by two-dimensional layers consisting of octahedra and tetrahedra in the *bc* plane separated by one-dimensional chains of trigonal prisms.

In the series of erbium chalcogenides (EuErCuCh_3 (Ch = S, Se, and Te)), a change in the space group from *Pnma* (EuErCuS$_3$) to *Cmcm* (EuErCuSe$_3$ and EuErCuTe$_3$) and a change in structural type from Eu$_2$CuS$_3$ to KZrCuS$_3$, respectively, were observed. As the chalcogen radius increased in the compounds (EuErCuCh_3 (Ch = S [15], Se [16], and Te (Table 1))), an increase in the unit cell parameters was observed, as follows:

$a_{Pnma}(c_{Cmcm})$ = 10.1005(2) Å ($a_{EuErCuS_3}$) → 10.4602(7) Å ($c_{EuErCuSe_3}$) → 11.1957(7) Å ($c_{EuErCuTe_3}$);
$b_{Pnma}(a_{Cmcm})$ = 3.91255(4) Å ($a_{EuErCuS_3}$) → 4.0555(3) Å ($c_{EuErCuSe_3}$) → 4.3086(3) Å ($c_{EuErCuTe_3}$);
$c_{Pnma}(b_{Cmcm})$ = 12.8480(2) Å ($a_{EuErCuS_3}$) → 13.3570(9) Å ($c_{EuErCuSe_3}$) → 14.3093(9) Å ($c_{EuErCuTe_3}$).

Correspondingly, the volume of the unit cell increased as follows: 507.737(14) Å^3 (EuErCuS$_3$) [15] → 566.62(6) Å^3 (EuErCuSe$_3$) [16] → 690.25(8) Å^3 (EuErCuTe$_3$) (Table 1).

A regular increase in the average metal-chalcogen bond lengths was also observed in the chalcogenide series. For example, we observed the following:

d(Eu–Ch): 3.060 Å (Ch = S) → 3.130 Å (Ch = Se) → 3.330 Å (Ch = Te);
d(Er–Ch): 2.723 Å (Ch = S) → 2.730 Å (Ch = Se) → 3.037 Å (Ch = Te);
d(Cu–Ch): 2.350 Å (Ch = S) → 2.468 Å (Ch = Se) → 2.654 Å (Ch = Te).

Thus, the increase in the chalcogen radius led to the transformation of the local geometry around the Eu^{2+}, resulting in a change in the type of its coordination polyhedron from a one-capped trigonal prism in EuErCuS$_3$ [15] to a trigonal prism in EuErCuSe$_3$ and EuErCuTe$_3$, a change in the structural type and space group in the series of EuErCuCh_3 compounds, and an increase in structural parameters.

When comparing the structural parameters of the EuErCuTe$_3$ compound with the already known europium tellurides EuGdCuTe$_3$ and EuLuCuTe$_3$ [42], their decrease was observed with a decrease in the ionic radius of Ln^{3+} (r_i(Gd^{3+}) = 0.938 Å > r_i(Er^{3+}) = 0.89 Å > r_i(Lu^{3+}) = 0.861 Å) [26]. The unit cell parameters of EuLnCuTe$_3$ (Ln = Gd [42], Er (Table 1), Lu) [42] changed as follows:

$a_{Pnma}(c_{Cmcm})$ = 11.3761(7) Å ($a_{EuGdCuTe_3}$) → 11.1957(7) Å ($c_{EuErCuTe_3}$) → 11.1174(7) Å ($c_{EuLuCuTe_3}$);
$b_{Pnma}(a_{Cmcm})$ = 4.3405(3) Å ($a_{EuGdCuTe_3}$) → 4.3086(3) Å ($c_{EuErCuTe_3}$) → 4.2937(9) Å ($c_{EuLuCuTe_3}$);
$c_{Pnma}(b_{Cmcm})$ = 14.3469(9) Å ($a_{EuGdCuTe_3}$) → 14.3093(9) Å ($c_{EuErCuTe_3}$) → 14.2876(9) Å ($c_{EuLuCuTe_3}$).

The volume of the unit cell decreased from 708.42(8) to 682.02(8) Å^3 in the EuLnCuTe$_3$ (Ln = Gd, Er, and Lu). Additionally, a consistent decrease in the average metal-tellurium bond length was observed in this series. Thus, we observed the following:

d(Eu–Te): 3.377 Å → 3.330 Å → 3.332 Å;
d(Ln–Te): 3.081 Å → 3.037 Å → 3.017 Å;
d(Cu–Te): 2.666 Å → 2.654 Å → 2.648 Å.

Thus, the decrease in the lanthanide radius in the series of EuLnCuTe$_3$ compounds (Ln = Gd, Er and Lu) was accompanied by a change in the structural type and space group, as well as decreases in the structural parameters.

3.2. Magnetic Properties of the EuErCuTe$_3$

The experimental field dependence of the magnetic moment of the sample at a temperature of 300 K had a linear form, characteristic of paramagnetic materials (Figure 3a). From this dependence, assuming the validity of the Curie law $m = HCT^{-1}$, the Curie constant C_{300K} = 0.232 m^3·K·kmol^{-1} and the corresponding effective magnetic moment μ_{300K} = 12.14 μ_B were calculated.

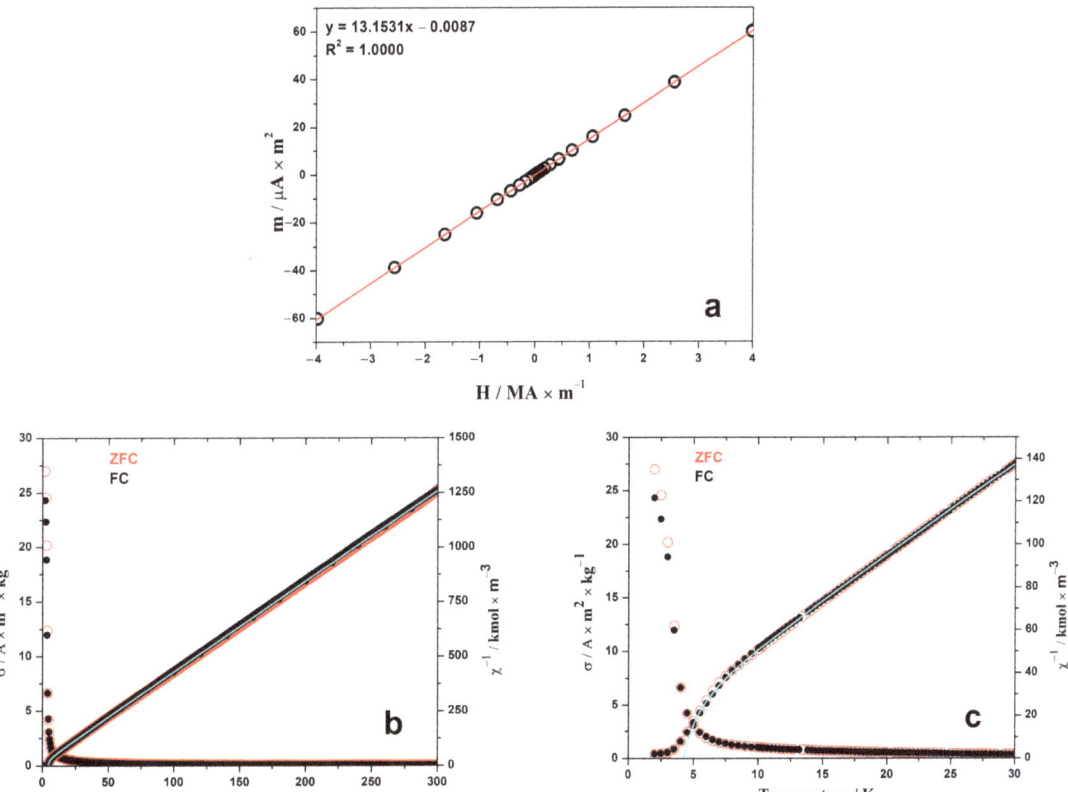

Figure 3. Field-dependent magnetization at 300 K (a) and temperature-dependent specific magnetization and reciprocal molar magnetic susceptibility at 40 MA m^{-1} (b,c) of EuErCuTe$_3$ sample. The measurements' temperature-dependent magnetizations were performed in the zero-field cooled (ZFC) and nonzero-field cooled (FC) modes.

Given the experimental data on the temperature-dependent magnetic moment of the EuErCuTe$_3$ sample, the temperature-dependent inverse molar magnetic susceptibility (Figure 3b,c) was calculated. Taking as a first approximation the Curie-Weiss law ($\chi^{-1} = C^{-1}(T - \theta_p)$) at temperatures from 50 to 300 K, the Curie constant $C_{50-300K}$ = 0.241 m^3·K·kmol^{-1} and the Weiss constant (Curie paramagnetic temperature) $\theta_p = -4.0$ K, as well as the effective magnetic moment $\mu_{50-300K}$ = 12.39 μ_B, were obtained.

The paramagnetic parameters of the EuErCuTe$_3$ compound corresponded well to the calculated parameters of the free ions ($\mu = \sqrt{7.94^2 + 9.58^2} = 12.44$ μ_B, C = 0.243 m^3·K·kmol^{-1}). However, the Curie paramagnetic temperature of this sample was negative, and the temperature dependence of the inverse susceptibility at low temperatures (Figure 3b,c) had a form characteristic of ferrimagnetic compounds. Therefore, to approximate this dependence in the temperature range 5 to 300 K, the Néel formula for a two-sublattice ferrimagnet model was used ($\chi^{-1} = T/C + \chi_0^{-1} - \sigma/(T - \theta)$). The calculations showed very good agreement between the experimental points and the theoretical model (Figure 3). The best fit was obtained with the following parameter values: C = 0.241 m^3·K·kmol^{-1}, χ_0^{-1} = 14.5 kmol·m^{-3}, σ = 37 kmol·K·m^{-1}, and θ = 3.1 K. Based on these data, the Néel temperature was determined at χ^{-1} = 0 using the following formula: $T_c = (\theta - C/\chi_0 + ((\theta - C/\chi_0)^2 + 4C(\theta/\chi_0 + \sigma))^{0.5})/2 = 4.2$ K. This value was close to the observed temperature at the divergence point of the magnetization curves for the FC and ZFC modes. Similar values for the phase transition points were obtained for EuErCuSe$_3$ (4.7 K [16]) and EuErCuS$_3$ (5.0 K [18]).

The magnetization in Bohr magnetons per formula unit plotted at 2 K (Figure 4) confirmed the conclusion regarding the ferrimagnetic structure of the moments in this compound. The shape of this curve exhibits the metamagnetic behavior of the moments. In the external fields of 0.03 and approximately 1 MA m^{-1}, jumps in susceptibility were observed, indicating a change in the magnetic structure. The saturation, which theoretically should approach the value of gS(Eu^{2+}) + g$_J$J(Er^{3+}) = 7 µ$_B$ + 9 µ$_B$ = 16 µ$_B$, was not achieved until reaching 4 MA·m^{-1}.

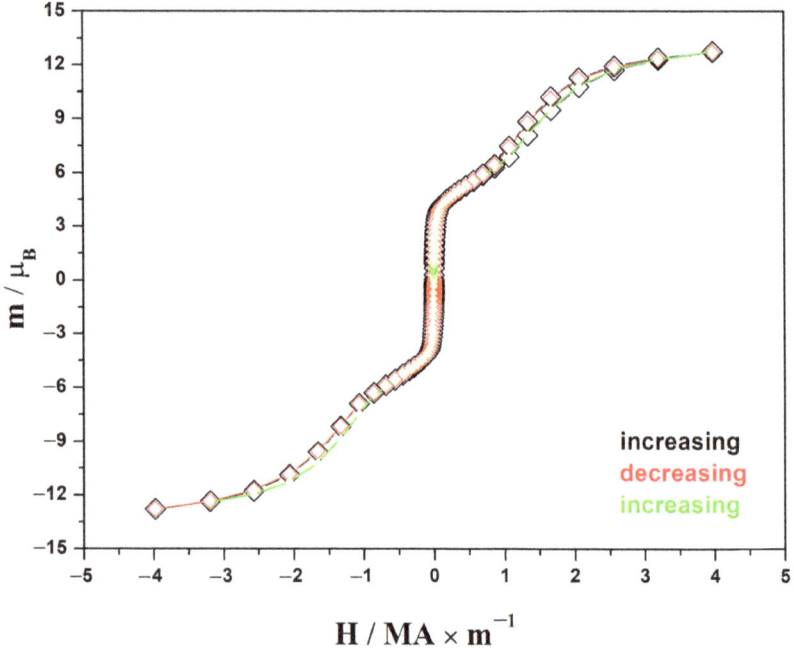

Figure 4. Magnetization curves of the EuErCuTe$_3$ sample at 2 K.

The magnetic properties of the EuErCuTe$_3$ compound were similar to those of EuGdCuTe$_3$. Both are ferrimagnets at low temperatures (lower than 4.2 K for the first and 7.9 K for the second), unlike EuLuCuTe$_3$, which is ferromagnetic under a temperature of 3.0 K.

3.3. Band Structure of EuErCuTe$_3$

We used the points Γ(0,0,0), Y($^1/_2$,$^1/_2$,0), T($^1/_2$,$^1/_2$,$^1/_2$), Z(0,0,$^1/_2$), S (0,$^1/_2$,0), and R(0,$^1/_2$,$^1/_2$) at the Brillouin zone of the space group *Cmcm*. The band structure (Figure 5) did not include the 4f states of erbium and europium since they were replaced by a pseudopotential.

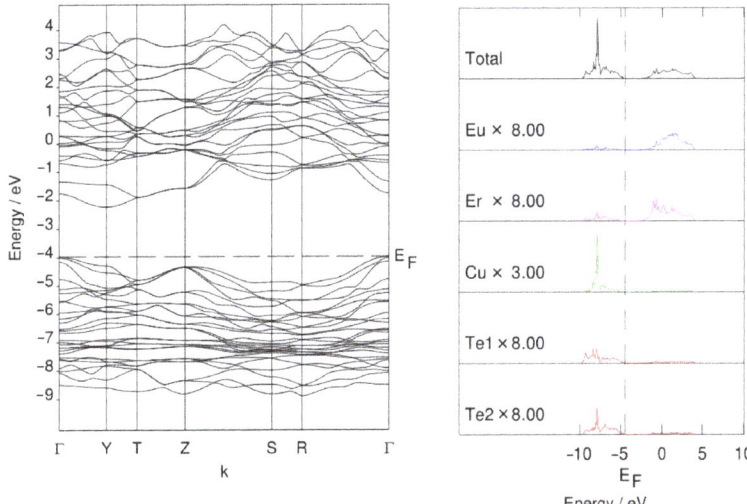

Figure 5. Electronic band structure of EuErCuTe$_3$.

As can be seen from the figure, copper and tellurium orbitals are the main contributions to the states near the top of the VB. Orbitals of erbium and europium are the main contributions to the bottom of the CB. The calculations predicted for EuErCuTe$_3$ an indirect band gap value of 1.75 eV (it was a HOMO–LUMO estimation). This value was close to the experimental data values for isostructural quaternary chalcogenides [16,18,43]. In the series EuErCuCh$_3$ (Ch = S, Se, and Te), decreases in the band gap widths of the compounds were observed (1.93 eV (EuErCuS$_3$) [18] → 1.79 eV (EuErCuSe$_3$) [16] → 1.75 eV (EuErCuTe$_3$)), which was consistent with the data on the narrowing of the band gap in the chalcogenide series [44]. Isostructural europium chalcogenides have lower band gap values compared to strontium chalcogenides in the space group $Cmcm$, for example, in SrRECuSe$_3$ (RE = Ho – Lu), the values range from 2.03–2.21 eV [43]. The narrower band gap of EuRECuCh$_3$ was explained by the presence of a $4f$–$5d$ transition in the Eu^{2+} ion [45].

3.4. Elastic Constants and Elastic Modulus

The elastic constants and the elastic modulus of the compound EuErCuTe$_3$ are presented in Table 5. This table presents the bulk module (B), shear module (G), Young's modulus (G), and Poisson's ratio. These are values for a polycrystal, and they are calculated by averaging the schemes of Voigt, Reuss, and Hill. The Voigt scheme assumes the uniformity of local strains. The Reuss scheme assumes the uniformity of local stresses. The Voigt scheme provides the upper bound, while the Reuss scheme provides the lower bound of the value. The Hill approximation provides the arithmetical average of the Voigt and Reuss values [46,47].

Table 5. The elastic constants and modulus, and also the Vickers hardness (GPa), of EuErCuTe$_3$.

C_{11}	C_{12}	C_{13}	C_{22}	C_{23}	C_{33}	C_{44}	C_{55}	C_{66}	Averaging Scheme	B	G	Y	Poisson's Ratio	H_V
122	52	40	95	47	113	10	31	45	Voigt	68	30	78	0.307	
									Reuss	67	23	61	0.349	3.2
									Hill	67	26	70	0.328	

The Voigt and Reuss estimates were very different (Table 5), which indicated the anisotropy of the elastic properties. The dependence of the Young's modulus on direction also illustrated the strong anisotropy of the elastic properties (Figure 6).

$$H_V = 0.92 \left(\frac{G}{B}\right)^{1.137} G^{0.708} \quad (1)$$

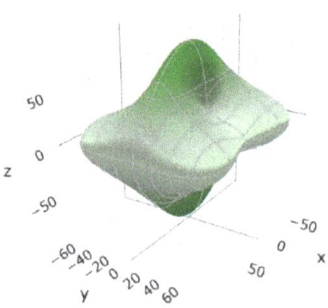

Figure 6. The Young's modulus (GPa) and its dependence on direction in the crystal EuErCuTe$_3$.

The empirical Formula (1) was used to calculate the hardness. According to [48] the formula was based on correlations between the Vickers hardness (H_V) and the ratio of the shear and bulk moduli. The parameters of the formula were determined from reproducing the hardness of more than forty compounds with ionic and covalent bonds [48]. In (1), the shear (G) and bulk (B) moduli were determined by the Hill estimate.

3.5. Raman, IR, and Phonon Spectra

From the DFT calculations, the wavenumbers and types of modes were determined (Table 6). From the calculations, displacement vectors were obtained. This made it possible to evaluate the participation of each ion in a particular mode. The values of the ion displacements characterized their participation in the modes (Figure 7).

Table 6. Phonons at the gamma point of the EuErCuTe$_3$.

Frequency, cm^{-1}	Type	IR Active/Inactive	IR Intensity IR (km·mol^{-1})	Raman Active/Inactive	Raman Intensity Raman (Arbitrary Units)	Involved Ions [1]
33	B$_{1u}$	A	10	I		Eu^S, Er^S, Cu^W, Te1^S, Te2^S
48	A$_u$	I	0	I		Er^S, Te2^S
54	B$_{1g}$	I	0	A	358	Eu^S, Cu^S, Te1^S, Te2^W
59	A$_g$	I	0	A	582	Eu^S, Cu^S, Te1^S, Te2
61	B$_{2g}$	I	0	A	287	Eu^S, Cu^W, Te2
67	B$_{2u}$	A	18	I		Eu, Er^S, Cu^S, Te1^S, Te2^W
71	B$_{1u}$	A	14	I		Eu^W, Er^S, Cu^S, Te1, Te2
81	B$_{3u}$	A	49	I		Eu, Er, Cu^S, Te1^W, Te2
82	B$_{2g}$	I	0	A	415	Eu, Cu^S, Te2
86	A$_g$	I	0	A	136	Eu^S, Cu^S, Te2
87	B$_{1g}$	I	0	A	202	Eu^S, Cu^S, Te1, Te2
88	B$_{1u}$	A	90	I		Eu^S, Er, Cu, Te2
90	B$_{2u}$	A	2	I		Eu^S, Er, Cu^S, Te1^W

Table 6. Cont.

Frequency, cm^{-1}	Type	IR Active/Inactive	IR Intensity IR (km·mol^{-1})	Raman Active/Inactive	Raman Intensity Raman (Arbitrary Units)	Involved Ions [1]
92	B$_{3u}$	A	115	I		Eu ^{S}, Er, Cu ^{W}, Te1 ^{W}, Te2
112	B$_{3u}$	A	16	I		Eu ^{W}, Er ^{S}, Cu ^{S}, Te1
113	B$_{3g}$	I	0	A	132	Te2 ^{S}
117	A$_u$	I	0	I		Er, Te2
119	B$_{1g}$	I	0	A	25	Eu ^{W}, Cu, Te1 ^{W}, Te2
120	B$_{1u}$	A	374	I		Er, Cu ^{S}, Te1 ^{W}, Te2
122.66	B$_{2u}$	A	495	I		Eu, Er, Te2
123.50	B$_{2g}$	I	0	A	1.5	Eu, Cu, Te1, Te2
131.92	B$_{3u}$	A	79	I		Eu, Er ^{W}, Cu, Te1 ^{S}, Te2
131.95	B$_{1g}$	I	0	A	47	Eu, Cu ^{S}, Te1 ^{S}, Te2 ^{W}
131.97	A$_g$	I	0	A	119	Eu, Te1, Te2
132.07	B$_{2u}$	A	135	I		Eu ^{W}, Cu ^{S}, Te1 ^{S}
138	A$_g$	I	0	A	51	Cu ^{S}, Te1, Te2 ^{W}
145	B$_{2g}$	I	0	A	116	Eu ^{W}, Cu ^{S}, Te1 ^{S}, Te2 ^{W}
146	A$_g$	I	0	A	1000	Cu, Te1, Te2
148	B$_{1u}$	A	5	I		Er, Cu, Te2
151	B$_{2g}$	I	0	A	59	Cu, Te1, Te2
154	B$_{1u}$	A	137	I		Er, Cu ^{W}, Te1 ^{S}, Te2 ^{W}
155	B$_{3u}$	A	185	I		Er, Cu, Te2
159	B$_{3u}$	A	2	I		Er, Cu, Te1, Te2

[1] The superscripts "S" and "W" denote strong and weak ion displacements in the modes, respectively.

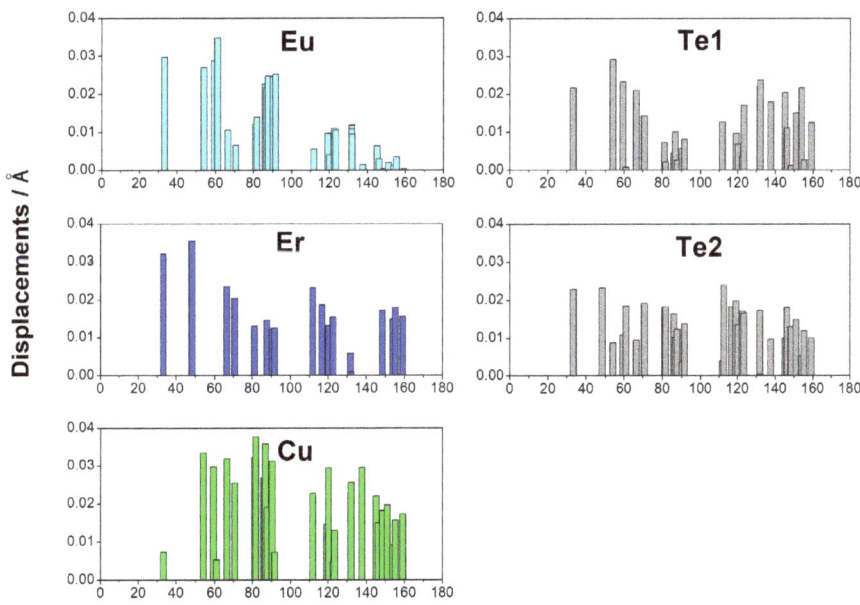

Figure 7. The values of the ion displacements at the phonon modes in the EuErCuTe$_3$ (space group: Cmcm).

According to the calculations, the phonon spectrum of the crystal at the gamma point lied in the frequency range of up to 170 cm^{-1}. In this frequency range, not only light copper ions are involved but also tellurium and erbium ions.

A strong mixing of vibrations in the structural units in the crystal EuErCuSe$_3$ could be noted. In crystal EuErCuTe$_3$, europium ions participate in the frequency range of up to ~95 cm^{-1}. The calculations predicted a gap in the phonon spectrum in the region ~95–110 cm^{-1} (Figure 7). The calculations predicted that in the crystal EuErCuSe$_3$, the most intense Raman mode had a frequency of approximately 146 cm^{-1} (A_{1g}) and the most intense infrared mode had a frequency of approximately 123 cm^{-1} (B_{2u}). These modes are illustrated in Figure 8.

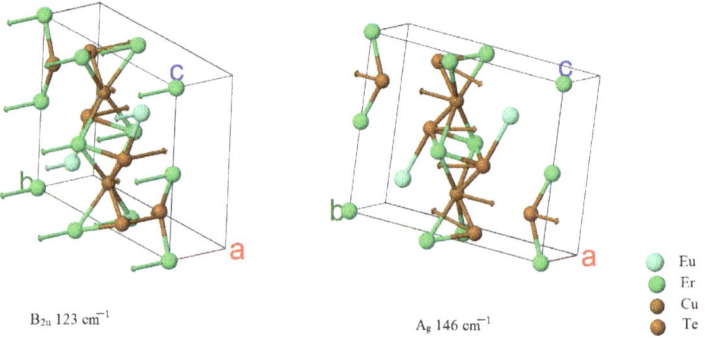

Figure 8. Ion displacements in the IR and Raman modes with maximum intensity.

The calculated Raman spectrum in comparison with the experimental one is shown in Figure 9. The results of calculating the phonon spectrum can be useful for interpreting IR and Raman spectra of the rare earth tellurides in EuLnCuTe$_3$.

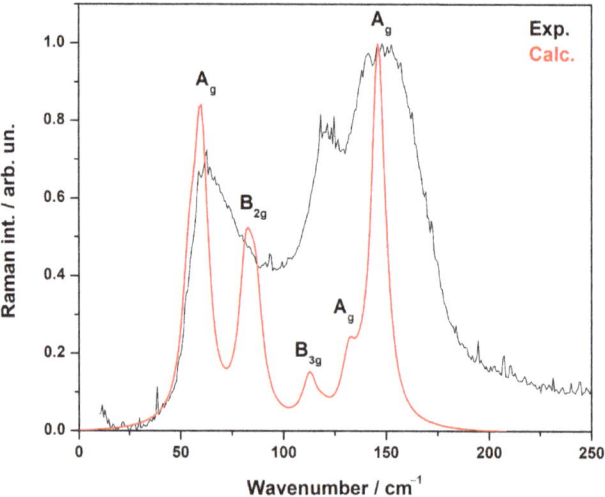

Figure 9. Raman spectrum modeling results. The calculations were carried out for the exciting laser wavelengths λ = 532 nm and T = 300 K.

The largest ion displacement was 0.038 Å. In the case when the displacement was greater than or equal to 0.02 Å, the displacement was indicated by "S". If the displacement did not exceed 0.01 Å, then the displacement was indicated by "W". If the value of the displacement was less than 0.005 Å, then the ion was not mentioned in the column "Participants".

4. Conclusions

This article discusses the synthesis, structure, and optical and magnetic properties of the new complex telluride EuErCuTe$_3$. The compound crystallizes in the KZrCuS$_3$ structure type. Its crystal structure is built from distorted [ErTe$_6$]$^{9-}$ octahedra and [CuTe$_4$]$^{7-}$ tetrahedra, forming two-dimensional layers. Trigonal prisms of [EuTe$_6$]$^{10-}$ are located between the layers. In the series of EuErCuCh_3 chalcogenides (Ch = S, Se, and Te), a change in the coordination polyhedron of Eu^{2+} was observed, along with a change in the structural type and space group and increases in the structural parameters. EuErCuTe$_3$ contains two magnetic ions, Eu^{2+} and Er^{3+}, and it undergoes a ferrimagnetic transition at 4.2 K. The obtained results correlate with the observed ferrimagnetic ordering in erbium sulfide and selenide EuErCuCh_3 (Ch = S and Se). EuErCuTe$_3$ is paramagnetic in the temperature range 300 K to 4.2 K. Within the framework of the DFT approach, the crystal structure and the IR, Raman, and "silent" modes were studied. The elastic constants and elastic moduli were calculated. The experimental Raman spectrum of the synthesized sample was interpreted using the calculated Raman spectra of the EuErCuTe$_3$. The theoretical calculations also allowed us to assign vibrational modes as well as to reveal the involved ions responsible for these modes.

Author Contributions: Conceptualization, A.V.R., M.V.G. and T.S.; software, A.V.R. and M.V.G.; validation, A.V.R.; formal analysis, M.V.G., A.A.G., V.A.C. and R.J.C.L.; data curation and DFT calculations, V.A.C., M.V.G., A.V.R. and T.S.; writing—original draft preparation, A.A.G., A.V.R., T.S., V.A.C. and M.V.G.; writing—review and editing, A.V.R., M.V.G., A.A.G., V.A.C. and T.S.; visualization, A.A.G. and M.V.G.; project administration, A.V.R.; funding acquisition, M.V.G. and A.V.R. All authors have read and agreed to the published version of the manuscript.

Funding: This research was funded by grant support from the Russian Science Foundation, grant number 24-23-00416 (https://rscf.ru/project/24-23-00416/ (accessed on 25 April 2024)).

Institutional Review Board Statement: Not applicable.

Informed Consent Statement: Not applicable.

Data Availability Statement: Data are available from the authors on request.

Conflicts of Interest: The authors declare no conflicts of interest.

References

1. Duczmal, M.; Pawlak, L. Magnetic properties of TlLnS$_2$ compounds (Ln = Nd, Gd, Dy, Er and Yb). *J. Alloys Compd.* **1994**, *209*, 271–274. [CrossRef]
2. Ahmed, N.; Nisar, J.; Kouser, R.; Nabi, A.G.; Mukhtar, S.; Saeed, Y.; Nasim, M.H. Study of electronic, magnetic and optical prop-erties of KMS_2 (M = Nd, Ho, Er and Lu): First principle calculations. *Mater. Res. Expr.* **2017**, *4*, 065903. [CrossRef]
3. Azarapin, N.O. Synthesis, Structure and Properties of Compounds BaRECuS$_3$ (RE = Rare Earth Element). Ph.D. Thesis, University of Tyumen, Tyumen, Russia, 2022.
4. Esmaeili, M.; Forbes, S.; Tseng, Y.C.; Mozharivskyj, Y. Crystal Structure, Electronic and Physical Properties of Monoclinic RECuTe$_2$ in Contrast to RECuSe$_2$ (RE: Pr, Sm, Gd, Dy and Er). *Solid State Sci.* **2014**, *36*, 89–93. [CrossRef]
5. Esmaeili, M.; Tseng, Y.-C.; Mozharivskyj, Y. Thermoelectric properties, crystal and electronic structure of semiconducting RECuSe$_2$ (RE = Pr, Sm, Gd, Dy and Er). *J. Alloys Compd.* **2014**, *610*, 555–560. [CrossRef]
6. Yao, J.; Deng, B.; Sherry, L.J.; McFarland, A.D.; Ellis, D.E.; van Duyne, R.P.; Ibers, J.A. Syntheses, structure, some band gaps, and electronic structures of CsLnZnTe$_3$ (Ln = La, Pr, Nd, Sm, Gd, Tb, Dy, Ho, Er, Tm, Y). *Inorg. Chem.* **2004**, *43*, 7735–7740. [CrossRef] [PubMed]
7. Mitchell, K.; Huang, F.Q.; Caspi, E.N.; McFarland, A.D.; Haynes, C.L.; Somers, R.C.; Jorgensen, J.D.; Van Duyne, R.P.; Ibers, J.A. Syntheses, structure, and selected physical properties of CsLnMnSe$_3$ (Ln = Sm, Gd, Tb, Dy, Ho, Er, Tm, Yb, Y) and AYbZnQ_3 (A = Rb, Cs; Q = S, Se, Te). *Inorg. Chem.* **2004**, *43*, 1082–1089. [CrossRef]
8. Yin, W.; Wang, W.; Bai, L.; Feng, K.; Shi, Y.; Hao, W.; Yao, J.; Wu, Y. Syntheses, structures, physical properties, and electronic structures of Ba$_2M Ln$Te$_5$ (M = Ga and Ln = Sm, Gd, Dy, Er, Y; M = In and Ln = Ce, Nd, Sm, Gd, Dy, Er, Y). *Inorg. Chem.* **2012**, *51*, 11736–11744. [CrossRef]
9. Yin, W.; Feng, K.; Wang, W.; Shi, Y.; Hao, W.; Yao, J.; Wu, Y. Syntheses, structures, optical and magnetic properties of Ba$_2MLn$Se$_5$ (M = Ga, In; Ln = Y, Nd, Sm, Gd, Dy, Er). *Inorg. Chem.* **2012**, *51*, 6860–6867. [CrossRef]

10. Ruseikina, A.V.; Andreev, O.V.; Galenko, E.O.; Koltsov, S.I. Trends in thermodynamic parameters of phase transitions of lan-thanide sulfides SrLnCuS$_3$ (Ln = La–Lu). *J. Therm. Anal. Calorim.* **2017**, *128*, 993–999. [CrossRef]
11. Ruseikina, A.V.; Solovyov, L.A. Grigoriev, M.V.; Andreev, O.V. Crystal structure variations in the series SrLnCuS$_3$ (Ln = La, Pr, Sm, Gd, Er and Lu). *Acta Crystallogr.* **2019**, *C75*, 584–588.
12. Ruseikina, A.V.; Soloviev, L.A.; Galenko, E.O.; Grigoriev, M.V. Refined Crystal Structures of SrLnCuS$_3$ (Ln = Er, Yb). *Russ. J. Inorg. Chem.* **2018**, *63*, 1225–1231. [CrossRef]
13. Huang, F.Q.; Ibers, J.A. Syntheses and structures of the quaternary copper tellurides K$_3Ln_4$Cu$_5$Te$_{10}$ (Ln = Sm, Gd, Er), Rb$_3Ln_4$Cu$_5$Te$_{10}$ (Ln = Nd, Gd), and Cs$_3$Gd$_4$Cu$_5$Te$_{10}$. *J. Solid State Chem.* **2001**, *160*, 409–414. [CrossRef]
14. Ikhioya, I.L.; Nkele, A.C.; Chigozirim, E.M.; Aisida, S.O.; Maaza, M.; Ezema, F.I. Effects of Erbium on the Properties of Electro-chemically-Deposited Zirconium Telluride Thin Films. *Nanoarchitectonics* **2020**, *2*, 110–118.
15. Ruseikina, A.V.; Chernyshev, V.A.; Velikanov, D.A.; Aleksandrovsky, A.S.; Shestakov, N.P.; Molokeev, M.S.; Grigoriev, M.V.; Andreev, O.V.; Garmonov, A.A.; Matigorov, A.V.; et al. Regularities of the property changes in the compounds EuLnCuS$_3$ (Ln = La–Lu). *J. Alloys Compd.* **2021**, *874*, 159968. [CrossRef]
16. Andreev, O.V.; Atuchin, V.V.; Aleksandrovsky, A.S.; Denisenko, Y.G.; Zakharov, B.A.; Tyutyunnik, A.P.; Habibullayev, N.N.; Velikanov, D.A.; Ulybin, D.A.; Shpindyuk, D.D. Synthesis, structure, and properties of EuLnCuSe$_3$ (Ln = Nd, Sm, Gd, Er). *Crystals* **2022**, *12*, 17. [CrossRef]
17. Koscielski, L.A.; Ibers, J.A. The Structural Chemistry of Quaternary Chalcogenides of the Type $AMM'Q_3$. *Z. Anorg. Allg. Chem.* **2012**, *638*, 2585–2593. [CrossRef]
18. Ruseikina, A.V.; Solovyov, L.A.; Chernyshev, V.A.; Aleksandrovsky, A.S.; Andreev, O.V.; Krylova, S.N.; Krylov, A.S.; Velikanov, D.A.; Molokeev, M.S.; Maximov, N.G.; et al. Synthesis, structure, and properties of EuErCuS$_3$. *J. Alloys Compd.* **2019**, *805*, 779–788. [CrossRef]
19. Boncher, W.; Dalafu, H.; Rosa, N.; Stoll, S. Europium chalcogenide magnetic semiconductor nanostructures. *Coord. Chem. Rev.* **2015**, *289*, 279–288. [CrossRef]
20. Wolf, M.; Sürgers, C.; Fischer, G.; Scherer, T.; Beckmann, D. Fabrication and magnetic characterization of nanometer-sized ellipses of the ferromagnetic insulator EuS. *J. Magn. Magn. Mater.* **2014**, *368*, 49–53. [CrossRef]
21. He, W.; Somarajan, S.; Koktysh, D.S.; Dickerson, J.H. Superantiferromagnetic EuTe nanoparticles: Room temperature colloidal synthesis, structural characterization, and magnetic properties. *Nanoscale* **2010**, *3*, 184–187. [CrossRef]
22. Oliveira, N.F., Jr.; Foner, S.; Shapira, Y.; Reed, T.B. EuTe. I. Magnetic Behavior of Insulating and Conducting Single Crystals. *Phys. Rev.* **1972**, *B5*, 2634. [CrossRef]
23. Chen, J.; Dresselhaus, G.; Dresselhaus, M.; Springholz, G.; Bauer, G. Magnetic Properties of Heisenberg Antiferromagnetic EuTe/PbTe Superlattices. *MRS Proc.* **1994**, *358*. [CrossRef]
24. Kępa, H.; Springholz, G.; Giebultowicz, T.M.; Goldman, K.I.; Majkrzak, C.F.; Kacman, P.; Blinowski, J.; Holl, S.; Krenn, H.; Bauer, G. Magnetic interactions in EuTe epitaxial layers and EuTe/PbTe superlattices. *Phys. Rev. B* **2003**, *68*, 024419. [CrossRef]
25. Li, Y.; Liu, J.; Zhang, P.; Jing, Q.; Liu, X.; Zhang, J.; Xiao, N.; Yu, L.; Niu, P. Electrical transport properties of EuTe under high pressure. *J. Mater. Chem. C* **2021**, *9*, 17371–17381. [CrossRef]
26. Shannon, R.D. Revised effective ionic radii and systematic studies of interatomic distances in halides and chalcogenides. *Acta Crystallogr.* **1976**, *A32*, 751–766. [CrossRef]
27. Pal, K.; Hua, X.; Xia, Y.; Wolverton, C. Unraveling the Structure-Valence-Property Relationships in $AMM'Q_3$ Chalcogenides with Promising Thermoelectric Performance. *ACS Appl. Energy Mater.* **2019**, *3*, 2110–2119. [CrossRef]
28. Pal, K.; Xia, Y.; Shen, J.; He, J.; Luo, Y.; Kanatzidis, M.G.; Wolverton, C. Accelerated discovery of a large family of quaternary chalcogenides with very low lattice thermal conductivity. *npj Comput. Mater.* **2021**, *7*, 82. [CrossRef]
29. Ruseikina, A.V.; Andreev, O.V.; Demchuk, Z.A. Preparation of polycrystalline samples of the EuLnCuS$_3$ (Ln = Gd, Lu) compounds. *Inorg. Mater.* **2016**, *52*, 537–542. [CrossRef]
30. Solovyeva, A.V. Regularities of phase equilibria in the systems A^{II}S–FeS, A^{II}S–FeS–Ln_2S$_3$, A^{II}S–Cu$_2$S–Ln_2S$_3$ (A^{II} = Mg, Ca, Sr, Ba; Ln = La–Lu). Ph.D. Thesis, University of Tyumen, Tyumen, Russia, 2012.
31. Sikerina, N.V. Regularities of Phase Equilibria in SrS–Ln_2S$_3$–Cu$_2$S Systems, Preparation and Structure of SrLnCuS$_3$ Compounds. Ph.D. Thesis, University of Tyumen, Tyumen, Russia, 2005.
32. Wakeshima, M.; Furuuchi, F.; Hinatsu, Y. Crystal structures and magnetic properties of novel rare-earth copper sulfides, EuRCuS$_3$ (R = Y, Gd–Lu). *J. Phys. Condens. Matter* **2004**, *16*, 5503–5518. [CrossRef]
33. Rare-Earth Metal Long Term Air Exposure Test, Metallium, Inc. Available online: https://www.elementsales.com/re_exp/ (accessed on 22 February 2024).
34. Keil, P.; Lutzenkirchen-Hecht, D.; Frahm, R. Investigation of room temperature oxidation of Cu in Air by YonedaXAFS. *AIP Conf. Proc.* **2007**, *882*, 490–492.
35. Sheldrick, G.M. A short history of SHELX. *Acta Crystallogr.* **2008**, *A64*, 112–122. [CrossRef]
36. Spek, A.L. *PLATON—A Multipurpose Crystallographic Tool*; Utrecht University: Utrecht, The Netherlands, 2008.
37. Crystal. Available online: http://www.crystal.unito.it/index.php (accessed on 10 January 2024).
38. Energy-Consistent Pseudopotentials of the Stuttgart/Cologne Group. Available online: http://www.tc.uni-koeln.de/PP/clickpse.en.html (accessed on 10 January 2024).

39. Towler, M. CRYSTAL Resourses Page. Available online: https://vallico.net/mike_towler/crystal.html (accessed on 10 January 2024).
40. Grigoriev, M.V.; Solovyov, L.A.; Ruseikina, A.V.; Aleksandrovsky, A.S.; Chernyshev, V.A.; Velikanov, D.A.; Garmonov, A.A.; Molokeev, M.S.; Oreshonkov, A.S.; Shestakov, N.P.; et al. Quaternary Selenides EuLnCuSe$_3$: Synthesis, Structures, Properties and In Silico Studies. *Int. J. Mol. Sci.* **2022**, *23*, 1503. [CrossRef] [PubMed]
41. Yang, L.; Powell, D.R.; Houser, R.P. Structural variation in copper(i) complexes with pyridylmethylamide ligands: Structural analysis with a new four-coordinate geometry index, τ_4. *Dalton Trans.* **2007**, *9*, 955–964. [CrossRef] [PubMed]
42. Ruseikina, A.V.; Grigoriev, M.V.; Garmonov, A.A.; Molokeev, M.S.; Schleid, T.; Safin, D.A. Synthesis, structures and magnetic properties of the Eu-based quaternary tellurides EuGdCuTe$_3$ and EuLuCuTe$_3$. *Cryst. Eng. Comm.* **2023**, *25*, 1716–1722. [CrossRef]
43. Ruseikina, A.V.; Grigoriev, M.V.; Solovyov, L.A.; Molokeev, M.S.; Garmonov, A.A.; Velikanov, D.A.; Safin, D.A. Unravelling the rare-earth (*RE*) element-induced magnetic and optical properties in the structures of quaternary selenides Sr*RE*CuSe$_3$. *Inorg. Chem. Commun.* **2023**, *156*. [CrossRef]
44. Pavlyuk, M.D. Detector Crystals Based on CdTe and Cd$_{1-x}$Zn$_x$Te for Direct Detection of X-ray and Gamma Quanta. Ph.D. Thesis, A.V. Shubnikov Institute of Crystallography RAS, Moscow, Russia, 2020.
45. Ruseikina, A.V.; Molokeev, M.S.; Chernyshev, V.A.; Aleksandrovsky, A.S.; Krylov, A.S.; Krylova, S.N.; Velikanov, D.A.; Grigoriev, M.V.; Maximov, N.G.; Shestakov, N.P.; et al. Synthesis, structure, and properties of EuScCuS$_3$ and SrScCuS$_3$. *J. Solid State Chem.* **2021**, *296*, 121926. [CrossRef]
46. Wu, J.; Zhao, E.J.; Xiang, H.P.; Hao, X.F.; Liu, X.J.; Meng, J. Crystal structures and elastic properties of superhard IrN$_2$ and IrN$_3$ from first principles. *Phys. Rev.* **2007**, *B76*, 54–115.
47. Korabelnikov, D.V.; Zhuravlev, Y.N. Ab-initio investigations of the elastic properties of chlorates and perchlorates. *Phys. Solid State* **2016**, *58*, 1166–1171. [CrossRef]
48. Tian, Y.; Xu, B.; Zhao, Z. Microscopic theory of hardness and design of novel superhard crystals. *Int. J. Refract. Met. Hard Mater.* **2012**, *33*, 93–106. [CrossRef]

Disclaimer/Publisher's Note: The statements, opinions and data contained in all publications are solely those of the individual author(s) and contributor(s) and not of MDPI and/or the editor(s). MDPI and/or the editor(s) disclaim responsibility for any injury to people or property resulting from any ideas, methods, instructions or products referred to in the content.

Article

Corrosion Resistance of Fe-Based Amorphous Films Prepared by the Radio Frequency Magnetron Sputter Method

Tai-Nan Lin [1], Pin-Hsun Liao [2], Cheng-Chin Wang [3], Hung-Bin Lee [2] and Leu-Wen Tsay [2,*]

[1] Department of Material Research, National Atomic Research Institute, Taoyuan 32546, Taiwan; tnlin@nari.org.tw
[2] Department of Optoelectronics and Materials Technology, National Taiwan Ocean University, Keelung 20224, Taiwan; 11089007@email.ntou.edu.tw (P.-H.L.); lhb6018@mail.ntou.edu.tw (H.-B.L.)
[3] Chung Yo Materials Co., Ltd., Kaohsiung 82059, Taiwan; kimwang@cymaterials.com.tw
* Correspondence: b0186@mail.ntou.edu.tw; Tel.: +886-2-2462-2192 (ext. 6405)

Abstract: Amorphous thin films can be applied to increase the anti-corrosion ability of critical components. Atomized FeCrNiMoCSiB powders were hot-pressed into a disc target for R. F. magnetron sputtering on a 316L substrate to upgrade its corrosion resistance. The XRD spectrum confirmed that the film deposited by R. F. magnetron sputtering was amorphous. The corrosion resistance of the amorphous film was evaluated in a 1 M HCl solution with potentiodynamic polarization tests, and the results were contrasted with those of a high-velocity oxy-fuel (HVOF) coating and 316L, IN 600, and C 276 alloys. The results indicated that the film hardness and elastic modulus, as measured using a nanoindenter, were 11.1 and 182 GPa, respectively. The principal stresses in two normal directions of the amorphous film were about 60 MPa and in tension. The corrosion resistance of the amorphous film was much greater than that of the other samples, which showed a broad passivation region, even in a 1 M HCl solution. Although the amorphous film showed high corrosion resistance, the original pinholes in the film were weak sites to initiate corrosion pits. After polarization tests, large, deep trenches were seen in the corroded 316L substrate; numerous fine patches in the IN 600 alloy and grain boundary corrosion in the C276 alloy were observed.

Keywords: R. F. magnetron sputtering; amorphous film; potentiodynamic polarization; XPS

Citation: Lin, T.-N.; Liao, P.-H.; Wang, C.-C.; Lee, H.-B.; Tsay, L.-W. Corrosion Resistance of Fe-Based Amorphous Films Prepared by the Radio Frequency Magnetron Sputter Method. *Materials* **2024**, *17*, 2071. https://doi.org/10.3390/ma17092071

Academic Editor: Costica Bejinariu

Received: 23 February 2024
Revised: 4 April 2024
Accepted: 26 April 2024
Published: 28 April 2024

Copyright: © 2024 by the authors. Licensee MDPI, Basel, Switzerland. This article is an open access article distributed under the terms and conditions of the Creative Commons Attribution (CC BY) license (https://creativecommons.org/licenses/by/4.0/).

1. Introduction

A variety of different amorphous alloys (AAs), including Ti- [1,2], Zr- [3,4], Ni- [5,6], and Cu-based [7,8] alloys, have been developed for distinct industries. The absence of microstructural heterogeneity, such as grain boundaries and precipitates, as well as mechanical homogeneity accounts for the remarkable mechanical and chemical properties of AAs [9,10]. Fe-based AAs have many superior characteristics, like excellent corrosion/wear resistances and low costs, among different AAs [11–13]. It is reported that FeCrMoCB-Tm AAs have a high glass-forming ability [14–19] but high brittleness at room temperature, which greatly restricts their engineering applications. Thus, Fe-based AAs have been provided as powders for thermal spraying [20–25] against corrosion and wear for industrial applications [26–28].

The chemical compositions, uniform structures, and stable passivation films account for the excellent corrosion resistance of AAs. The addition of Cr has an obviously positive effect on increasing the corrosion resistance of the FeCrNiB AA with a stable passivation film [29]. Forming the more insoluble Cr_2O_3 oxide will stabilize and enhance the formation of dense passivation films on FeCrMoCBY coatings [30]. Compared with the FeCrBCP amorphous coating, the addition of a minor amount of Mo (5 at.%) to feedstock powders is advantageous to upgrade the corrosion and wear resistances of coatings [31]. The corrosion resistance of the FeCrMoPCB AA in seawater improves with increasing Cr/Mo content [32]. Furthermore, the addition of C (1–3 wt.%) to FeCuNbSiB amorphous ribbons can significantly improve the corrosion resistance in a 0.1 M H_2SO_4 solution [33]. In a 0.5 M KNO_3 solution, the corrosion potential increases, and the passivation current density decreases with increasing Ni content up to 4 at.% in FeNiCuNbSiB amorphous ribbons [34]. Increasing the crystalline phase in the FeCrMoCBY amorphous coating accounts for a decreased corrosion resistance [35]. The carbide precipitates result in forming a Cr- and Mo-depleted zone and degrade the stability of the passivation film of the FeMoCrYCB amorphous coating [36].

Austenitic stainless steels (SSs) are known to be susceptible to chloride stress corrosion cracking (SCC) despite their high general corrosion resistance. Amorphous thin films can be deposited on the critical components against harsh corrosion attack. According to the literature, the characteristics of Fe-based amorphous thin films have rarely been investigated. An Fe-based (FeMoCrYCB) thin film in an artificial sweat solution shows much superior passivation stability relative to 304 SS [37]. In prior work [38], an Fe-based AA (FeCrNiMoCSiB) was used as the feedstock powder for HVOF spraying on a 316L plate. The corrosion rate of the high-velocity oxy-fuel (HVOF) coating is identical to that of the 316L SS in seawater and a 1 M HCl solution, but much higher corrosion-wear resistance was observed in seawater [38]. In this work, an Fe-based AA (FeCrNiMoCSiB) acts as the target material for R. F. magnetron sputtering on 316L SS to upgrade the corrosion resistance of 316L components. The corrosion characteristics of the amorphous film were tested in a 1 M HCl solution with potentiodynamic polarization tests, and the results were contrasted with those of the HVOF coating [38] and 316L [38], IN 600, and C276 alloys. The corroded features of distinct specimens were inspected using a scanning electron microscope (SEM). Moreover, the surface passivation film was analyzed using X-ray photoelectron spectroscopy (XPS).

2. Materials and Experiments

In this work, Ar-gas atomized amorphous powders were used for the target material for subsequent sputtering, which Chung Yo Materials Co. offered. The chemical compositions, in weight percentages of the feedstock powders, were 13.65 Ni, 14.41 Mo, 21.53 Cr, 2.30 B, 2.07 C, and 2.73 Si, and the balance was Fe. A 404 F3 high-temperature differential scanning calorimeter (HT-DSC, Netzsch, Selb, Germany) was applied to determine the transformation temperature of the powder. The composition of the C276 alloy was 15.7 Cr, 15.8 Mo, 3.2 W, 6.0 Fe, 0.008 C, 0.6 Mn, and residual Ni. The nominal composition of the IN 600 alloy was the same as that of a Ni alloy with 16 Cr and 8 Fe. The feedstock powders were hot-pressed and sintered into a 6.0 mm plate and then wire-cut to be a 75 mm disc as a target for sputtering. Prior to sputtering, the 316L plate was ground using up to 3000# abrasive paper and then subjected to polishing to a mirror finish. Thin films were deposited on a Si wafer and a mirror-finished 316L plate using an R. F. magnetron sputter system under high vacuum. The sputtering variables included a sputtering power of 120 W, a target-to-substrate distance of 60 mm, a working pressure of 0.7 Pa, and a rotating speed of 5 rpm. Distinct film thicknesses up to 1500 nm could be achieved by altering the deposition time. The film compositions were determined using an electron probe microanalyzer (EPMA, JEOL JXA 8200) equipped with a wavelength-dispersive spectroscope. The phases of the target and the film were detected with a D2 X-ray diffractometer (XRD, Bruker Billerica, MA, USA). The microstructures of the prepared specimens were inspected with a 3400 SEM (Hitachi, Tokyo, Japan). A Hysitron TI 980 nanoindenter (Bruker, Billerica, MA, USA) loaded at 2000 μN was applied to determine the nano-hardness and the Young's modulus of the prepared film. An optical profiler was used to examine the surface metrology of the film, which provided non-contact surface measurements. A scratch tester (Revetest Scratch Tester: RST^3) was used to evaluate the thin film's adhesion property.

The residual stresses of the film deposited on the Si substrate were calculated according to Stoney's equation [39] as follows:

$$\sigma r = EH/6(1 - v_S)T_f R_f, \qquad (1)$$

where σr is the planar stress in the film; E, H, and v_S are the Young's modulus (130.2 GPa), the thickness (525 μm), as well as the Poisson's ratio (0.279) of the Si substrate, respectively; T_f is the film's thickness; and R_f is the radius of the film's curvature.

The potentiodynamic polarization curve of the film was investigated using a standard three-electrode cell system and contrasted with those of the HVOF coating and 316L, IN 600, and C276 alloys. A saturated calomel electrode (SCE) acted as the counter electrode, and the reference electrode was a platinum plate. The potentiodynamic polarization test was carried out in a 1 M HCl solution from −1.0 V to +2.0 V and at a 1 mV/s scanning rate. The elemental depth profile and composition valence of the amorphous film before and after the corrosion test were determined using X-ray photoelectron spectroscopy (XPS, ULVAC-PHI, PHI 5000 Versa Probe) assembled with an Ar$^+$ ion etching gun.

3. Results and Discussion

The inherent characteristics of the feedstock are shown in Figure 1, which reveals the appearance and the differential scanning calorimetry (DSC) curve of the powder. The atomized powder used in this work (Figure 1a) predominantly comprised smooth and round particles without a dendritic structure after solidification. As shown in Figure 1b, the DSC curve of the feedstock powder revealed a glass transition temperature (T_g) of about 555 °C, a peak crystallization temperature (T_p) of 659 °C, solidus temperature (T_S) of 1111 °C, and a liquidus temperature (T_L) of 1168 °C. The solidus temperature of the powder was below 1120 °C. To be used as the target for sputtering, the feedstock powders were pre-vacuumed and hot-pressed at 1080 °C, then the compact powders were sintered for 8 hours into a plate of a 6.0 mm thickness. A 75 mm diameter disc was wire-cut from the sintered plate to be the target for R. F. sputtering (Figure 1c).

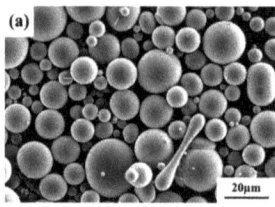
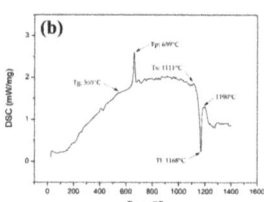
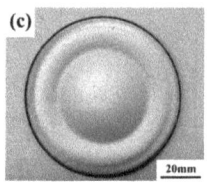

Figure 1. (a) The appearance, (b) the DSC curve of the feedstock powder, and (c) the appearance of the target used for sputtering.

Figure 2 displays the XRD spectra of the sputtered film and the target. To avoid the interference of the sub-surface layer with the XRD spectrum, the film was deposited directly on the Si wafer. The XRD pattern of the film showed wide and broad peaks in the spectrum (Figure 2a), which indicated the very highly amorphous constituent of the film. By contrast, the XRD pattern of the target (Figure 2b) comprised many sharp peaks, which were identified to be attributed to complicated carbides, borides, and intermetallics formed in the target owing to the slow cooling of the sintered target after hot-pressing the feedstock powders. Although the target was crystalline, the film prepared by R. F. magnetron sputtering was amorphous. Metallic glass films can be fabricated by controlling the sputtering parameters, such as the power and working pressure, so that the as-deposited films inherit the composition and amorphous structure of the target.

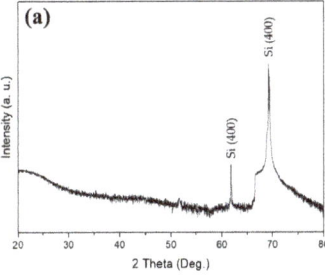
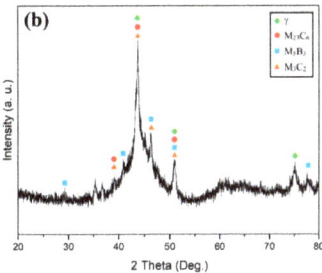

Figure 2. XRD spectra of (**a**) the film (The former is from Si (400) Cu Kβ (1.392 Å), and the latter is from Si (400) Cu Kα (1.54 Å); (**b**) the target.

It was found that peeling was more likely to occur when directly depositing the amorphous film on the 316L substrate, but peeling and/or cracking did not occur on the Si wafer. To avoid the peeling and cracking of the amorphous film, an about 300 nm thick layer of pure Ti was pre-sputtered on the 316L substrate. Figure 3 displays the SEM morphologies of the amorphous film. The amorphous film deposited on the Si wafer, in the cross-sectional view shown in Figure 3a, consisted of a Ti film as the intermediate layer. The chemical compositions of the film, as determined by EPMA, in weight percentages were 16.88 Ni, 13.00 Mo, 21.80 Cr, 1.47 B, 2.38 C, 2.65 Si, and 0.24 Ti, and Fe was the balance. Overall, the deposited film had the about same constituents as the target material. The minor Ti content of the film meant that the yielding volume of the electron beam during EPMA analysis was above the film thickness and reached the intermediate layer. Figure 3b displays the top surface morphology of the amorphous film deposited on the 316L substrate. It was observed that cracking was not found in the film, but some fine particles and pinholes were present in the examined samples (Figure 3b). Those fine particles on the film surface could be from the contaminants present in the chamber. The chemical compositions of those fine particles were inspected using SEM/EDX. The results indicated that most of those fine particles in the examined sample had the about same constituents as the target. It was deduced that those fine particles would be the residues from the target during sputtering. Moreover, those pinholes would be critical defects and would deteriorate the corrosion resistance of the film. The surface texture of the film deposited on the polished 316L substrate was investigated by a 3D contour profiler, as shown in Figure 3c. The Sa, Sp, and Sv values of the film were 20.96 nm, 0.85 μm, and 2.21 μm, respectively. Overall, the amorphous film had a very low surface roughness.

A nanoindenter was applied to measure the hardness and elastic modulus of the amorphous film, using the sample with the film deposited on the 316 SS. The hardness of the film fell in the range between 10.80 and 11.12 GPa after eight measurements. The elastic modulus of the film was about 181 GPa. The nano-hardness of the film was roughly converted to micro-Vickers hardness values, i.e., from about HV 1102 to 1135. This result revealed that the amorphous film would provide very high wear resistance as compared with traditional engineering alloys. The residual stresses of the amorphous film deposited on the Si substrate were calculated according to Stoney's equation [39]. Ten measurements were performed at different sites in the film. The principal stresses in two normal directions were calculated according to Stoney's equation. The two perpendicular residual stresses were in tension and around 60 MPa. The tensile residual stress of the film was low, as compared with the expected strength of the amorphous film [12]. Scratch test experiments determine the practical adhesion strength and mechanical failure modes of hard (HV = 5 GPa or higher) thin (≤30 μm) coatings on metal and ceramic substrates at ambient temperatures according to ASTM C1624-5. In our study, the measured adhesion strength for the as-deposited amorphous film on the stainless-steel substrate was around 25 N at a 5 mm scratch distance, with the scratch load ranging from 50 to 200 N. This result indicates the amorphous film adheres well on the stainless-steel substrate, as expected.

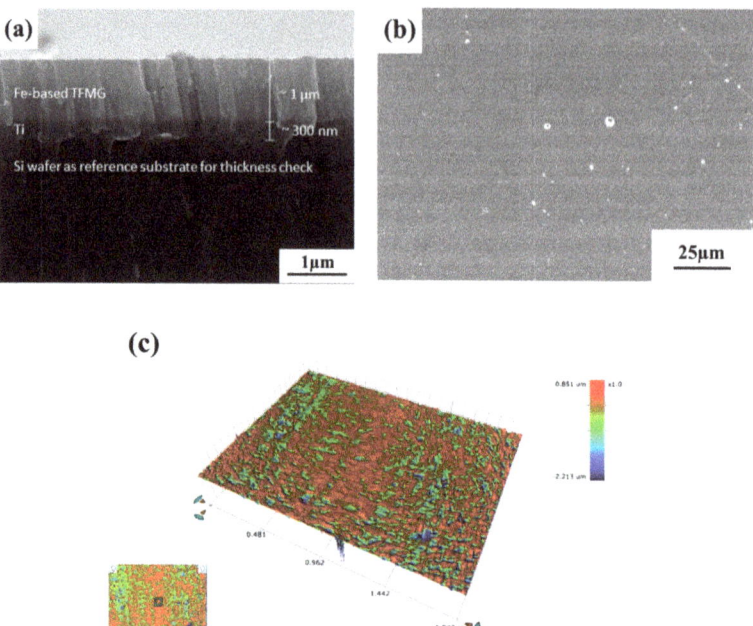

Figure 3. (**a**) The microstructure in the cross-sectional view, (**b**) the surface morphology, and (**c**) the 3D contour profile of the amorphous film.

Potentiodynamic polarization tests of various specimens were performed at room temperature in a 1 M HCl solution, and the polarization curves are shown in Figure 4. Table 1 lists the corrosion potential (E_{Corr}), corrosion current density (i_{Corr}), and pitting potential (E_{Pit}) of the tested samples. In our prior work [38] studying the corrosion performance of the HVOF coating using the same amorphous powder as used in this work, a 1 M HCl solution was a much more severe corrosion condition relative to seawater and 0.5 M H_2SO_4. To distinguish the corrosion resistances among the IN 600, C276, and amorphous film, a 1 M HCl solution is used in this work. Figure 4a reveals the polarization curves of the 316L, IN 600, and C276 alloys and amorphous coating in the 1 M HCl solution. In such a harsh environment, none of the three alloys had an evident passivation region. Among the three alloys, the C276 alloy had a higher E_{Corr} and a lower i_{Corr} than the other two alloys in the 1 M HCl solution. The better corrosion resistance of the C276 alloy was the result of the high alloy contents relative to the other two alloys. The E_{Corr} and i_{Corr} values of 316L were about −0.34 V and 58.2 µA/cm², respectively [38], and around −0.28 V and 29.2 µA/cm² for the IN 600 alloy. This indicated that the IN 600 alloy showed slightly higher corrosion resistance than 316L. Figure 4b displays the potentiodynamic polarization curves of the C276 and 316L alloys and amorphous film in the 1 M HCl solution. This revealed that the amorphous film possessed a marked passivation region and a pitting potential near 1.0 V, whereas the C276 alloy showed an increased current density with increasing potential. Obviously, a passivation zone was not found in the polarization curve of the C276 alloy (Figure 4b). Among the tested samples, only the amorphous film showed a broad passivation zone during corrosion in the 1 M HCl solution. Undoubtedly, the amorphous film provided excellent protection against the harsh HCl attack.

Table 1. The E_{Corr}, i_{Corr}, and E_{Pit} values of distinct samples tested in 1 M HCl solution.

Sample	316L [38]	IN 600	Amorphous Coating [38]	C276	Amorphous Film
E_{Corr} (V)	−0.34	−0.28	−0.29	−0.22	−0.3
i_{Corr} (μA/cm^2)	58.2	29.2	39.2	20.5	58.3
E_{Pit} (V)	0.0	0.25	--	1.04	0.85

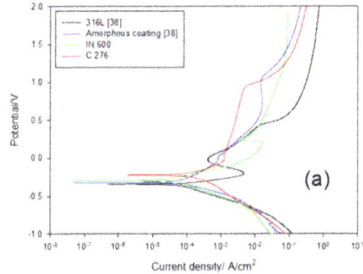 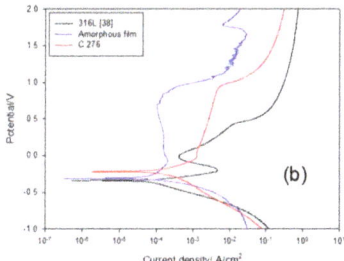

Figure 4. The potentiodynamic polarization curves of the (**a**) IN 600 and C276 alloys in comparison with the 316L and amorphous coating and (**b**) the C276 and amorphous film in comparison with 316L tested in 1 M HCl solution.

The surface morphologies of the distinct test pieces after the polarization tests are displayed in Figure 5. After testing in the 1 M HCl solution, the 316L sample presented severe corrosion owing to the link between fine corrosion pits and corroded trenches (Figure 5a). As compared with the 316L SS, a completely different corroded appearance was seen for the IN 600 alloy tested in the 1 M HCl solution (Figure 5b). The IN 600 alloy showed extensive surface corrosion, which was caused by the extensive dissolution of fine patches in the 1 M HCl solution. By contrast, grain boundary corrosion was more likely to occur for the C276 alloy under the same test conditions (Figure 5c). In addition, the amorphous coating showed fine pits initiated at the interface between residual powders and the molten zone; then, those pits connected to form fine ditches after the polarization test [38]. It was noticed that the corrosion feature of the amorphous film was obviously different from that of the other samples (Figure 5d). Fine pits penetrated the film through pinholes to the Ti buffer layer and grew into a more significant pore, which accounted for the failure of the amorphous film in the 1 M HCl solution (Figure 5d). The original pinholes led to the degraded corrosion resistance of the amorphous film in the harsh HCl environment.

An Fe-based AA (FeCrNiMoCSiB) was used as the target material and was deposited on 316L SS using R. F. magnetron sputtering. The XRD pattern of the target was composed of peaks attributed to complex carbides, borides, and intermetallics, whereas the deposited film was in the amorphous state. The presence of complex precipitates in the target could be attributed to the slow cooling of the sintered target after hot-pressing. Although the target was crystalline, the film prepared by R. F. magnetron sputtering was amorphous. Metallic glass films can be fabricated by controlling the sputtering parameters, such as the power and working pressure so that the as-deposited films inherit the composition and amorphous structure of the target. In addition, the high cooling rate of the deposited film after sputtering was helpful to form the amorphous film. The elastic modulus of the amorphous film, as determined using a nanoindenter, was about 181 GPa, which agrees with those of Fe-based amorphous alloys [12]. In addition, the amorphous film possessed a nano-hardness from around 10.80 to 11.12 GPa, which meant that the investigated film was expected to have a very high wear resistance. The principal residual stresses were in tension and around 60 MPa. The low principal residual stress could be partly attributed to comparable elastic moduli and thermal conductivities between the 316L and amorphous film. The polarization curve of the amorphous film in the 1 M HCl solution exhibited a marked passivation region and a pitting potential near 1.0 V. In such a harsh solution, only the amorphous film showed a wide passivation region among the tested samples. Although a few pinholes in the film were the cause of the failure in the 1 M HCl solution, the anti-corrosion ability of the amorphous film could be associated with the inherent alloy composition.

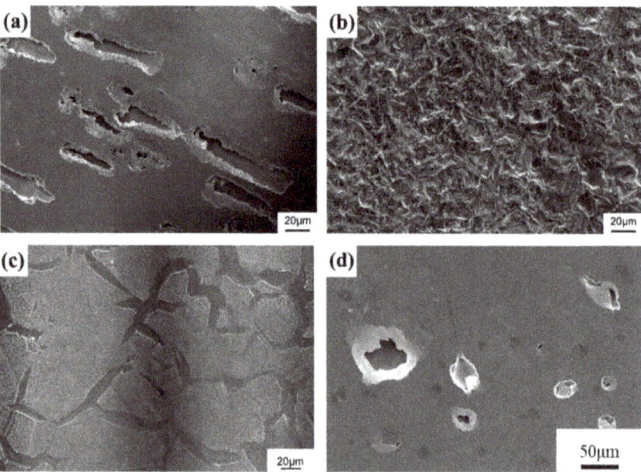

Figure 5. The surface morphologies of the (**a**) 316L, (**b**) IN 600, (**c**) and C276 alloys and (**d**) the amorphous film after polarization tests in the 1 M HCl solution.

As displayed in Figure 6 and Table 2, the XPS depth profile (d) was achieved using an Ar$^+$ ion beam accelerated at 3 keV and a sputtering area of approximately 2 × 2 mm^2, and the binding energies of the elements, e.g., Fe, Ni, Cr, Mo, Si, Ti, and O, were measured. The etching rate of the film was about 8.16 nm/min for SiO$_2$. The charging effect was amended using the C1s peak at 284.6 eV as the baseline for the investigated species [40]. In addition, all the peaks were distinguished based on the NIST database and the results reported in the available literature [41]. As shown in Figure 6a, the O content decreased from 60 at.% to 11 at.% as the depth increased from the surface to the depth after about 1.5 min of sputtering, when the low O content was detected after sputtering. However, after the film was attacked in the 1 M HCl solution, the contents of elements like Fe, Cr, Mo, and Si showed less change, but the Ti content increased with increasing sputtering time. At the depth corresponding to a sputtering time of 3.5 min (Figure 6b), the amorphous film was totally demolished.

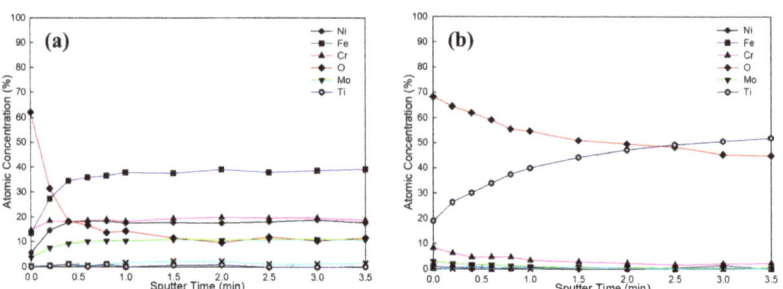

Figure 6. XPS depth profiles of different elements in the amorphous film (**a**) before and (**b**) after corrosion in the 1 M HCl solution.

The commercial software XPS Peak version 4.1 was applied to identify the specific peaks, and the binding energy of each species was referenced according to the NIST database, as previous reported. In this sputtered film, elements like Fe, Ni, Cr, Mo, Si, and Ti were analyzed as the primary elements. Table 2 lists the binding energies and spectral peaks corresponding to the relative valences. Figure 7 displays the resolved spectra of the Fe, Ni, Cr, Mo, Si, and Ti elements. Figure 7a reveals the Fe spectra, showing that all the Fe peaks were located at binding energies of approximately 720.0 eV, 707.0 eV, 710.3 eV, and 724.1 eV [42–45]. Figure 7b shows the Ni spectra, which indicate that all the Ni peaks were located at binding energies of approximately 852.8 eV, 870.1 eV, and 873.6 eV, as well as an Ni^{2+} peak located at 860.4 eV [46–48]. As presented in Figure 7c, initially, the Cr peak was located at a binding energy of 583.3 eV, while a Cr peak appeared at 574.1 eV, Cr^{3+} peaks appeared at 575.9 eV and 586.4 eV, and a Cr^{6+} peak appeared at 577.7 eV [42,43,49,50]. Figure 7d displays the initial Mo spectral peaks; they indicate that initially, the Mo in the film was composed of Mo (according to the peak at 228.6 eV) as well as Mo^{4+} (according to the peak at 231.8 eV) [43,49,51,52]. Based on these results, it was deduced that the target used in this work might be contaminated by oxygen during hot-pressing at elevated temperatures and was found to have an O content of 10 at.% (Figure 6a). Figure 7e reveals the Si spectra, showing a Si^0 peak at 99.4 eV and Si^{4+} peaks at 101.7 eV and 103.2 eV [53,54]. The intensity of the Ti peak was so low that it could be ignored, as shown in Figure 7f. Figure 8 reveals the XPS spectra of the amorphous film after corrosion in the 1 M HCl solution, and the corresponding data are also listed and referenced in Table 2. Because of the failure of the film, the XPS depth profile contained peaks attributed to Ti and residual Cr and Mo. As shown in Figure 8f, the Ti spectra included Ti_{2p} and Ti^0 peaks at 453.9 eV and 460.1 eV, respectively, as well as Ti^{2+} peaks located at 455.3 eV and 461.0 eV, Ti^{3+} peaks located at 457.1 eV and 462.7 eV, and Ti^{4+} peaks located at 458.7 eV and 464.4 eV [43,55]. Figure 8c shows the Cr core-level spectra and the spectra of the Cr band; most of the Cr remained in the metallic chromium state, with Cr^{+3} peaks at 575.9 eV and 586.4 eV. Chromium might transform to its oxide according to the Cr peak at 574.1 eV and the Cr^{+3} peaks at 575.9 eV and 586.4 eV [42,43], which showed the difference before and after the corrosion test. By contrast, the Cr and Cr^{+3} peaks at 575.9 eV and 586.4 eV, respectively, increased in intensity, which implied the formation of condensed and insoluble Cr_2O_3 in the film. Moreover, the Cr_2O_3 oxide could effectively prevent Cl- from penetrating the substrate. Figure 8d presents the XPS spectra of the Mo element. According to a study by Hashimoto et al. [56], MoO_2 can provide substantial corrosion protection because Mo can form a very dense and insoluble oxide relative to the Cr oxide. According to Tian et al. [43], Mo^{+6} can dissolve in water to form MoO_4^{-2}, which can adhere to the surface and prevent Cl- from penetrating the film. In addition, Mo^{+6} effectively avoids the further pitting attack of the corroded iron, which prevents the film from corroding and being demolished. Therefore, the high Cr (above 20%) and Mo (above 14%) contents of the amorphous film ensured the formation of dense and stable Cr and Mo oxides [29–31], which were beneficial to form a protective layer, even in the 1 M HCl solution. As revealed in a prior study, the addition of 1–3 wt.% of C to Fe-based AAs could also upgrade their corrosion resistance [33]. In this work, the C content (2.07%) of the investigated film fell within the suggested range. In addition, the 14% Ni content of the film would assist in the formation of Ni oxides, which were also helpful to increase the corrosion resistance of the film. With the addition of 2.73% Si, Ni is reported to increase the Si content in the SiO_2 passivation film and make the film more protective [13]. As mentioned previously, only a few pinholes were present in the amorphous film, and there was a lack of other defects, like grain boundaries and precipitates, in the film. Therefore, the uniform composition/structure and advantageous alloying made this Fe-based AA film show excellent anti-corrosion ability in a harsh environment.

Table 2. Measured and reference binding energies of the XPS spectra generated for amorphous film (a) before and (b) after corrosion.

		Before Corrosion	
		Binding Energy (eV)	References
Cr $2p_{3/2}$	Cr^0	574.1	[42,51]
	Cr^{3+}	575.9	[43]
Cr $2p_{1/2}$	Cr^0	583.3	[42]
	Cr^{3+}	586.4	
Fe $2p_{3/2}$	Fe^0	707.0	[42–44]
	Fe^{2+}	710.3	[45]
Fe $2p_{1/2}$	Fe^0	720.0	[42–44]
	Fe^{2+}	724.1	[45]
Ni $2p_{3/2}$	Ni^0	870.1	[48]
		873.6	[47]
Ni $2p_{1/2}$	Ni^0	852.8	[46,47,50]
	Ni^{2+}	860.4	
Mo 3d	Mo^0	228.0	[43,50,53]
	Mo^{4+}	231.1	[55]
Si $2p_{3/2}$	Si^{4+}	103.2	
Si $2p_{1/2}$	Si^0	99.4	[54]
	Si^{4+}	101.7	
		After corrosion	
		Binding Energy (eV)	References
Cr $2p_{3/2}$	Cr^0	574.1	[50,51]
	Cr^{3+}	575.9	[43]
	Cr^{6+}	577.7	[51]
Cr $2p_{1/2}$	Cr^0	583.3	[42]
	Cr^{3+}	586.4	
Mo 3d	Mo^0	228.6	[43,50,52]
	Mo^{4+}	231.8	
Ti $2p_{2/3}$	Ti^0	453.9	
	Ti^{2+}	455.3	
	Ti^{3+}	457.1	
	Ti^{4+}	458.7	[43,56]
Ti $2p_{1/3}$	Ti^0	460.0	
	Ti^{2+}	461.0	
	Ti^{3+}	462.7	
	Ti^{4+}	464.4	

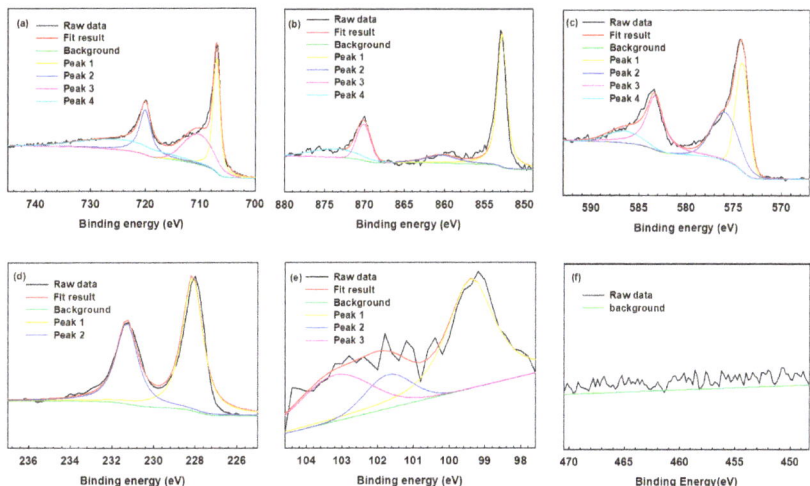

Figure 7. The deconvolution curve-fitting analyses of the binding-energy peaks of the (**a**) Fe, (**b**) Ni, (**c**) Cr, (**d**) Mo, (**e**) Si, and (**f**) Ti elements in the measured XPS spectra of the amorphous film.

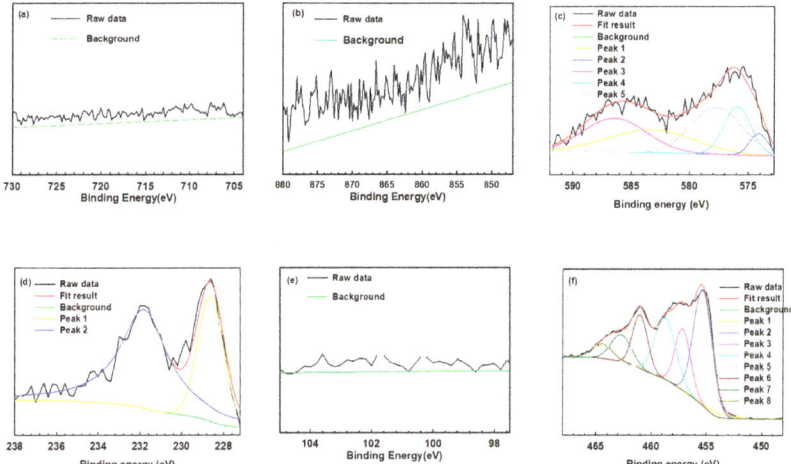

Figure 8. The deconvolution curve-fitting analyses of the binding-energy peaks of the (**a**) Fe, (**b**) Ni, (**c**) Cr, (**d**) Mo, (**e**) Si, and (**f**) Ti elements in the measured XPS spectra of the film after testing in the 1 M HCl solution.

4. Conclusions

(1) With the Ti buffer layer, the amorphous film was successfully deposited on the 316L substrate using R. F. magnetron sputtering, although the FeCrNiMoCSiB target was composed of complex carbides, borides, and intermetallics. The elastic modulus and the nano-hardness of the amorphous film, as measured using a nanoindenter, were about 181 GPa and 11 GPa, respectively;

(2) The polarization curve of the amorphous film in the 1 M HCl solution indicated a broad passivation region and a pitting potential near 1.0 V. The pinholes in the film were the cause of the failure in the 1 M HCl solution after the polarization tests. The XPS spectra of the amorphous film revealed that the dense, insoluble Cr and Mo oxides were beneficial to form a protective layer, even in the 1 M HCl solution. As

mentioned previously, only a few pinholes were present in the amorphous film, and there was a lack of other defects, like grain boundaries and precipitates, in the film. Therefore, the uniform composition/structure and advantageous alloying made this Fe-based AA film show excellent anti-corrosion ability in a harsh environment.

Author Contributions: All the authors, writing—review and editing; T.-N.L., P.-H.L. and C.-C.W., investigation and visualization; H.-B.L., methodology and supervision; L.-W.T., conceptualization, resources, project administration, and funding acquisition. All the authors have read and agreed to the published version of the manuscript. All authors have read and agreed to the published version of the manuscript.

Funding: This research was funded under grant number MOST 110-2622-E-019-008.

Institutional Review Board Statement: Not applicable.

Informed Consent Statement: Not applicable.

Data Availability Statement: The data are available on request because of restrictions.

Acknowledgments: The authors are grateful for the HT-DSC data from the Thermal Analysis System at the Instrumentation Center, National Taiwan University. The authors are grateful to the Ministry of Science and Technology (National Taiwan University) for the assistance in EPMA analysis (EPMA000300). The authors are grateful to the Core Facility Center of National Cheng Kung University (OTHER002200) for granting access to the 3D optical profiler.

Conflicts of Interest: Author Cheng-Chin Wang was employed by the company Chung Yo Materials Co., Ltd. The remaining authors declare that the research was conducted in the absence of any commercial or financial relationships that could be construed as a potential conflict of interest.

References

1. Tantavisut, S.; Lohwongwatana, B.; Khamkongkaeo, A.; Tanavalee, A.; Tangpornprasert, P.; Ittiravivong, P. The novel toxic free titanium-based amorphous alloy for biomedical application. *J. Mater. Res.* **2018**, *7*, 248–253. [CrossRef]
2. Kuball, A.; Gross, O.; Bochtler, B.; Adam, B.; Ruschel, L.; Zamanzade, M.; Busch, R. Development and characterization of titanium-based bulk metallic glasses. *J. Alloys Compd.* **2019**, *790*, 337–346. [CrossRef]
3. Inoue, A.; Yokoyama, Y.; Shinohara, Y.; Masumoto, T. Preparation of Bulky Zr-Based Amorphous Alloys by a Zone Melting Method. *Mater. Trans.* **1994**, *35*, 923–926. [CrossRef]
4. Wang, F.; Yin, D.; Lv, J.; Zhang, S.; Ma, M.; Zhang, X.; Liu, R. Effect on microstructure and plastic deformation behavior of a Zr-based amorphous alloy by cooling rate control. *J. Mater. Sci. Technol.* **2021**, *82*, 1–9. [CrossRef]
5. Xu, D.; Duan, G.; Johnson, W.L.; Garland, C. Formation and properties of new Ni-based amorphous alloys with critical casting thickness up to 5 mm. *Acta Mater.* **2004**, *52*, 3493–3497. [CrossRef]
6. Wang, A.P.; Chang, X.C.; Hou, W.L.; Wang, J.Q. Corrosion behavior of Ni-based amorphous alloys and their crystalline counterparts. *Corros. Sci.* **2007**, *49*, 2628–2635. [CrossRef]
7. Xu, D.; Duan, G.; Johnson, W.L. Unusual Glass-Forming Ability of Bulk Amorphous Alloys Based on Ordinary Metal Copper. *Phys. Rev. Lett.* **2004**, *92*, 245504. [CrossRef] [PubMed]
8. Fu, H.M.; Zhang, H.F.; Wang, H.; Hu, Z.Q. Cu-based bulk amorphous alloy with larger glass-forming ability and supercooled liquid region. *J. Alloys Compd.* **2008**, *458*, 390–393. [CrossRef]
9. Li, H.X.; Lu, Z.C.; Wang, S.L.; Wu, Y.; Lu, Z.P. Fe-based bulk metallic glasses: Glass formation, fabrication, properties and applications. *Prog. Mater. Sci.* **2019**, *103*, 235–318. [CrossRef]
10. Gostin, P.F.; Gebert, A.; Schultz, L. Comparison of the corrosion of bulk amorphous steel with conventional steel. *Corros. Sci.* **2010**, *52*, 273–281. [CrossRef]
11. Huang, D.; Li, R.; Huang, L.; Ji, V.; Zhang, T. Fretting wear behavior of bulk amorphous steel. *Intermetallics* **2011**, *19*, 1385–1389. [CrossRef]
12. Suryanarayana, C.; Inoue, A. Iron-based bulk metallic glasses. *Int. Mater. Rev.* **2013**, *58*, 131–166. [CrossRef]
13. Souza, C.A.C.; Ribeiro, D.V.; Kiminami, C.S. Corrosion resistance of Fe–Cr-based amorphous alloys: An overview. *J. Non-Cryst. Solids* **2016**, *442*, 56–65. [CrossRef]
14. Pang, S.J.; Zhang, T.; Asami, K.; Inoue, A. New Fe-Cr-Mo-(Nb, Ta)CB Glassy Alloys with High Glass-Forming Abilit and Good Corrosion Resistance. *Mater. Trans.* **2001**, *42*, 376–379. [CrossRef]
15. Pang, S.J.; Zhang, T.; Asami, K.; Inoue, A. Synthesis of Fe–Cr–Mo–C–B–P bulk metallic glasses with high corrosion resistance. *Acta Mater.* **2002**, *50*, 489–497.
16. Pang, S.J.; Zhang, T.; Asami, K.; Inoue, A. Bulk glassy Fe–Cr–Mo–C–B alloys with high corrosion resistance. *Corros. Sci.* **2002**, *44*, 1847–1856. [CrossRef]

17. Shen, J.; Chen, Q.; Sun, J.; Fan, H.; Wang, G. Exceptionally high glass-forming ability of an FeCoCrMoCBY alloy. *Appl. Phys. Lett.* **2005**, *86*, 151907. [CrossRef]
18. Amiya, K.; Inoue, A. Fe-(Cr,Mo)-(C,B)-Tm Bulk Metallic Glasses with High Strength and High Glass-Forming Ability. *Mater. Trans.* **2006**, *47*, 1615–1618. [CrossRef]
19. Hirata, A.; Hirotsu, Y.; Amiya, K.; Inoue, A. Crystallization process and glass stability of an $Fe_{48}Cr_{15}Mo_{14}C_{15}B_6Tm_2$ bulk metallic glass. *Phys. Rev. B* **2008**, *78*, 144205. [CrossRef]
20. Liu, X.Q.; Zheng, Y.G.; Chang, X.C.; Hou, W.L.; Wang, J.Q.; Tang, Z.; Burgess, A. Microstructure and properties of Fe-based amorphous metallic coating produced by high velocity axial plasma spraying. *J. Alloys Compd.* **2009**, *484*, 300–307. [CrossRef]
21. Zhou, Z.; Wang, L.; Wang, F.C.; Zhang, H.F.; Liu, Y.B.; Xu, S.H. Formation and corrosion behavior of Fe-based amorphous metallic coatings by HVOF thermal spraying. *Surf. Coat. Technol.* **2009**, *204*, 563–570. [CrossRef]
22. Zhang, C.; Liu, L.; Chan, K.C.; Chen, Q.; Tany, C.Y. Wear behavior of HVOF-sprayed Fe-based amorphous coatings. *Intermetallics* **2012**, *29*, 80–85. [CrossRef]
23. Wang, Y.; Zheng, Y.G.; Ke, W.; Sun, W.H.; Hou, W.L.; Chang, X.C.; Wang, J.Q. Slurry erosion–corrosion behaviour of high-velocity oxy-fuel (HVOF) sprayed Fe-based amorphous metallic coatings for marine pump in sand-containing NaCl solutions. *Corros. Sci.* **2011**, *53*, 3177–3185. [CrossRef]
24. Huang, Y.; Guo, Y.; Fan, H.; Shen, J. Synthesis of Fe–Cr–Mo–C–B amorphous coating with high corrosion resistance. *Mater. Lett.* **2012**, *89*, 229–232. [CrossRef]
25. Tului, M.; Bartuli, C.; Bezzon, A.; Marino, A.L.; Marra, F.; Matera, S.; Pulci, G. Amorphous Steel Coatings Deposited by Cold-Gas Spraying. *Metals* **2019**, *9*, 678. [CrossRef]
26. Wang, Y.; Zheng, Y.G.; Ke, W.; Sun, W.H.; Wang, J.Q. Corrosion of high-velocity oxy-fuel (HVOF) sprayed iron-based amorphous metallic coatings for marine pump in sodium chloride solutions. *Mater. Corros.* **2011**, *63*, 685–694. [CrossRef]
27. Farmer, J.C.; Choi, J.S.; Saw, C.; Haslem, J.; Day, D.; Hailey, P.; Lian, T.; Rebak, R.; Perepezko, J.; Payer, J. Iron-Based Amorphous-Metals: High-Performance Corrosion-Resistant Materials (HPCRM) Development. *Metall. Mater. Trans.* **2009**, *40*, 1289–1305. [CrossRef]
28. Bolelli, G.; Bonferroni, B.; Laurila, J.; Lusvarghi, L.; Milanti, A.; Niemi, K.; Vuoristo, P. Micromechanical properties and sliding wear behaviour of HVOF-sprayed Fe-based alloy coatings. *Wear* **2012**, *276–277*, 29–47. [CrossRef]
29. Botta, W.J.; Berger, J.E.; Kiminami, C.S.; Roche, V.; Nogueira, R.P.; Bolfarini, C. Corrosion resistance of Fe-based amorphous alloys. *J. Alloys Compd.* **2014**, *586*, S105–S110. [CrossRef]
30. Wang, M.; Zhou, Z.; Wang, Q.; Wang, Z.; Zhang, X.; Liu, Y. Role of passive film in dominating the electrochemical corrosion behavior of FeCrMoCBY amorphous coating. *J. Alloys Compd.* **2019**, *811*, 151962. [CrossRef]
31. Nayak, S.K.; Faridi, M.A.; Gopi, M.; Kumar, A.; Laha, T. Fe-based metallic glass composite coatings by HVOF spraying: Influence of Mo on phase evolution, wear and corrosion resistance. *Mater. Charact.* **2022**, *191*, 112149. [CrossRef]
32. Zhang, C.; Li, Q.; Xie, L.; Zhang, G.; Mu, B.; Chang, C.; Li, H.; Ma, X. Development of novel Fe-based bulk metallic glasses with excellent wear and corrosion resistance by adjusting the Cr and Mo contents. *Intermetallics* **2023**, *153*, 107801. [CrossRef]
33. Li, X.; Zhao, X.; Liu, F.; Yu, L.; Wang, Y. Effect of C addition on the corrosion properties of amorphous Fe-based alloys. *Int. J. Mod. Phys. B* **2019**, *33*, 1940006. [CrossRef]
34. Li, X.; Zhao, X.; Lv, F.; Liu, F.; Wang, Y. Improved corrosion resistance of new Fe-based amorphous alloys. *Int. J. Mod. Phys. B* **2017**, *31*, 1744010. [CrossRef]
35. Yang, Y.; Zhang, C.; Peng, Y.; Yu, Y.; Liu, L. Effects of crystallization on the corrosion resistance of Fe-based amorphous coatings. *Corros. Sci.* **2012**, *59*, 10–19. [CrossRef]
36. Zhang, C.; Zhang, Z.-W.; Chen, Q.; Liu, L. Effect of hydrostatic pressure on the corrosion behavior of HVOF-sprayed Fe-based amorphous coating. *J. Alloys Compd.* **2018**, *758*, 108–115. [CrossRef]
37. Li, Z.; Zhang, C.; Liu, L. Wear behavior and corrosion properties of Fe-based thin film metallic glasses. *J. Alloys Compd.* **2015**, *650*, 127–135. [CrossRef]
38. Liao, P.-H.; Jian, J.-W.; Tsay, L.-W. The Corrosion and Wear-Corrosion of the Iron-Base Amorphous Coating Prepared by the HVOF Spraying. *Metals* **2023**, *13*, 1137. [CrossRef]
39. Janssen, G.C.A.M.; Abdalla, M.M.; Keulen, F.V.; Pujada, B.R.; Venrooy, B.V. Celebrating the 100th anniversary of the Stoney equation for film stress: Developments from polycrystalline steel strips to single crystal silicon wafers. *Thin Solid Films* **2009**, *517*, 1858–1867. [CrossRef]
40. Wagner, C.D.; Riggs, W.M.; Davis, L.E.; Moulder, J.F.; Muilenberg, G.E. *Handbook of X-ray Photoelectron Spectroscopy*; Perkin-Elmer Corporation: Eden Prairie, MN, USA, 1979; pp. 190–195.
41. Naumkin, A.V.; Kraut-Vass, A.; Powell, C.J. *NIST X-ray Photoelectron Spectroscopy Database*; Measurement Services Division of the National Institute of Standards and Technology (NIST) Technology Services: Gaithersburg, MD, USA, 2008.
42. Xia, H.; Chen, Q.; Wang, C. Evaluating corrosion resistances of Fe-based amorphous alloys by YCr/Mo values. *J. Rare Earths* **2017**, *35*, 406–411. [CrossRef]
43. Tian, W.P.; Yang, H.W.; Zhang, S.D. Synergistic Effect of Mo, W, Mn and Cr on the Passivation Behavior of a Fe-Based Amorphous Alloy Coating. *Acta Metall. Sin.* **2018**, *31*, 308–320. [CrossRef]
44. Jiang, F.X.; Chen, D.; Zhou, G.W.; Wang, Y.N.; Xu, X.H. The dramatic enhancement of ferromagnetism and band gap in Fe-doped In_2O_3 nanodot arrays. *Sci. Rep.* **2018**, *8*, 2417. [CrossRef] [PubMed]

45. Lin, H.E.; Kubota, Y.; Katayanagi, Y.; Kishi, T.; Yano, T.; Matsushita, N. Solution-processed $Cu_{2-x}O$-Fe_2O_3 composites as novel supercapacitor anodic materials. *Electrochim. Acta* **2019**, *323*, 134794. [CrossRef]
46. Luo, C.; Li, D.; Wu, W.; Yu, C.; Li, W.; Pan, C. Preparation of 3D reticulated ZnO/CNF/NiO heteroarchitecture for high-performance photocatalysis. *Appl. Catal. B Environ.* **2015**, *166–167*, 217–223. [CrossRef]
47. Prietoa, P.; Nistor, V.; Nounehb, K.; Oyamac, M.; Abd-Lefdil, M.; Díaz, R. XPS study of silver, nickel and bimetallic silver-nickel nanoparticles prepared by seed-mediated growth. *Appl. Surf. Sci.* **2012**, *258*, 8807–8813. [CrossRef]
48. Mansour, A.N. Nickel Monochromated Al Kα XPS Spectra from the Physical Electronics Model 5400 Spectrometer. *Surf. Sci. Spectra* **1994**, *3*, 221–230. [CrossRef]
49. Lee, C.Y.; Lin, T.J.; Sheu, H.H.; Lee, H.B. Study on corrosion and corrosion-wear behavior of Fe-based amorphous alloy coating prepared by high velocity oxygen fuel method. *J. Mater. Res. Technol.* **2021**, *15*, 4880–4895. [CrossRef]
50. Ning, W.; Zhai, H.; Xiao, R.; He, D.; Liang, G.; Wu, Y.; Li, W.; Li, X. The Corrosion Resistance Mechanism of Fe-Based Amorphous Coatings Synthesised by Detonation Gun Spraying. *J. Mater. Eng. Perform.* **2020**, *29*, 3921–3929. [CrossRef]
51. Huang, Y.; Ge, J.; Hu, J.; Zhang, J.; Hao, J.; Wei, Y. Nitrogen-Doped Porous Molybdenum Carbide and Phosphide Hybrids on a Carbon Matrix as Highly Effective Electrocatalysts for the Hydrogen Evolution Reaction. *Adv. Energy Mater.* **2018**, *8*, 1701601. [CrossRef]
52. Vasquez, R.P. X-ray photoelectron spectroscopy study of Sr and Ba compounds. *J. Electron Spectrosc. Relat. Phenom.* **1991**, *56*, 217–240. [CrossRef]
53. He, J.W.; Xu, X.; Corneille, J.S.; Goodman, D.W. X-ray photoelectron spectroscopic characterization of ultra-thin silicon oxide films on a Mo(100) surface. *Surf. Sci.* **1992**, *279*, 119–126. [CrossRef]
54. Óvári, L.; Kiss, J. Angle-resolved XPS investigations of the interaction between O_2 and Mo_2C/Mo(100). *Vacuum* **2005**, *80*, 204–207. [CrossRef]
55. Dahle, S.; Gustus, R.; Viöl, W.; Maus-Friedrichs, W. DBD Plasma Treatment of Titanium in O_2, N_2 and Air. *Plasma Chem. Plasma Process.* **2012**, *32*, 1109–1125. [CrossRef]
56. Hashimoto, K. 2002 W.R. Whitney Award Lecture: In Pursuit of New Corrosion-Resistant Alloys. *Corrosion* **2002**, *58*, 715–722. [CrossRef]

Disclaimer/Publisher's Note: The statements, opinions and data contained in all publications are solely those of the individual author(s) and contributor(s) and not of MDPI and/or the editor(s). MDPI and/or the editor(s) disclaim responsibility for any injury to people or property resulting from any ideas, methods, instructions or products referred to in the content.

Article

Characterization of Interfacial Corrosion Behavior of Hybrid Laminate EN AW-6082 ∪ CFRP

Alexander Delp [1,*], Shuang Wu [2], Jonathan Freund [3], Ronja Scholz [1], Miriam Löbbecke [3], Thomas Tröster [2], Jan Haubrich [3] and Frank Walther [1]

[1] Chair of Materials Test Engineering (WPT), TU Dortmund University, Baroper Str. 303, D-44227 Dortmund, Germany
[2] Automotive Lightweight Design (LiA), Paderborn University, Warburger Str. 100, D-33098 Paderborn, Germany
[3] German Aerospace Center (DLR e.V. Deutsches Zentrum für Luft-und Raumfahrt), Institute of Materials Research, Linder Höhe, D-51147 Cologne, Germany
* Correspondence: alexander.delp@tu-dortmund.de; Tel.: +49-231-755-90173

Citation: Delp, A.; Wu, S.; Freund, J.; Scholz, R.; Löbbecke, M.; Tröster, T.; Haubrich, J.; Walther, F. Characterization of Interfacial Corrosion Behavior of Hybrid Laminate EN AW-6082 ∪ CFRP. *Materials* **2024**, *17*, 1907. https://doi.org/10.3390/ma17081907

Academic Editor: Andrea Bernasconi

Received: 28 February 2024
Revised: 15 April 2024
Accepted: 17 April 2024
Published: 19 April 2024

Copyright: © 2024 by the authors. Licensee MDPI, Basel, Switzerland. This article is an open access article distributed under the terms and conditions of the Creative Commons Attribution (CC BY) license (https://creativecommons.org/licenses/by/4.0/).

Abstract: The corrosion behavior of a hybrid laminate consisting of laser-structured aluminum EN AW-6082 ∪ carbon fiber-reinforced polymer was investigated. Specimens were corroded in aqueous NaCl electrolyte (0.1 mol/L) over a period of up to 31 days and characterized continuously by means of scanning electron and light microscopy, supplemented by energy dispersive X-ray spectroscopy. Comparative linear sweep voltammetry was employed on the first and seventh day of the corrosion experiment. The influence of different laser morphologies and production process parameters on corrosion behavior was compared. The corrosion reaction mainly arises from the aluminum component and shows distinct differences in long-term corrosion morphology between pure EN AW-6082 and the hybrid laminate. Compared to short-term investigations, a strong influence of galvanic corrosion on the interface is assumed. No distinct influences of different laser structuring and process parameters on the corrosion behavior were detected. Weight measurements suggest a continuous loss of mass attributed to the detachment of corrosion products.

Keywords: CFRP; corrosion exposure; EN AW-6082; galvanic corrosion; hybrid laminate; intrinsic bonding; laser structuring; linear sweep voltammetry; materials engineering; surface modification

1. Introduction

The improvement of lightweight structures for the automotive sector is an often-considered approach for reducing energy demand during operation. Sector-specific materials, e.g., carbon fiber-reinforced polymers (CFRP) with high specific strength, and light metals, such as aluminum with high impact strength and excellent formability, are predestined for synergetic hybridization. This combination of the individual advantages can be achieved by means of multi-material systems, which are also termed hybrid structures [1–4]. The joining technology is of utmost importance as it defines the load transmission within the hybrid structure. Common techniques are bolting, blind riveting, and welding, which lead to defects or thermal disruptions [5,6]. An alternative technique is adhesive bonding [7]. When adhesively bonded materials originating from different material groups, surface properties are important regarding joining technology, while different electrochemical potentials are challenging with respect to corrosion protection [8,9].

A commonly used aluminum alloy that inherits beneficial corrosion resistance and mechanical strength is EN AW-6082-T6 [10,11]. The surface functionalization of this alloy, which includes cleaning and surface structuring, is possible via laser pre-treatment [12,13]. Depending on the pretreatment parameters, this functionalization can increase the joint strength [14–16]. The combination of laser structuring of the metal component and co-curing of the metal sheet and CFRP allows a performant hybrid system made of EN

AW-6082 ∪ CFRP [17]. Nevertheless, galvanic insulation of the carbon fibers and the aluminum alloy surface cannot be ensured. The enlarged surface roughness due to laser structuring enhances this effect by producing a larger surface area and material tips, which lead to a greater probability of direct fiber–metal contact. Therefore, it is necessary to characterize the corrosion behavior by means of short-term corrosion testing procedures, i.e., linear sweep voltammetry (lsv) [18,19], and long-term corrosion testing, i.e., corrosion exposure testing with a focus on laser structure. The corrosion reaction is triggered by the CFRP component, which itself usually remains undamaged, while the metal component dissolves [20]. EN AW-6082 is susceptible to pitting corrosion and fragmentation of the oxide layer in NaCl solutions [21]. The combination of both leads to enhanced corrosion processes [20]. This study aims to understand the short- and long-term corrosion morphology evolution of a hybrid laminate consisting of EN AW-6082 ∪ CFRP when considering different laser structuring parameters at the surface of the Al component. Excitation linear sweep voltammetry and corrosion exposure tests were performed. The mass changes were evaluated. The corrosion morphology evolution was characterized by means of light microscopy. The corrosion products were classified according to their appearance and composition via SEM and EDX.

2. Materials and Methods

2.1. Material

The hybrid laminate consists of laser structured EN AW-6082 T6 sheet (Al), t = 2 mm, intrinsically bonded to five unidirectional layers CFRP Sigrapreg C (U230 0/NF E20/39%; SGL Carbon SE, Wiesbaden, Germany) in a P200S hot press (VOGT Labormaschinen GmbH, Berlin, Germany), as shown in Figure 1a. Laser structuring was realized with a Nd:YAG-Laser CL20 (Clean Lasersysteme GmbH, Herzogenrath, Germany), wavelength λ = 1064 nm, using three sets of parameters (lp), varying laser frequency f, laser power P, laser spot overlap o, and number of scans N, as listed in Table 1.

Figure 1. (a) EN AW-6082 ∪ CFRP embedded in epoxy resin and polished up to grain size of 1 mm, (b) micrograph of the three different laser parameter lp1–3 and as-rolled condition. CFRP component on top with fiber orientation horizontal and Al component (lp0) on the bottom, (c) experimental setup for corrosion exposure tests in aqueous solution of H_2O and NaCl consisting of eight single epoxy-embedded specimens of EN AW-6082 ∪ CFRP and one epoxy-embedded specimen of three sheets of EN AW-6082.

Table 1. Parameter of laser structuring (lp) on EN AW-6082 T6 sheet consisting of laser frequency f, laser Power P, laser spot overlap o, and number of scans N [22].

Laser Parameter	f (kHz)	P (W)	o (10^{-2})	N
lp0	N/A	N/A	N/A	N/A
lp1	60	20	10	5
lp2	40	20	50	1
lp3	60	15	50	1

After laser structuring, CFRP was bonded to Al via co-curing. The hybrid systems, tested via corrosion exposure tests, were produced with a temperature T = 150 °C and pressure p = 0.5 MPa (for lp2) or p = 0.8 MPa (for lp1) for t = 5 min. Additional specimens with the following bonding parameters for lsv investigation of the influence of laser parameters were used: T = 160 °C, p = 0.8 MPa, t = 20 min. The resulting morphology of the laser structure can be seen in Figure 1b. Details of all used specimens are listed in Table 2, including the total exposure time Σt_{exp} and reference specimen.

Table 2. Prepared specimen configuration and experimental procedure for corrosion exposure tests of EN AW-6082 ∪ CFRP in aqueous solution of $0.1\ mol_{NaCl}/L_{H2O}$, including an indication of whether lsv was performed.

Parameter Set	Fiber Direction (°)	Number	lsv	Total Exposure Time (h)
lp1	90	K1lp1	yes	168
lp1	0	K2lp1	yes	168
lp1	0	K3lp1	no	744
lp1	90	K4lp1	no	744
lp2	0	K1lp2	no	168
lp2	90	K2lp2	no	168
lp2	90	K3lp2	no	744
lp2	0	K4lp2	no	744
EN AW-6082	N/A	ref	no	744
lp0	0	lp0	yes	N/A
lp1	0	lp1	yes	N/A
lp2	0	lp2	yes	N/A
lp3	0	lp3	yes	N/A
EN AW-6082	N/A	Al	yes	N/A
CFRP	N/A	CFRP	yes	N/A

2.2. Specimen Preparation

Eight specimens, four pretreated with lp1 and four pretreated with lp2, of EN AW-6082 ∪ CFRP laminate were prepared with two different fiber orientations, each (90° and 0° with regard to longitudinal fiber direction) for corrosion exposure testing (initial letter "K"). Two different fiber orientations were chosen to take the different distribution of carbon fiber volume content at the interface into account. For lsv, four specimens with lp0 to lp3 were prepared. Additionally, two EN AW-6082 specimens and one specimen of pure CFRP were provided. All specimens were embedded in epoxy resin EpoFix (Struers GmbH, Willich, Germany), ground, and polished up to particle size s = 1 µm. A prepared specimen is shown in Figure 1a. For characterization of the epoxy water absorption, a total of thirty cylindrical specimens of pure epoxy were prepared, using two diameters (d_1 = 40 mm; d_2 = 30 mm) and five filling levels (5/5 to 1/5), with a coverage of three specimens for each diameter. This resulted in six sets of specimens, three per diameter and five filling levels each, i.e., ten different masses. To ensure conductivity, the back side of specimens, used for lsv, were exposed by grinding and coated with conductive silver paint.

2.3. Experimental Setup

All weight measurements were performed using an analytical scale AUW220D (Shimadzu Corp., Kyoto, Japan), e = 1 mg; d = 0.01 mg. For iterative condition monitoring, a digital light microscope VHX-7000 (Keyence Corp., Osaka, Japan) was employed. Corrosion products were characterized via a scanning electron microscope (SEM) Mira 3 XMU (Tescan, Dortmund, Germany) and energy-dispersive X-ray spectroscopy (EDX) with Octane Elect

Plus detector (Ametek GmbH, Meerbusch, Germany). Surface proportions of Al and CFRP were measured by means of ImageJ 1.53r.

For linear-sweep voltammetry, a three-electrode setup in a customized PMMA cell, described in [18], and a Gamry Interface 1000 potentiostat (Gamry Instruments Inc., Warminster, PA, USA) with Ag/AgCl as reference inside of a Luggin capillary (RE-1CP, ALS Co., Ltd., Tokyo, Japan), with a constant potential of 45.3 mV against saturated calomel electrode (RE-2BP, ALS Co., Ltd., Tokyo, Japan; +0.242 V vs. standard hydrogen electrode [23]) was employed.

2.4. Testing Method

Specimens for corrosion exposure testing were weighed, characterized via light microscopy, and placed in a beaker, as shown in Figure 1c, filled with 0.1 mol/L NaCl (purity > 99.8%, Ca < 0.01%, Mg < 0.002%, abherents < 0.0015%, batch 073196635; Carl Roth GmbH, Karlsruhe, Germany) in deionized H_2O (conductivity κ < 2.5 µS/cm) equal to 11.688 $g_{NaCl}/2\ L_{H2O}$. Conductivity of the mixed solution was measured (InLab 742 Mettler Toledo, DC, USA). All specimens were investigated via light microscopy after 24 h, 48 h, 72 h, 96 h, and 168 h. Additionally, four selected specimens were investigated after 336 h, 504 h, and 744 h to characterize the long-term corrosion evolution at the surface. Before light microscopy analysis, specimens were dipped into deionized H_2O, air-dried, and weighed. All exposed specimens were replaced in the beaker after light microscopy measurements. The medium was replaced every 168 h. To prevent movement, evaporated water was not replaced. The short-term corrosion behavior of K1lp1 and K2lp2 (see Table 1) was investigated by means of lsv in accordance with ISO 17475 [24] before corrosion exposure tests and after t_{exp} = 168 h. Further specimens, no. lp0 to lp3, CFRP, and Al (see Table 1) were investigated by means of lsv. After a 1 h set-up time to reach open circuit potential (OCP), the measurement was realized with a potential feed $\Delta \dot{E}$ = 1 mVs^{-1} and a potential range ΔE of ±300 mV vs. OCP. After the first lsv of K1lp1 and K2lp1, before corrosion exposure, the specimens were polished again. The second measurement was conducted on the corroded surface. Specimens lp0 to lp3 were tested three times each. After each lsv, the surface was polished again.

3. Results

3.1. Specimen Dimensions

The diameter and height of embedded specimens, as well as the area of the CFRP and Al fractions, were measured and are listed in Table 3. Due to the geometries of the initial EN AW-6082 ∪ CFRP with regard to fiber orientation, the volume of specimens with 90° orientation is higher, with areas of approximately 11–12 mm² CFRP and 18–19 mm² Al, while for 0° orientation, the areas are approximately 29–32 mm² CFRP and 19–20 mm².

Table 3. Specimen dimensions and partial surfaces of CFRP and Al, as well as the total volume.

Number	Diameter (mm)	Height (mm)	Area CFRP (mm²)	Area Al (mm²)	Volume (mm³)
K1lp1	29.8	17.3	11.26	18.02	12,054
K2lp1	30.0	11.3	19.25	30.18	7988
K3lp1	29.8	12.5	19.70	31.11	8703
K4lp1	29.8	17.4	18.23	11.84	12,091
K1lp2	30.1	12.2	19.17	31.53	8660
K2lp2	29.8	17.1	10.47	18.64	11,935
K3lp2	30.0	17.1	10.94	18.58	12,080
K4lp2	29.9	12.7	17.82	29.63	8927
ref	30.0	11.4	00.00	00.00	8062

3.2. Weight Measurement

The weights of the specimens continuously increased in a range of 0.04 g during the testing period. Detailed records and are shown in Table 4 for over a period of 744 h, i.e., 4 weeks. The percentage increases were calculated with regard to initial weights of 0.25–0.27%, while the mass increase for reference EN AW-6082 reached $\Delta m/m_0 = 0.38\%$. This is shown in Figure 2a. The weight increase in reference specimens was distinctly higher than the weight increases in hybrid specimens. Investigations with pure epoxy indicate a continuous solution absorption during a time period of 31 days. The percentage increases were calculated with regard to initial weights for the arithmetic mean and fluctuated between $\Delta m/m_0 = 0.2\%$ for the highest initial weight of approximately m = 42 g to $\Delta m/m_0 = 0.5\%$ for an initial weight of m = 6 g, as shown in Figure 2b. The percentage weight increase is generally higher for specimens with lower initial weights. The blue area in Figure 2b represents the maximum percentage of weight increase, which was measured for hybrid specimens K4lp2 in corrosion exposure tests.

Table 4. Results of weight measurements during corrosion exposure testing.

Weight (g)	0 h (D0)	24 h (D1)	48 h (D2)	72 h (D3)	96 h (D4)	168 h (W1)	336 h (W2)	504 h (W3)	744 h (W4)
K1lp1	14.6750	14.6802	14.6844	14.6855	14.6880	14.6934	n.a.	n.a.	n.a.
K2lp1	10.5495	10.5519	10.5555	10.5567	10.5591	10.5637	n.a.	n.a.	n.a.
K3lp1	10.8513	10.8587	10.862	10.8629	10.8651	10.8706	10.8759	10.8792	10.8794
K4lp1	14.6666	14.6734	14.6788	14.6791	14.6817	14.6886	14.6959	14.7008	14.7012
K1lp2	10.5432	10.5462	10.5493	10.5505	10.5526	10.5568	n.a.	n.a.	n.a.
K2lp2	10.6009	10.6085	10.6130	10.6136	10.6166	10.6217	n.a.	n.a.	n.a.
K3lp2	13.6981	13.7044	13.709	13.7094	13.7129	13.7174	13.7248	13.7296	13.7301
K4lp2	11.1066	11.1121	11.1150	11.1162	11.1190	11.1258	11.1324	11.1376	11.1379
ref	10.1690	10.1740	10.1783	10.1797	10.1828	10.1893	10.1976	10.2035	10.2055

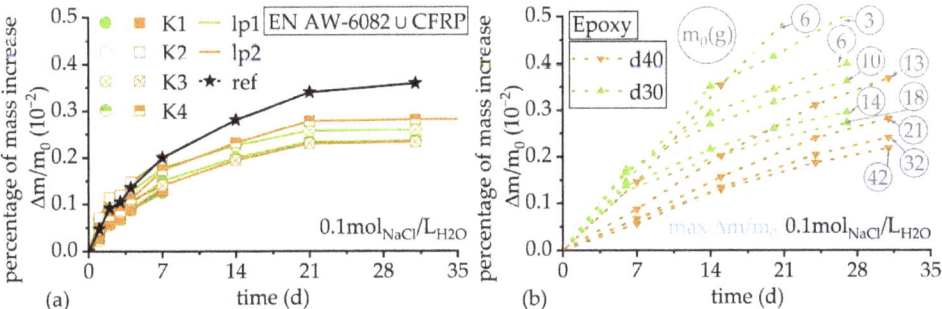

Figure 2. (a) Percentage increase in corrosion exposure specimen mass with regard to initial weight over time. The lines between measurements describe linear approximations for enhanced visualization. (b) Percentage increase in pure epoxy specimens with regard to initial weight over time. The initial weight m_0 is listed in grey, inside a circle. The maximum percentage of mass increase in corrosion exposure specimens is plotted as blue area. Each triangle describes one measurement. The lines between triangles describe linear approximations for enhanced visualization. Two different diameters, d40 = 40 mm and d30 = 30 mm, were considered. The conductivity of the solution continuously increased, while the water volume decreased due to evaporation. This led to an increase in NaCl concentration from $c_0 = 0.10\ mol_{NaCl}/L_{H2O}$ to $c_0 = 0.11\ mol_{NaCl}/L_{H2O}$ within one week, which is shown in Table 5.

Table 5. Evaporated water over a time period of one week and the solution (s) conductivities κ at the start and end of solution usage with s1: t = 0 h...168 h; s2 t = 168 h...336 h; s3 t = 336 h...504 h, and s4 t = 504 h...744 h. Resulting c is calculated by c_0 divided by remaining water volume.

Batch	Initial κ (mS/cm)	Final κ (mS/cm)	Remaining Water Volume (mL)	Resulting c (mol/L)
s1	10.50	11.24	1791	0.1117
s2	10.58	10.89	1817	0.1101
s3	10.29	11.76	1803	0.1109
s4	10.35	11.12	1763	0.1134

3.3. Linear Sweep Voltammetry

The OCP for both fiber orientation showed comparable OCP and current density i_{corr} before exposure testing. This can be seen in Figure 3a. After one week of exposure testing, both Tafel plots, i.e., i_{corr} and OCP, reached comparable trajectories.

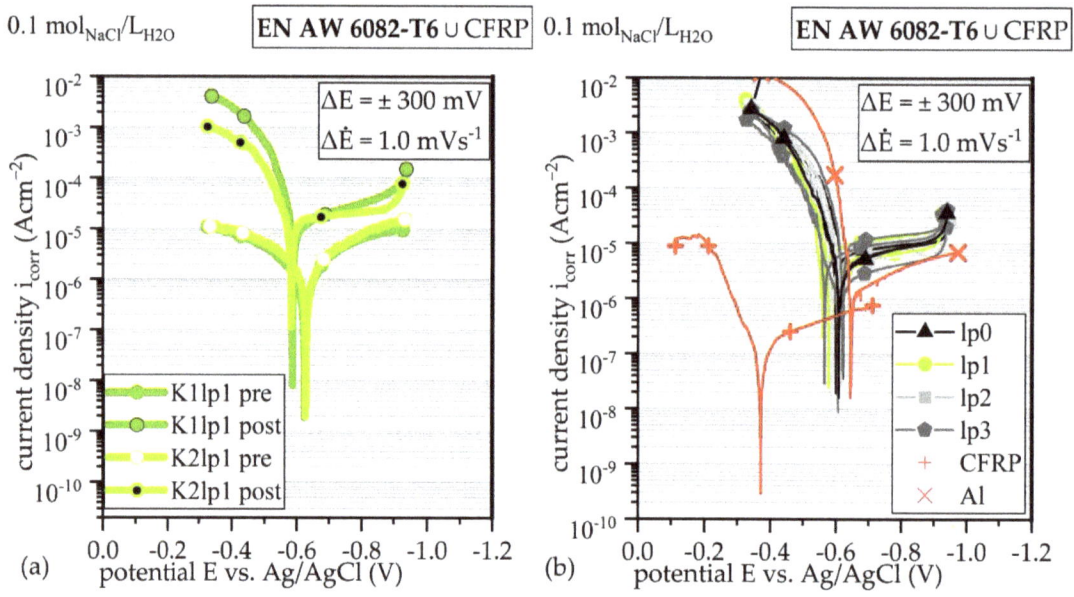

Figure 3. Tafel plots of lsv with logarithmic ordinate for comparison of short-term corrosion behavior: (a) before (pre) and after (post) corrosion exposure testing and (b) short-term corrosion behavior with regard to different lp, as well as pure CFRP and Al.

Comparing lp, lsv detects no differences within a statistical scatter, as shown via Tafel plots in Figure 3b. Pure Al shows higher OCP, while i_{corr} was within a comparable range to the hybrid material. Lsv on CFRP shows distinct differences, as shown in [18]. When comparing the different production parameters between K-series and lp-series (compare Figure 3a,b), no remarkable differences were detectable. Due to the strong oscillation of the OCP in the presence of carbon fibers and the passivation effects of Al, a reliable determination of the corrosion current density, and therefore determination of the mass loss based on Faraday's law, was not possible.

3.4. Microscopic Analyses

The corrosion evolution during exposure testing showed the continuous progress of corrosion development on the surface morphology over the first seven days until week one (W1) for all specimens. Between W1 and week two (W2), and between W2 and week three (W3), there is no distinct change of surface pit morphology, but an increase of corrosion products at EN AW 6082 ∪ CFRP. The evolution of the surface morphology at the reference specimens, as shown in Figure 4, starts at random corrosion nuclei and forms randomly distributed corrosion pits over the whole surface after 24 h (D1). With further corrosion progress, the corrosion pit diameters increase until W1. White, salt-like corrosion products develop at the rims of the corrosion pits, until the entire surface is covered at W1; see Figure 4. In this state, few new corrosion pits emerge (Figure 4 (A)) and the development of new corrosion products and morphology at the surface stagnates.

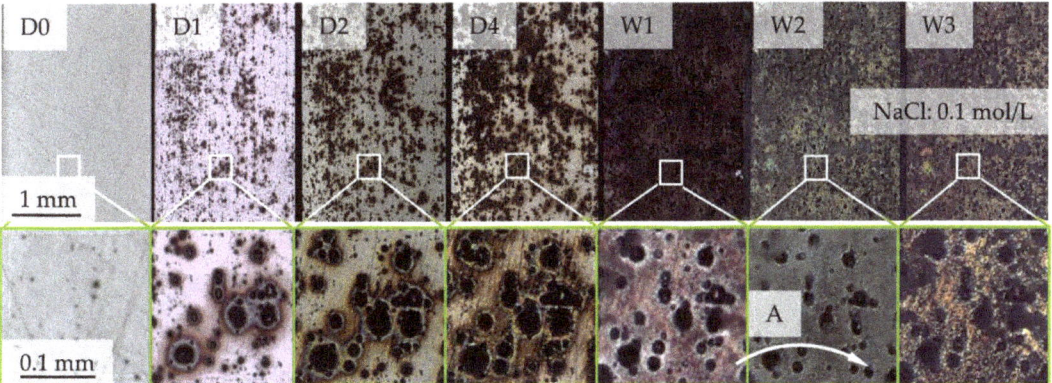

Figure 4. Depiction of corrosion evolution at EN AW-6082 over a time period of three weeks with recordings before exposure (D0), after one, two, and four days (D1, D2, D4) and after one, two, and three weeks (W1, W2, W3). Recorded with confocal light source. The evolution of one example area is marked via in a white box. (A) Traces the appearance of one pitting hole after two weeks via arrow.

SEM investigations revealed that the surface structure of CFRP at the interface remains intact after a corrosion exposure time of t − 168 h, as shown in Figure 5. Despite no visible direct contact between carbon fibers and Al, the Al component is peeled off and forms a trench at the interface (see Figure 5a). The oxide layer shows cracks and aluminum oxides adhere randomly at the surface. Corrosion products, which appear white under light microscopy, are shown in Figure 5b and can be identified as aluminum oxides.

In the case of EN AW-6082 ∪ CFRP hybrids, the number of corrosion pits at the Al–overall surface is distinctly lower, but the diameter is higher. A representative overview for two different fiber orientations and lp is shown in Figure 6.

From D1, salt-like, white corrosion products develop around the interface, independent from fiber orientation or lp. The diameter of the appearing corrosion products is distinctly higher than the pitting hole itself. From D1 to D2, the quantity of detectable single corrosion pits next to the interface is higher when testing in the 0° fiber orientation. From D3, the distribution of corrosion products at the surface is similar, regardless of the lp and fiber orientations. From W1 to W3, the amount of corrosion products increases while the amount of corrosion pits remains similar. Only a minor growth in diameter, as visible in Figure 4, is observed. Due to the drying process (air-drying between the extraction actions), the corrosion products are U-shaped with streamlined distribution around the corrosion pits.

After lsv, Al exhibited no increased corrosion at the interface, as shown in Figure 7. Corrosion pits are distributed with a quantity comparable to corrosion exposure tests at

EN AW-6082 ∪ CFRP, while no corrosion products adhere at the surface. Surfaces of pure Al show the same morphology as the surfaces of EN AW-6082 ∪ CFRP. EDX on corrosion products at the interface areas of EN AW-6082 ∪ CFRP revealed Al, C, and O, as well as small signals for Mo and Cl, but not Na, as shown in Figure 8.

Figure 5. Representative SEM micrographs of surface morphology and corrosion products after an exposure time of t = 168 h on EN AW-6082 ∪ CFRP at specimen K2lp2. (**a**) Overview of the interface with visible trench at the Al component at the interface and corrosion products at the surface; (**b**) detailed overview over salt-like white corrosion products.

Figure 6. Depiction of corrosion evolution at EN AW-6082 ∪ CFRP over a time period of three weeks with recordings before exposure (D0), after one, two, three, and four days (D1, D2, D3, D4), and after one, two, and three weeks (W1, W2, W3) under consideration of fiber orientation and lp, using confocal microscopy for observation. The fiber orientation in length direction is marked with an arrow, the fiber orientation in view direction is marked via dot.

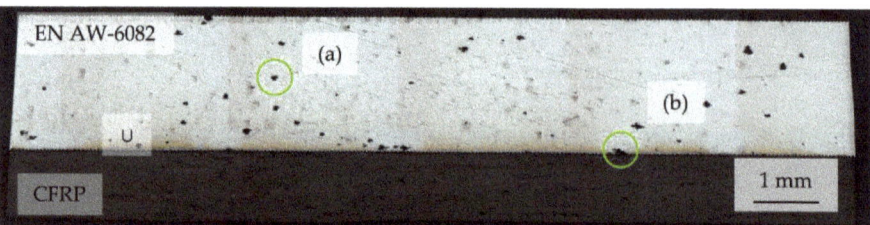

Figure 7. Example micrograph of EN AW-6082 ∪ CFRP after lsv measurement with (**a**) corrosion pits at the Al component, and (**b**) corrosion pits at the interface.

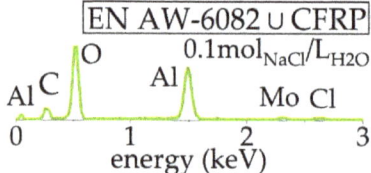

Figure 8. Representative results from EDX on corrosion products after corrosion exposure testing.

4. Discussion

lsv and corrosion exposure testing did not show differences in the corrosion behavior for specimens with different laser pretreatments and production parameters. The ratio of the local surface increase at the interface to the total proportion of the Al component on the overall surface is small. Therefore, there is no quantifiable acceleration of corrosion processes during lsv and exposure testing with regard to different laser structuring after joining. Furthermore, it cannot be ensured that all embedded CFRP fibers are conductively connected, despite the conductive silver coating. Especially when orientated in a transverse direction, the inter-fiber contacts are responsible for electric conductivity. The investigations of Zappalorto et al. [25] and Zhao et al. [26] have proven the distinctly higher electric resistivity in out-of-plane orientation for CFRP. The reason for this is the local insulation of fibers against each other and against the Al component. When orientated in the transverse direction (the fiber's longitudinal axis is transverse to specimen height), the fibers are not connected to the counter-side and are insulated against each other via the matrix. When orientated in the longitudinal direction (the fiber's longitudinal axis is parallel to the specimen height), it is assumed that the higher conductivity of aluminum leads to the conduction of current via Al. As already stated in a previous publication [18], the OCP for EN AW-6082 ∪ CFRP is decreased compared to pure EN AW-6082, while the corrosion current density (y-axis shift, Figure 3) reaches a similar range. Comparable results for CFRP and EN AW-5754-O were observed by Li et al. [6] under the influence of a more aggressive medium (3.5 wt% NaCl). Additionally, it must be considered that it is not possible to distinguish between the conductive and non-conductive areas of the CFRP component after hybridization, i.e., the areas that were actually tested and those that remained non-conductively enclosed by the epoxy resin or are epoxy resin. It has to be assumed that the real surface area of EN AW 6082 ∪ CFRP exposed to corrosion processes and the medium was smaller than the total surface of EN AW-6082 ∪ CFRP, which was used to calculate i_{cor}. Based on those considerations, it is assumed that the real corrosion current density has to be calculated on basis of the EN AW-6082 component of the EN AW-6082 ∪ CFRP hybrid. The narrowed surface area leads to a locally higher i_{cor} than assumed.

Weight measurements indicate a mass decrease by rinsing before microscopy. The epoxy resin of the mount absorbs more solution than the material of the embedded hybrid specimens' increase in mass, indicating a more stable adhesion of corrosion products on EN AW 6082 compared to the surface of hybrid specimens. Therefore, continuous mass loss from EN AW-6082 ∪ CFRP is evident. This implies continuous corrosion processes at the aluminums surface at the interface.

The light microscopic analyses show a strong influence of the CFRP component on the corrosion evolution: large amounts of corrosion products continuously agglomerate at the interface, while the corrosion products at the reference material, pure Al, are evenly distributed. Additionally, only randomly distributed, individual points of the Al component with no direct contact to the interface develop corrosion pits. This is observed through lsv and corrosion exposure testing. SEM investigations confirm that the Al component dissolves while the CFRP component remains intact. The greatest loss of material occurs directly at the interface, which indicates a strong influence of galvanic corrosion. The lsv measurement of EN AW-6082 ∪ CFRP causes no agglomeration of corrosion products at the interface and generates a similar corrosion pit pattern to exposure testing. Therefore, it can again be assumed that the conductivity of the carbon fibers is restricted. Overall, there is a strong influence of galvanic corrosion. Due to the roughness peaks, the surface texture of EN AW-6082 provides randomly distributed contact points to carbon fibers (compare [5]). The corrosion morphology of the reference material during corrosion exposure testing, when compared to the morphology of EN AW-6082 ∪ CFRP, indicates that the interface acts like a sacrificial anode for the aluminum component without direct contact to CFRP at EN AW-6082 ∪ CFRP. Although the surface areas at the EN AW-6082 ∪ CFRP interface will have a different degree of direct connection to CFRP fibers, there is no global difference, as the PDP measurements have shown. Therefore, it can be assumed that the laser structuring has no effect on the global corrosion properties of the EN AW-6082 ∪ CFRP hybrid.

The EDX measurements suggest that the formed corrosion products consist of aluminum hydroxide. Traces of Cl further indicate the conversion of aluminum chlorohydrate [27]. Unlike in [6], without contact to galvanize steel, dawsonite could not be detected. The results also indicate that there are no remaining traces of NaCl after rinsing.

5. Conclusions

The results of this investigation of the corrosion evolution of a hybrid laminate consisting of laser-structured EN AW-6082 ∪ CFRP under the influence of NaCl electrolyte (0.1 mol/L) can be summarized as follows:

1. Galvanic coupling and passivation of the Al component, limited conductivity of the carbon fibers, and the random distribution of exposed fibers at specimen cut leads to high statistical scatter of the lsv measurement as well as uncertainties in the determination of the true surface and therefore limits the applicability of lsv for the hybrid material to qualitative comparisons.
2. A continuous mass loss was detected during corrosion exposure tests and could be allocated to the direct contact region of the interface, proving the dominance of galvanic corrosion on the long-term corrosion evolution of EN AW-6082 ∪ CFRP, while the corrosion mechanism of pure EN AW-6082 under same condition was identified as pitting corrosion. The interface acts comparably to a sacrificial anode for the Al base material.
3. Corrosion products were identified as aluminum oxides.

Author Contributions: Conceptualization, T.T., J.H. and F.W.; methodology, A.D.; validation, R.S. and M.L.; formal analysis, A.D., S.W. and J.F.; investigation, A.D.; resources, T.T., J.H. and F.W.; data curation, A.D., S.W. and J.F.; writing—original draft preparation, A.D., S.W. and J.F.; writing—review and editing, R.S., T.T., J.H., M.L. and F.W.; visualization, A.D., S.W. and J.F.; supervision, T.T., J.H. and F.W.; project administration, A.D., S.W. and J.F.; funding acquisition, T.T., J.H. and F.W. All authors have read and agreed to the published version of the manuscript.

Funding: The research project is funded by the German Research Foundation (Deutsche Forschungsgemeinschaft, DFG), project number 426499947, "Energy-efficient manufacturing and mechanism-based corrosion fatigue characterization of laser-structured hybrid structures". The authors thank the German Research Foundation and the Ministry of Culture and Science of North Rhine-Westphalia

(Ministerium fuer Kultur und Wissenschaft des Landes Nordrhein-Westfalen, MKW NRW) for their financial support within the major research instrumentation program for the in situ atomic force microscope (445052562), including the digital microscope VHX-7000.

Institutional Review Board Statement: Not applicable.

Informed Consent Statement: Not applicable.

Data Availability Statement: The data presented in this study are available on request from the corresponding author. The data are not publicly available due to being part of an ongoing study.

Acknowledgments: The authors thank Jan Philipp Schwalke for his support.

Conflicts of Interest: The authors declare no conflicts of interest.

References

1. Bader, B.; Türck, E.; Vietor, T. Multi material design. A current overview of the used potential in automotive industries. In *Technologies for Economical and Functional Lightweight Design. Zukunftstechnologien für den Multifunktionalen Leichtbau*; Springer: Berlin/Heidelberg, Germany, 2019; pp. 3–13. [CrossRef]
2. Min, J.; Li, Y.; Li, J.; Carlson, B.E.; Lin, J. Friction stir blind riveting of carbon fiber reinforced polymer composite and aluminum alloy sheets. *Int. J. Adv. Manuf. Technol.* **2015**, *76*, 1403–1410. [CrossRef]
3. Vermeeren, C.A.J.R. An historic overview of the development of fibre metal laminates. *Appl. Compos. Mater.* **2003**, *10*, 189–205. [CrossRef]
4. Malekinejad, H.; Carbas, R.J.C.; Akhavan-Safar, A.; Marques, E.A.S.; Castro Sousa, F.; da Silva, L.F.M. Enhancing fatigue life and strength of adhesively bonded composite joints: A comprehensive review. *Materials* **2023**, *16*, 6468. [CrossRef] [PubMed]
5. Pramanik, A.; Basak, A.K.; Dong, Y.; Sarker, P.K.; Uddin, M.S.; Littlefair, G.; Dixit, A.R.; Chattopadhyaya, S. Jointing of carbon fibre reinforced polymer (CFRP) composites and aluminium alloys—A review. *Compos. Part A Appl. Sci. Manuf.* **2017**, *101*, 1–29. [CrossRef]
6. Li, S.; Khan, H.A.; Hihara, L.H.; Cong, H.; Li, J. Corrosion behavior of friction stir blind riveted AL/CFRP and Mg/CFRP joints exposed to a marine environment. *Corros. Sci.* **2018**, *132*, 303–309. [CrossRef]
7. Park, S.Y.; Choi, W.J.; Choi, H.S.; Kwon, H.; Kim, S.H. Recent trends in surface treatment technologies for airframe adhesive bonding processing: A review (1995–2008). *J. Adhes.* **2010**, *86*, 192–221. [CrossRef]
8. Kwon, D.S.; Yoon, S.H.; Hwang, H.Y. Effects of residual oils on the adhesion characteristics of metal-CFRP adhesive joints. *Compos. Struct.* **2019**, *207*, 240–254. [CrossRef]
9. Ireland, R.; Arronche, L.; Saponara, V.L. Electrochemical investigation of galvanic corrosion between aluminum 7075 and glass fibre/epoxy composites modified with carbon nanotubes. *Compos. Part B Eng.* **2012**, *43*, 183–194. [CrossRef]
10. Xhanari, K.; Finšgar, M. The corrosion inhibition of AA6082 aluminium alloy by certain azoles in chloride solution: Electrochemistry and surface analysis. *Coatings* **2019**, *9*, 380. [CrossRef]
11. Lumley, R. *Fundamentals of Aluminium Metallurgy*; Elsevier: Amsterdam, The Netherlands, 2018; ISBN 9780081020647.
12. Rechner, R.; Jansen, I.; Beyer, E. Influence on the strength and aging resistance of aluminium joints by laser pre-treatment and surface modification. *Int. J. Adhes.* **2010**, *30*, 595–601. [CrossRef]
13. Heckert, A.; Zaeh, M.F. Laser surface pre-treatment of aluminium for hybrid joints with glass fibre reinforced thermoplastics. *Phys. Procedia* **2014**, *56*, 1171–1181. [CrossRef]
14. Steinert, P.; Dittes, A.; Schimmelpfennig, R.; Scharf, I.; Lampke, T.; Schubert, A. Design of high strength polymer metal interfaces by laser microstructured surfaces. *IOP Conf. Ser. Mater. Sci. Eng.* **2018**, *373*, 012015. [CrossRef]
15. Zhang, Z.; Shan, J.; Tan, X.; Zhang, J. Improvement of the laser joining of CFRP and aluminum via laser pre-treatment. *Int. J. Adv. Manuf. Technol.* **2017**, *90*, 3465–3472. [CrossRef]
16. Schanz, J.; Meinhard, D.; Dostal, I.; Riegel, H.; De Silva, A.K.M.; Harrison, D.K.; Knoblauch, V. Comprehensive study on the influence of different pretreatment methods and structural adhesives on the shear strength of hybrid CFRP/aluminum joints. *J. Adhes.* **2022**, *98*, 1772–1800. [CrossRef]
17. Wu, S.; Delp, A.; Freund, J.; Walther, F.; Haubrich, J.; Löbbecke, M.; Tröster, T. Adhesion properties of the hybrid system made of laser-structured aluminium EN AW 6082 and CFRP by co-bonding-pressing. *J. Adhes.* **2023**, *100*, 1–29. [CrossRef]
18. Delp, A.; Freund, J.; Wu, S.; Scholz, R.; Löbbecke, M.; Haubrich, J.; Tröster, T.; Walther, F. Influence of laser-generated surface micro-structuring on the intrinsically bonded hybrid system CFRP-EN AW 6082-T6 on its corrosion properties. *Compos Struct.* **2022**, *285*, 115238. [CrossRef]
19. Narsimhachary, D.; Rai, P.K.; Shariff, S.M.; Padmanabham, G.; Mondal, K.; Basu, A. Corrosion behavior of laser-brazed surface made by joining of AA6082 and galvanized steel. *J. Mater. Eng. Perform.* **2019**, *28*, 2115–2127. [CrossRef]
20. Song, G.-L.; Zhang, C.; Chen, X.; Zheng, D. Galvanic activity of carbon fiber reinforced polymers and electrochemical behavior of carbon fiber. *Corros. Commun.* **2021**, *1*, 26–39. [CrossRef]
21. Li, Z.; Peng, M.; Wei, H.; Zhang, W.; Lv, Q.; Zhang, F.; Shan, Q. First-principles study on surface corrosion of 6082 aluminum alloy in H+ and Cl medium. *J. Mol. Struct.* **2023**, *1294*, 136570. [CrossRef]

22. Freund, J.; Lützenkirchen, I.; Löbbecke, M.; Delp, A.; Walther, F.; Wu, S.; Tröster, T.; Haubrich, J. Transferability of the structure-property relationships from laser-pretreated metal-polymer joints to aluminum-CFRP hybrid joints. *J. Compos. Sci.* **2023**, *7*, 427. [CrossRef]
23. McCafferty, E. *Introduction to Corrosion Science*; Springer: New York, NY, USA, 2010. [CrossRef]
24. ISO 17475:2005+Cor.1:2006; German version EN ISO 17475:2008: Corrosion of Metals and Alloys—Electrochemical Test Methods—Guidelines for Conducting Potentiostatic and Potentiodynamic Polarization Measurements. DIN e.V.: Beuth Verlag, Berlin, 2006.
25. Zappalorto, M.; Panozzo, F.; Carraro, P.A.; Quaresimin, M. Electrical response of laminate with a delamination: Modelling and experiments. *Compos. Sci. Technol.* **2017**, *143*, 31–45. [CrossRef]
26. Zhang, D.-Q.; Li, J.; Joo, H.G.; Lee, K.Y. Corrosion properties of Nd:YAG laser-GMA hybrid welded AA6061 Al alloy and its microstructure. *Corros. Sci.* **2009**, *51*, 1399–1404. [CrossRef]
27. Zhao, Q.; Zhang, K.; Zhu, S.; Xu, H.; Cao, D.; Zhao, L.; Zhang, R.; Yin, W. Review on the electrical resistance/conductivity of carbon fiber reinforced polymer. *Appl. Sci.* **2019**, *9*, 2390. [CrossRef]

Disclaimer/Publisher's Note: The statements, opinions and data contained in all publications are solely those of the individual author(s) and contributor(s) and not of MDPI and/or the editor(s). MDPI and/or the editor(s) disclaim responsibility for any injury to people or property resulting from any ideas, methods, instructions or products referred to in the content.

Article

Application of Alcohols to Inhibit the Formation of Ca(II) Dodecyl Sulfate Precipitate in Aqueous Solutions

Csaba Bús [1], Marianna Kocsis [1], Áron Ágoston [2], Ákos Kukovecz [3], Zoltán Kónya [3] and Pál Sipos [1,*]

[1] Department of Molecular and Analytical Chemistry, University of Szeged, Dóm Square 7-8, 6720 Szeged, Hungary; bus.csaba@szte.hu (C.B.); mkocsis@chem.u-szeged.hu (M.K.)
[2] Department of Physical Chemistry and Material Science, University of Szeged, Rerrich Béla Square 1, 6720 Szeged, Hungary
[3] Department of Applied and Environmental Chemistry, University of Szeged, Rerrich Béla Square 1, 6720 Szeged, Hungary; konya@chem.u-szeged.hu (Z.K.)
* Correspondence: sipos@chem.u-szeged.hu

Abstract: The presence of alkaline earth cations, in particular, Ca^{2+} and Mg^{2+} ions in brine, causes undesired effects in solutions containing anionic surfactants because of precipitate formation. In the present study, an anionic surfactant, sodium dodecyl sulfate (SDS), was investigated, focusing on the determination of various properties (surface tension, critical micelle concentration, micelle size, turbidity) in the presence of alcohols and, in particular, the inhibition of the precipitation of SDS with calcium ions. The calcium ions were added to the surfactant in increasing concentrations (3.0–10.0 g/L), and short-carbon-chain alcohols (methanol, ethanol, *n*-propanol and *n*-butanol) were used to shift the onset of precipitate formation. The critical micelle concentration (CMC) of SDS in the presence of alcohols was also determined. It was established that among these alcohols, methanol and ethanol did not exert significant effects on the solubility of the $Ca(DS)_2$ precipitate, while *n*-propanol and *n*-butanol were found to be much more efficient inhibitors. In addition, all the alcohols in the applied concentration range (up to 20 V/V%) were found to decrease the critical micelle concentration of SDS.

Keywords: surfactant; precipitation; tensiometry; critical micelle concentration; sodium dodecyl sulfate

Citation: Bús, C.; Kocsis, M.; Ágoston, Á.; Kukovecz, Á.; Kónya, Z.; Sipos, P. Application of Alcohols to Inhibit the Formation of Ca(II) Dodecyl Sulfate Precipitate in Aqueous Solutions. *Materials* **2024**, *17*, 1806. https://doi.org/10.3390/ma17081806

Academic Editor: Andrei Victor Sandu

Received: 27 February 2024
Revised: 10 April 2024
Accepted: 11 April 2024
Published: 15 April 2024

Copyright: © 2024 by the authors. Licensee MDPI, Basel, Switzerland. This article is an open access article distributed under the terms and conditions of the Creative Commons Attribution (CC BY) license (https://creativecommons.org/licenses/by/4.0/).

1. Introduction

Long-chain alkyl sulfonates and sulfates are a commonly applied anionic surfactants [1]. The main functions of surfactants include reducing interfacial tension (IFT) and the alteration of wettability. In some cases, the IFT between water and oil can be reduced from 20–30 mN/m to a value in the order of 10^{-3} mN/m as a result of surfactant addition [2], but generally, this decrease is lower (i.e., 10 mN/m [3]). Numerous parameters have been described to affect IFT, e.g., temperature, pressure, salinity, surfactant type and concentration, and solute type and concentration [4,5].

Divalent cations, most importantly, alkaline earth metal ions (i.e., Ca(II) and Mg(II)) are known to adversely influence the applicability of surfactants. These cations may be bound to the polar head of the anionic surfactants, which affects surface tension (ST) reduction. In waterflooding precipitation, dynamic conditions need to be considered [6]. Furthermore, anionic surfactants may form precipitates with these divalent cations if the latter are present in sufficient concentrations. For example, precipitation with these divalent cations is not expected in brine water with a divalent cation concentration of less than 15–900 ppm [7,8].

Several studies aimed at the inhibition of such precipitate formation have been published in the open literature. Zhang et al. demonstrated that $[B(OH)_4]^-$ ions are effective complexing agents of bivalent cations, like Ca^{2+} and Mg^{2+}, that increase their effective solubility in the aqueous phase. According to their results, at first, an amorphous precipitate is formed which dissolves as the divalent cation concentration is increased [9].

Khaled et al. applied in situ-generated sodium acrylate to inhibit precipitation caused by divalent cations [10]. They claimed that a delicate interplay among three parallel effects is responsible for the efficiency of the inhibition: the direct complexation of sodium acrylate with the ions present in the solution; the adsorption of sodium acrylate on the crystal surface or at the active growth sites of the solid phase present; and a change in the ionic strength of the solution as a result of sodium acrylate addition, and hence, an increase in the effective solubilities of the calcium and magnesium compounds [10]. The advantage of the in situ production of the inhibitor is the use of hard brines without the need for softening the injection water [11].

Tri-sodium citrate was used to sequester calcium in a study published by Miyazaki et al. The experimental data confirmed that tri-sodium citrate prevents the precipitation caused by calcium ions up to 900 ppm in solutions containing surfactin (a class of biosurfactant with an uncommon cyclic lipopeptide as a head group) [12].

Different electrolytes were reported to reduce the CMC of both anionic and non-ionic surfactants [13]. The added salts exert a shielding effect in solutions of anionic surfactants, which suppresses the dissociation of the ions from the micelle, resulting in a decrease in the repulsive forces between the polar head groups [14].

Alcohols are the most commonly used cosolvents of surfactants to improve their properties. Alcohols were also described as cosurfactants which can stabilize microemulsions in mixtures with oil. Some studies are concerned with the interactions between surfactants and alcohols. The effects of alcohols on CMC depend on the chain length of the alcohols [15]. Alcohols with shorter carbon chains are miscible with water, split between the micelles and the bulk phase [16]. According to Rao et al., in solutions of anionic surfactants the C_4–C_7 alcohols decrease the CMC, which is associated with an increase in the distance between the head groups through solubilization and coordination with the micelle surface. Furthermore, the interaction between the alkyl chains of the surfactants and those of the alcohols results in an increase in entropy [17,18]. Methanol and ethanol are reported to increase the CMC value in solutions of polyoxyethylene lauryl ethers according to the facilitation of solvation by water, while alcohols with longer carbon chains (C_3–C_7) are reported to be able to intercalate into the micellar structure, forming mixed micelles, which also decreases the CMC [19,20].

The CMC of SDS in aqueous solutions has been determined in a large variety of studies by using various experimental techniques (including dynamic light scattering, fluorescent microscopy, UV-Vis spectroscopy, tensiometry and conductometry). The different experimental techniques provide various CMC values of SDS ranging from 3.46 mmol/L to 8.03 mmol/L. These concentration values are equal to a 1.0–2.32 g/L SDS concentration [21–28].

The present paper is concerned with the effect of alcohols on the surfactant precipitation induced by the addition of calcium ions. The results indicate that under well-defined conditions, alcohols with carbon chains longer than C_2 are capable of inhibiting precipitation. The inhibition effects of n-propanol and n-butanol on the calcium ion-induced precipitation in SDS solutions are also presented in this paper.

2. Materials and Methods

Calcium chloride dihydrate ($CaCl_2 \times 2H_2O$), methanol, ethanol, n-propanol and n-butanol were purchased from VWR Chemicals (Debrecen, Hungary). Sodium dodecyl sulfate ($C_{12}H_{25}NaO_4S$, SDS) was purchased from Sigma-Aldrich (St. Louis, MO, USA). All chemicals were used without further purification. All solutions were prepared using MilliQ-MilliPore water, which was produced by reverse osmosis and was further purified by UV irradiation, using a Puranity TU3 UV/UF+ system (VWR, Debrecen, Hungary).

Mass spectrometric measurements were processed with a 1260 Infinity II HPLC setup coupled to a G6125B LC-MSD (mass-selective detection) from Agilent (Santa Clara, CF, USA), applying electrospray ionization (ESI). For analysis, a ~0.075 g/L (75 ppm) aqueous surfactant solution was prepared, and ultrapure water was used as an eluent. The mass spectrometric measurements were carried out in negative-ion mode, scanning the m/z region from 150 to 400.

The exact composition of the SDS was determined by using LC-MS. The mass spectrum of the SDS is shown in Figure 1.

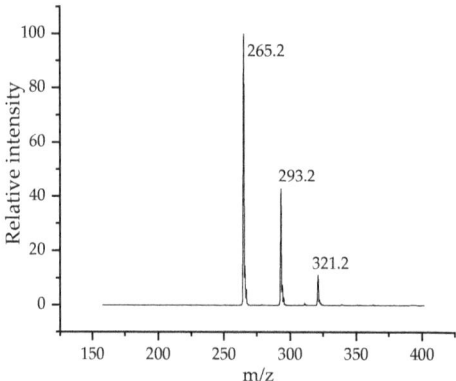

Figure 1. Mass spectrum of the as-received sodium dodecyl sulfate.

The molar mass of the dodecyl sulfate (DS^-) anion was 265.2 g/mol. The recorded mass spectra indicated the DS^- anion to be the main surfactant compound, but two other surfactant homologues were also detected; the 293.2 m/z value indicated a tetradecil sulfate, while the third compound was a hexadecil sulfate with a value of 321.2 m/z. The relative intensity values showed that the investigated surfactant was a mixture of the C_{12} homologue in a proportion of 67% (w/w), C_{14} at 27% (w/w), and C_{16} at 6% (w/w), respectively. From these data, an average molar mass of 299.12 was calculated and used to calculate the molar concentration of SDS.

The turbidity values were measured by using an HI-98703 Precision Turbidity Portable Meter (Hanna Instruments, Keysborough, Australia). The instrument was calibrated according to the calibration standards provided by the supplier (0.1, 15, 100 and 750 NTU). The reported results are based on three independent measurements.

The surface tensions at the air–water interface and the CMC values of the surfactant solutions were determined at 25.0 ± 0.1 °C using a tensiometer (type K100; Krüss, Hamburg, Germany). The measurements were carried out using a Wilhelmy plate.

Dynamic light scattering (DLS) measurements were carried out by using a Malvern Zetasizer Nano ZS instrument (Malvern, UK) operating with a 4 mW helium–neon laser light source (λ = 633 nm) for the determination of micelle size values. The measurements were carried out at room temperature using quartz cuvettes. The measurements were made in back-scattering mode at an incident angle of 173°. At the start of the measurement, a 120 s equilibration time was employed. The reported data were calculated from three independent measurements. The data evaluation was carried out by using the Zetasizer software (7.11). To minimize the effect of the presence of solid contaminants (for example, airborne dust), the solutions were filtered through a 0.1 μm diameter filter before measurements. The refractive index of the solvents were measured by a Mettler Toledo RM50 refractometer (Mettler Toledo, Columbus, OH, USA) at room temperature, and the viscosity values were measured using and Ostwald-type viscosimeter (VWR Chemicals, Debrecen, Hungary) at room temperature (Supporting information, Table S1).

For the measurements, SDS solutions with 0.1, 0.5, 1.0, 2.5 and 5.0 g/L concentrations were prepared using MilliQ water. The effects of the investigated alcohols on the SDS solutions were also studied; to the aqueous SDS solutions, 5.0 V/V%, 10.0 V/V% and 20.0 V/V% methanol, ethanol or n-propanol were added. In the case of n-butanol, the maximum amount of alcohol was 5.0 V/V% because at higher concentrations, *n*-butanol was found to be immiscible with the aqueous surfactant solutions.

From calcium chloride dehydrate, stock solutions with concentrations of 3.0, 5.0 and 10.0 g/L (with respect to calcium ions) were prepared using MilliQ water. The effects of calcium ions on the SDS solutions were determined by using the solution with the highest (5.0 g/L) SDS concentration. The calcium ions were then added to the SDS-containing samples in the form of an aqueous solution. Upon adding alcohols to the samples, their concentrations were made up to 10.0 V/V% (except for *n*-butanol; see above); here, we attempted to take the high environmental load of this process into consideration [29]. For samples where the precipitate was visibly not dissolved, the amount of alcohol was increased to 20.0 V/V%.

During the experiments, the alcohols were added to the surfactant solutions first, before adding the stock solutions containing Ca^{2+}.

3. Results and Discussion

3.1. SDS in Various Alcohol–Water Mixtures

3.1.1. Determination of CMC of SDS in Alcohol–Water Mixtures

The CMC values were determined via surface tension measurements. To the SDS solutions, 5.0, 10.0 and 20.0 V/V% alcohol–water mixture was added as a titrant. The titrated stock solution was prepared with a starting surfactant content of 10.0 g/L. To determine the CMC value, the measured surface tension values were plotted as a function of the logarithm of the surfactant concentration. The titration curve of the aqueous and alcohol-containing SDS solutions are illustrated in Figure 2.

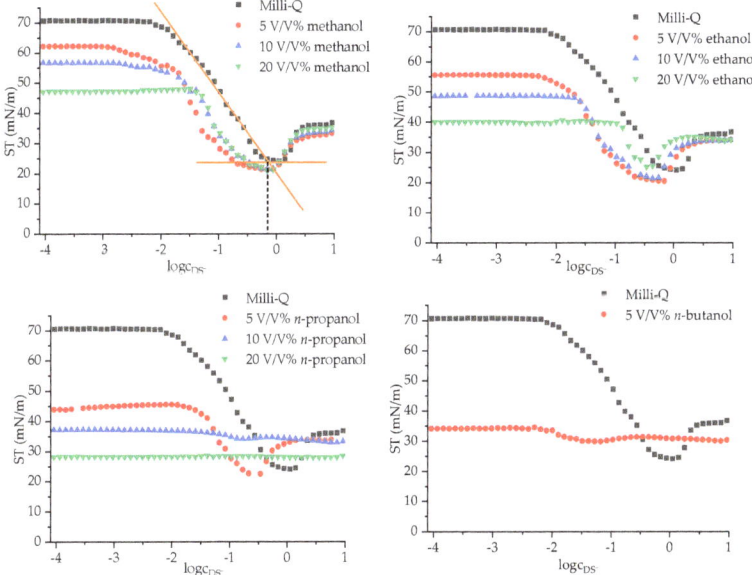

Figure 2. Tensiometric curves of the aqueous and alcohol-containing SDS solutions (c_{DS^-} expressed in g/L).

The critical micelle concentrations obtained from the tensiometric curves are listed in Table 1. The way the CMC was extracted from the curves shown in Figure 1 is demonstrated in the upper left figure. For CMC, the lower intersection point was always chosen; there is another section on the titration curves where the ST becomes constant; according to the literature [30], the second breakpoint at higher SDS concentrations is likely to correspond to a second CMC, with micelles exhibiting structures different from those corresponding to the first one.

Table 1. Critical micelle concentrations (CMC) of SDS in the presence of alcohols.

Alcohol V/V%	Solvent			
	Methanol	Ethanol	n-Propanol	n-Butanol
0		0.63 g/L (2.4 mM)		
5.0	0.17 g/L (0.7 mM)	0.16 g/L (0.6 mM)	0.17 g/L (0.7 mM)	0.03 g/L (0.1 mM)
10.0	0.34 g/L (1.3 mM)	0.21 g/L (0.8 M)	0.14 g/L (0.5 mM)	–
20.0	0.35 g/L (1.3 mM)	0.29 g/L (1.1 mM)		–

On the basis of the tensiometric measurements, the aqueous SDS solution has a somewhat lower CMC than the values reported in the literature; this may be associated with the way of extracting the CMC from the tensiometric curve. Short-carbon-chain-length alcohols (methanol and ethanol) were reported to increase the CMC of SDS [16,31] relative to the CMC in pure water. In our case, the addition of any alcohol resulted in CMC values which increase with increasing alcohol concentration but are always smaller than that in pure water, even at the largest alcohol concentrations. The reason for this most probably lies in the alcohol concentrations being much larger in our systems than in those reported in the literature. The addition of propanol had considerable effects on the tensiometric behavior of SDS: in the presence of 5.0 V/V% and 10.0 V/V% propanol, the CMC decreased, but in the presence of 20.0 V/V%, the titration curve did not have an inflexion point; thus, CMC determination was not possible. This curve indicates that higher n-propanol content inhibits micelle formation. The CMC values measured in n-propanol-containing solvents are consistent with the literature values [31]. The addition of 5.0 V/V% n-butanol considerably decreased the CMC from 0.66 g/L to 0.04 g/L. These results were also in line with data published in the literature [31].

3.1.2. Measurement of Micelle Size Values in the Presence of Alcohols

Dynamic light scattering (DLS) is a conventional technique used to measure micelle size values [32,33]. For the DLS measurements in this study, SDS solutions were prepared in a broad concentration range both below and over the critical micelle concentration determined via tensiometry. The alcohol-containing solutions were prepared with the same composition as the aqueous ones.

SDS was previously reported to form micelles in water with diameters between 3.5 and 4.0 nm [34,35]. The micelle size values obtained by us are listed in Table 2.

Table 2. Micelle size values of SDS in aqueous solutions, obtained from DLS measurements (values marked by an asterisk are supposedly artefacts).

Sample Composition	Particle Size/nm	Polydispersity Index (PdI)
0.1 g/L SDS	167 *	0.3
0.5 g/L SDS	181 *	0.2
1.0 g/L SDS	6.6	0.5
2.5 g/L SDS	4.6	0.5
5.0 g/L SDS	2.9	0.5

The obtained micelle diameter values do not indicate micelle formation below a 1.0 g/L SDS concentrations. Under a 1.0 g/L surfactant concentration, larger aggregates were extracted from the experimental data; they are supposedly fitting artefacts which are associated with the small number of inhomogeneities remaining in the solution (in spite of the careful filtration). Hereafter, micelle diameters over 100 nm should be considered artefacts as well (Supporting Information, Figures S1–S29). In solutions with 1.0–5.0 g/l surfactant concentrations, micelles with 3–7 nm diameters were detected, which corresponds to the micelle size found in the literature [34,35].

The addition of alcohols did not considerably change the micelle size. The corresponding data are shown in Table 3.

Table 3. Micelle size and polydispersity index values (in parentheses) of SDS in the presence of methanol and ethanol.

c (SDS)/g/L	Solvent					
	5.0 V/V% Methanol	10.0 V/V% Methanol	20.0 V/V% Methanol	5.0 V/V% Ethanol	10.0 V/V% Ethanol	20.0 V/V% Ethanol
0.5	123.4 (0.2)	17.8 (0.2)	24.4 (0.2)	191.4 (0.4)	83.8 (0.21)	21.8 (0.05)
1.0	4.9 (0.3)	5.9 (0.4)	6.5 (0.5)	5.0 (0.6)	4.5 (0.49)	3.6 (0.10)
2.5	3.5 (0.4)	3.7 (0.3)	18.0 (0.2)	3.0 (0.4)	2.7 (0.35)	58.0 (0.7)
5.0	2.3 (0.6)	2.3 (0.5)	2.0 (0.1)	2.0 (0.4)	1.7 (0.23)	89.3 (0.5)

During the dissolution of the surfactant in a medium containing methanol or ethanol, a higher alcohol fraction remains in the intermicellar phase and a smaller part is solubilized [16]. Alcohols with carbon chains below C_3 were reported to decrease micelle size [36]. Our experimental results confirmed the CMC-decreasing effect of alcohols obtained from the tensiometric measurements. The micelle-size-decreasing effects of methanol and ethanol were confirmed by these results until 10.0 V/V% alcohol content was reached. However, in the presence of 20.0 V/V% ethanol-containing media, the micelle size values did not follow the same trends; with an increase in the surfactant concentration, the micelle size values increased. The micelle size values determined in n-propanol and n-butanol containing media are shown in Table 4.

Table 4. Micelle size and polydispersity index values of SDS in the presence of n-propanol and n-butanol.

c (g/L)	Solvent			
	5.0 V/V% n-Propanol	10.0 V/V% n-Propanol	20.0 V/V% n-Propanol	5.0 V/V% n-Butanol
0.1	139.9 (0.7)	149.0 (0.6)	156.6 (0.3)	0.70 (0.8)
0.5	51.0 (0.5)	25.6 (0.5)	56.5 (0.8)	0.9 (0.5)
1.0	3.1 (0.5)	53.8 (0.5)	23.2 (0.6)	1.4 (0.42)
2.5	1.7 (0.4)	1.5 (0.7)	1.5 (0.6)	1.3 (0.4)
5.0	1.2 (0.3)	1.0 (0.4)	0.9 (0.4)	1.0 (0.3)

The results obtained for n-propanol- and n-butanol-containing solutions follows trends similar to those obtained for methanol. Increasing the surfactant concentration did not decrease the micelle size until 1.0 g/L-containing SDS solutions were reached; moreover, the CMC of SDS was found to be lower than 0.5 g/L in the presence of n-propanol. The values recorded in 20.0 V/V% n-propanol indicate the inhibition of micelle formation (Supporting Information Figures S30–S41), or even the dissolution of the micelles. These results confirm the results extracted from the tensiometric curves.

n-butanol was determined to be highly solubilized in the micelles and to cause a rapid decrease in micelle size [37]. Our results did not indicate this considerable decrease until a 2.5 g/L surfactant concentration. Moreover, in the case of n-butanol, all the studied samples contained SDS in concentrations above the CMC. According to the observed particle size values recorded in n-butanol-containing media and comparing these with the tensiometric results, it is indicated that in the presence of n-butanol, a molecularly dispersed solution is formed instead of a colloid solution with a micellar structure (Supporting Information Figures S42–S46).

3.2. Inhibition Effect of Alcohols on the Precipitation of DS⁻ with Calcium Ions

3.2.1. Determination the Inhibition Effects of Alcohols on the Precipitation Using Turbidimetry

The precipitation of DS^- with calcium ions was investigated in solutions containing 5.0 g/L SDS. During the first measurement, the calcium ion concentration was varied from 3.0 g/L to 10.0 g/L. First, the alcohols were added to the samples in a 10.0 V/V% concentration (in the case of n-butanol, 5.0 V/V%). In samples where large amounts of precipitate occurred (ethanol and methanol), the amounts of the added alcohols were increased to 20.0 V/V%. The precipitation was characterized by using turbidimetry, which is an effective method to make distinctions between transparent and precipitated solutions.

The turbidity values of the calcium- and alcohol-containing samples are shown in Figure 3.

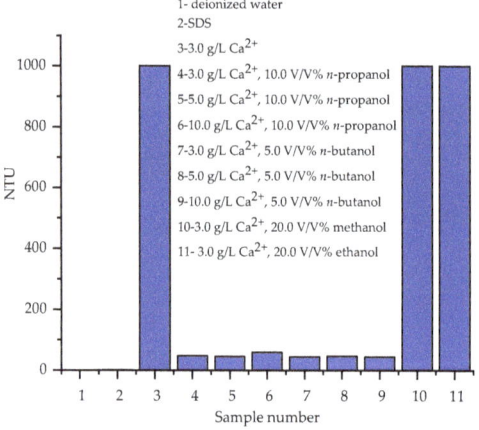

Figure 3. Turbidity values (in arbitrary unit) of Ca^{2+}-containing SDS solutions (c_{SDS} = 5.0 g/L, $c_{Ca^{2+}}$ = 3.0–10.0 g/L) in the presence of various alcohols.

The turbidity values approximately represent the inhibition effects of alcohols on the precipitation of DS^- with Ca^{2+} ions. The addition of 10.0 V/V% n-propanol enhanced the solubility of $Ca(DS)_2$ precipitates and successfully inhibited precipitation at all investigated Ca^{2+} concentrations. The addition of 5.0 V/V% n-butanol to the calcium ion-containing SDS solutions resulted in similar turbidity values compared to the n-propanol-containing ones, these turbidity values confirmed the inhibition effects of n-butanol on calcium ion-induced precipitation in SDS solutions. In the presence of 20.0 V/V% methanol and ethanol, considerably higher (>1000) NTU values were measured. Such high turbidity values showed the presence of high amounts of solid precipitates in the samples. Therefore, methanol and ethanol were found to have no considerable effect on the solubility of the precipitates, even at the highest 20.0 V/V% concentration.

The inhibition effects of n-propanol and n-butanol were further studied by increasing the calcium ion concentration in the samples. Solutions with 10.0 V/V% n-propanol and 5.0 V/V% n-butanol containing 5.0 g/L SDS were prepared by adding the calcium ion-containing stock solutions. Of these solutions, the n-propanol-containing ones were found to be "calcium resistant" until the above-mentioned 10.0 g/L Ca^{2+} concentration was reached, while the n-butanol-containing samples remained transparent even after reaching a 100.0 g/L Ca^{2+} concentration. Although the metal ion concentration was not increased above this concentration, this experiment indicated that the addition of n-butanol results in the collapse of the SDS's micellar structure and therefore inhibits the precipitation of $Ca(DS)_2$.

The methanol- and ethanol-containing solvents were also further investigated to determine the minimum amount of calcium which induces precipitation in the presence of these alcohols. During the experiments, 5.0 g/L SDS stock solutions with 10.0 V/V% and 20.0 V/V% methanol and ethanol were prepared. A total of 6.0 mL of the alcohol-containing solution was added to the sample holder of the turbidity meter. Ca^{2+} titrant solution (1.5 g/L) was added to the surfactant solution in 100 µL increments and the turbidity was measured after each step. The calcium solution was added to the alcohol-containing solution until precipitation was detected, and the surfactant–calcium ion molar ratios were calculated.

The experimental results are illustrated in Figure 4.

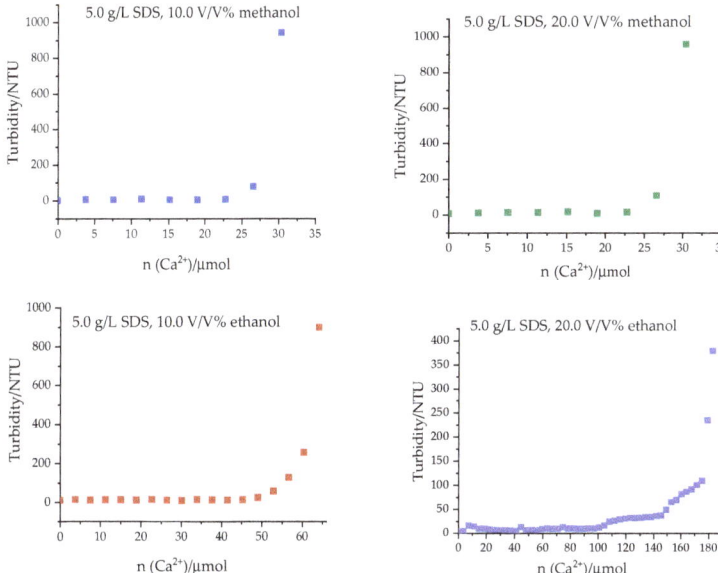

Figure 4. Turbidity of the methanol- and ethanol-containing SDS solutions as a function of the amount of added Ca(II).

The sudden increase in the turbidity values indicates the onset of the formation of precipitates. From the volume and concentration of the calcium solution, the corresponding molar ratios were calculated. The measurements indicated that the solutions containing methanol become turbid at an SDS/Ca^{2+} = 4.67 molar ratio, independently of the amount of methanol present. In ethanol–water mixtures, this value was found to be SDS/Ca^{2+} = 2.06 at 10.0 V/V% and 0.68 at 20.0 V/V%. Accordingly, in comparison, the $Ca(DS)_2$ precipitate is more soluble in methanol–water mixtures than in ethanol–water ones.

3.2.2. Measurement of Micelle Diameters in SDS Solutions Containing Calcium and Alcohol

As n-propanol and n-butanol did, while methanol and ethanol did not, enhance the solubility of $Ca(DS)_2$ precipitates, the micelle size values were determined only for the n-propanol- and n-butanol-containing solutions. The micelle size values of Ca^{2+}-containing samples in the presence of these alcohols are shown in Figure 5.

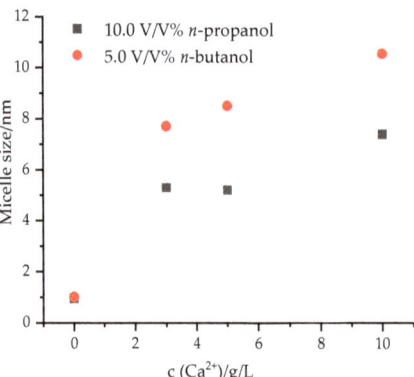

Figure 5. Micelle size values of calcium ion-containing SDS solutions in the presence of alcohols (c_{SDS} = 5.0 g/L).

In the presence of 10.0 V/V% n-propanol and 5.0 V/V% n-butanol, SDS in a 5.0 g/L concentration was found to be present in molecular form (0.96 nm diameter in 10 V/V% n-propanol solution and 1.01 nm diameter in 5 V/V% n-butanol). With the addition of calcium to these systems, micelles are indeed formed. The micelle size values increased upon increasing the metal ion concentration in the presence of both alcohols. In n-propanol-containing media approximately 5.5–7.0 nm micelle sizes were detected. The presence of n-butanol resulted in the formation of larger micelles measuring 8.0–11.0 nm diameter. According to these experimental results, it can be concluded that higher micelle sizes facilitate the solubilization of SDS in the presence of calcium ions.

3.2.3. Measurement of Surface Tension Samples of Calcium- and Alcohol-Containing SDS Samples

The surface tension was measured in calcium ion-containing SDS samples in the presence of n-propanol and n-butanol. The results are presented in Figure 6.

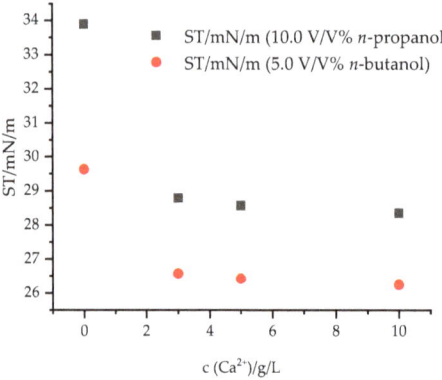

Figure 6. Surface tension of calcium ion- and alcohol-containing SDS solutions (c_{SDS} = 5.0 g/L).

Generally, the addition of alcohols resulted in a decrease in the surface tension of solutions containing SDS and calcium ions. Upon comparing the alcohol-containing samples, it can be stated that in the presence of n-propanol, higher surface tensions were measured; therefore, the larger the carbon chain length, the larger the decrease in the surface tension both in the presence and in the absence of calcium ions.

4. Conclusions

In the present study, the effects of alcohols on the inhibition of the precipitation of sodium dodecyl sulfate (SDS) with calcium ions were studied. The samples were characterized by using tensiometry, turbidimetry and dynamic light scattering. All the investigated alcohols (methanol, ethanol and n-propanol) decreased the CMC of the SDS-containing solutions (depending on the type and added amount of alcohol) from 0.66 g/L to 0.148–0.346 g/L, while in the presence of 5.0 V/V% n-butanol, 0.03 g/L CMC was obtained. Among the alcohols, n-propanol and n-butanol inhibited the formation of Ca(DS)$_2$ precipitation in 5.0 g/L SDS solutions until a 10.0 g/L Ca^{2+} concentration was reached. The addition of calcium ions increased the micelle size values to 7.4 nm in n-propanol- and to 10.5 nm in n-butanol-containing solutions.

In the presence of 10.0 V/V% propanol and butanol, the calcium ion concentration could be increased above 10.0 g/L without the formation of Ca(DS)$_2$ precipitates. In the presence of n-propanol, the solutions were precipitated when the calcium ion concentration surpassed 10.0 g/L, while the n-butanol-containing ones remained transparent even at 100 g/L. This phenomenon strongly indicates the dissolution of micelles under these specific experimental conditions.

According to these results, the use of alcohols as cosolvents may inhibit precipitation caused by divalent ions in SDS solutions. Therefore, the addition of alcohols can be an appropriate method to achieve such inhibition; however, the applicable types of alcohols are limited. The inhibition effects of n-propanol and n-butanol on calcium ion-induced precipitation were described for the first time in this paper.

Supplementary Materials: The following supporting information can be downloaded at: https://www.mdpi.com/article/10.3390/ma17081806/s1. Table S1. Viscosity (cP) and refractive index of the applied alcohol-water solvents, Figures S1–S46. Correlation diagrams of dynamic light measurements processed in alcohol containing media.

Author Contributions: Conceptualization, P.S. and C.B.; methodology, M.K., C.B. and Á.Á.; investigation, C.B. and M.K.; resources, P.S., Z.K. and Á.K.; writing—original draft preparation, C.B. and M.K.; writing—review and editing, P.S. All authors have read and agreed to the published version of the manuscript.

Funding: The financial support provided by the Ministry of Human Resources Hungary (Grant No. GINOP-2.3.4-15-2020-00006) is gratefully acknowledged. This paper was also supported by the National Research, Development and Innovation Fund from the UNKP-23-5 New National Excellence Program of the Ministry for Innovation and Technology as well as by the János Bolyai Research Scholarship of the Hungarian Academy of Sciences.

Institutional Review Board Statement: Not applicable.

Informed Consent Statement: Not applicable.

Data Availability Statement: Data are contained within the article.

Conflicts of Interest: The authors declare no conflicts of interest.

References

1. Negin, C.; Ali, S.; Xie, Q. Most common surfactants employed in chemical enhanced oil recovery. *Petroleum* **2007**, *3*, 197–211. [CrossRef]
2. Sheng, J.J. Comparison of the effects of wettability alteration and IFT reduction on oil recovery in carbonate reservoirs. *Asia-Pac. J. Chem. Eng.* **2013**, *8*, 154–161. [CrossRef]
3. de Aguiar, H.B.; de Beer, A.G.F.; Strader, M.L.; Roke, S. The interfacial tension of nanoscopic oil droplets in water is hardly affected by SDS surfactant. *J. Am. Chem. Soc.* **2010**, *132*, 2122–2123. [CrossRef] [PubMed]
4. Sheng, J.J. Status of surfactant EOR technology. *Petroleum* **2015**, *1*, 97–105. [CrossRef]
5. Sheng, J.J. Optimum phase type and optimum salinity profile in surfactant flooding. *J. Pet. Sci. Eng.* **2010**, *75*, 143–153. [CrossRef]
6. Khormali, A.; Ahmadi, S. Prediction of barium sulfate precipitation in dynamic tube blocking tests and its inhibition for waterflooding application using response surface methodology. *J. Pet. Explor. Prod. Technol.* **2023**, *13*, 2267–2281. [CrossRef]

7. Mukherjee, S.; Dac, P.; Sivapathasekaran, C.; Sen, R. Antimicrobial biosurfactants from marine Bacillus circulans: Extracellular synthesis and purification. *Lett. Appl. Microbiol.* **2009**, *48*, 281–288. [CrossRef]
8. Jakubowska, A. Interactions of univalent counterions with headgroups of monomers and dimers of an anionic surfactant. *Langmuir* **2015**, *31*, 3293–3300. [CrossRef] [PubMed]
9. Zhang, J.; Nguyen, Q.P.; Flaaten, A.K.; Pope, A.G. Mechanisms of Enhanced Natural Imbibition with Novel Chemicals. *SPE Reserv. Eval. Eng.* **2009**, *12*, 912–920. [CrossRef]
10. Elraies, K.A. The effect of a new in situ precipitation inhibitor on chemical EOR. *J. Pet. Explr. Prod. Technol.* **2013**, *3*, 133–137. [CrossRef]
11. Amjad, Z. Effect of precipitation inhibitors on calcium phosphate scale formation. *Can. J. Chem.* **1989**, *67*, 850–856. [CrossRef]
12. Miyazaki, N.; Sugai, Y.; Sasaki, K.; Okamoto, Y.; Yanagisawa, S. Screening of the Effective Additive to Inhibit Surfactin from Forming Precipitation with Divalent Cations for Surfactin Enhanced Oil Recovery. *Energies* **2020**, *13*, 2430. [CrossRef]
13. Shinoda, K. The Effect of Alcohols on the Critical Micelle Concentrations of Fatty Acid Soaps and the Critical Micelle Concentration of Soap Mixtures. *J. Phys. Chem.* **1954**, *58*, 1136–1141. [CrossRef]
14. Páhi, A.B.; Varga, D.; Király, Z.; Mastalir, Á. Thermodynamics of micelle formation of the ephedrine-based chiral cationic surfactant DMEB in water, and the intercalation of DMEB in montmorillonite. *Colloid Surf. A* **2008**, *319*, 77–83. [CrossRef]
15. Zana, R.; Eljebari, M.J. Fluorescence probing investigation of the self-association of alcohols in aqueous solution. *J. Phys. Chem.* **1993**, *97*, 11134–11136. [CrossRef]
16. Zana, R. Aqueous surfactant-alcohol systems: A review. *Adv. Colloid Interface Sci.* **1995**, *57*, 1–65. [CrossRef]
17. Rao, I.V.; Ruckenstein, E. Micellization Behavior in the Presence of Alcohols. *Colloid Interface Sci.* **1986**, *113*, 375–388. [CrossRef]
18. Flockhart, B.D. The critical micelle concentration of sodium dodecyl sulfate in ethanol-water mixtures. *J. Colloid Sci.* **1957**, *12*, 557–565. [CrossRef]
19. Ray, A.; Nemethy, G. Micelle formation by nonionic detergents in water-ethylene glycol mixtures. *J. Phys. Chem.* **1971**, *75*, 809–815.
20. Kalyanasundaram, K.; Thoma, J.K. Environmental effects on vibronic band intensities in pyrene monomer fluorescence and their application in studies of micellar systems. *J. Am. Chem. Soc.* **1997**, *99*, 2039–2044. [CrossRef]
21. Scholz, N.; Behnke, T.; Resch-Genger, U. Determination of the Critical Micelle Concentration of Neutral and Ionic Surfactants with Fluorometry, Conductometry, and Surface Tension-A Method Comparison. *J. Fluoresc.* **2018**, *28*, 465–476. [CrossRef]
22. Pérez-Rodríguez, M.; Prieto, G.; Rega, C.; Varela, L.M.; Sarmiento, F.; Mosquera, V. A Comparative Study of the Determination of the Critical Micelle Concentration by Conductivity and Dielectric Constant Measurements. *Langmuir* **1998**, *14*, 4422–4426. [CrossRef]
23. Paillet, S.; Grassl, B.; Desbriéres, J. Rapid and quantitative determination of critical micelle concentration by automatic continuous mixing and static light scattering. *Anal. Chim. Acta* **2009**, *636*, 236–241. [CrossRef]
24. Aguiar, J.; Carpena, P.; Molina-Bolívar, A.; Carnero Ruiz, C. On the determination of the critical micelle concentration by the pyrene 1:3 ratio method. *J. Colloid Interface Sci.* **2003**, *258*, 116–122. [CrossRef]
25. Cai, L.; Gochin, M.; Liu, K. A facile surfactant critical micelle concentration determination. *Chem. Commun.* **2011**, *47*, 5527. [CrossRef]
26. Fluksman, A.; Benny, O. A robust method for critical micelle concentration determination using coumarin-6 as a fluorescent probe. *Anal. Methods* **2019**, *11*, 3810–3818. [CrossRef]
27. Tanhaei, B.; Seghatoleslami, N.; Chenar, M.P.; Ayati, A.; Hesampour, M.; Mänttäri, M. Experimental Study of CMC Evaluation in Single and Mixed Surfactant Systems, Using the UV-Vis Spectroscopic Method. *J. Surfact. Deterg.* **2013**, *16*, 357–362. [CrossRef]
28. Zhou, W.; Zhu, L. Solubilization of pyrene by anionicnonionic mixed surfactants. *J. Hazard. Mater.* **2004**, *109*, 213–220. [CrossRef]
29. Austad, T.; Matre, B.; Milter, J.; Saevareid, A.; Øyno, L. Chemical flooding of oil reservoirs 8. Spontaneous oil expulsion from oil- and water-wet low permeable chalk material by imbibition of aqueous surfactant solutions. *Colloids Surfaces A Physicochem. Eng. Aspects* **1998**, *137*, 117–129. [CrossRef]
30. Ádám, A.A.; Ziegenheim, S.; Janovák, L.; Szabados, M.; Bús, C.; Kukovecz, Á.; Kónya, Z.; Dékány, I.; Sipos, P.; Kutus, B. Binding of Ca^{2+} Ions to Alkylbenzene Sulfonates: Micelle Formation, Second Critical Concentration and Precipitation. *Materials* **2023**, *16*, 494. [CrossRef]
31. Shirahama, K.; Kashiwabara, T.J. The CMC-decreasing effects of some added alcohols on the aqueous sodium dodecyl sulfate solutions. *Colloid Interface Sci.* **1971**, *36*, 65. [CrossRef]
32. Yoshimura, T.; Ohno, A.; Esumi, K. Mixed micellar properties of cationic trimeric-type quaternary ammonium salts and anionic sodium n-octyl sulfate surfactants. *J. Colloid Interface Sci.* **2004**, *272*, 191–196. [CrossRef] [PubMed]
33. Kaushik, P.; Vaidya, S.; Ahmad, T.; Ganguli, A.K. Optimizing the hydrodynamic radii and polydispersity of reverse micelles in the Triton X-100/water/cyclohexane system using dynamic light scattering and other studies. *Colloids and Surfaces A Physicochem. Eng. Aspects* **2007**, *293*, 162–166. [CrossRef]
34. Chun, B.J.; Choi, J.I.; Jang, S.S. Molecular dynamics simulation study of sodium dodecyl sulfate micelle: Water penetration and sodium dodecyl sulfate dissociation. *Colloids and Surfaces A Physicochem. Eng. Aspects* **2015**, *474*, 36–43. [CrossRef]
35. Mirgorod, Y.; Chekadanov, A.; Dolenko, T. Structure of micelles of sodium dodecyl sulphate in water: X-ray and dynamic light scattering study. *Chem. J. Mold.* **2019**, *14*, 107–119. [CrossRef]

36. Candau, S.; Zana, R. Effect of alcohols on the properties of micellar systems: III. Elastic and quasielastic light scattering study. *J. Colloid Interface Sci.* **1981**, *84*, 206–219. [CrossRef]
37. Almgrem, M.; Swarup, S. Size of sodium dodecyl sulfate micelles in the presence of additives i. alcohols and other polar compounds. *J. Colloid Interface Sci.* **1983**, *91*, 256–266. [CrossRef]

Disclaimer/Publisher's Note: The statements, opinions and data contained in all publications are solely those of the individual author(s) and contributor(s) and not of MDPI and/or the editor(s). MDPI and/or the editor(s) disclaim responsibility for any injury to people or property resulting from any ideas, methods, instructions or products referred to in the content.

Article

Rapid Synthesis of Noble Metal Colloids by Plasma–Liquid Interactions

Yuanwen Pang [1], Hong Li [1,*], Yue Hua [1], Xiuling Zhang [1] and Lanbo Di [1,2,3,*]

1. College of Physical Science and Technology, Dalian University, Dalian 116622, China; pangyuanwen2021@163.com (Y.P.); huayuexx@163.com (Y.H.); xiulz@sina.com (X.Z.)
2. State Key Laboratory of Structural Analysis for Industrial Equipment, Dalian University of Technology, Dalian 116024, China
3. Key Laboratory of Advanced Technology for Aerospace Vehicles of Liaoning Province, Dalian University of Technology, Dalian 116024, China
* Correspondence: lihong10@dlu.edu.cn (H.L.); dilanbo@163.com (L.D.)

Abstract: The interactions between plasma and liquids cause complex physical and chemical reactions at the gas–liquid contact surface, producing numerous chemically active particles that can rapidly reduce noble metal ions. This study uses atmospheric-pressure surface dielectric barrier discharge (DBD) plasma to treat ethanol aqueous solutions containing noble metal precursors, and stable gold, platinum, and palladium colloids are obtained within a few minutes. To evaluate the mechanism of the reduction of noble metal precursors by atmospheric-pressure surface DBD plasma, the corresponding metal colloids are prepared first by activating an ethanol aqueous solution with plasma and then adding noble metal precursors. It is found that the long-lived active species hydrogen peroxide (H_2O_2) plays a dominant role in the synthesis process, which has distinct effects on different metal ions. When $HAuCl_4$ and H_2PdCl_4 are used as precursors, H_2O_2 acts as a reducing agent, and $AuCl_4^-$ and $PdCl_4^{2-}$ ions can be reduced to metallic Au and Pd. However, when $AgNO_3$ is the precursor, H_2O_2 acts as an oxidising agent, and Ag^+ ions cannot be reduced to obtain metal colloids because metallic Ag can be dissolved in H_2O_2 under acidic conditions. A similar phenomenon was also observed for the preparation of Pd colloid-PA with a plasma-activated ethanol aqueous solution using $Pd(NO_3)_2$ as a Pd precursor.

Keywords: noble metal colloids; surface DBD plasma; plasma–liquid interactions; hydrogen peroxide

Citation: Pang, Y.; Li, H.; Hua, Y.; Zhang, X.; Di, L. Rapid Synthesis of Noble Metal Colloids by Plasma–Liquid Interactions. *Materials* 2024, 17, 987. https://doi.org/10.3390/ma17050987

Academic Editors: Daniela Caschera and Norbert Robert Radek

Received: 20 December 2023
Revised: 3 February 2024
Accepted: 19 February 2024
Published: 21 February 2024

Copyright: © 2024 by the authors. Licensee MDPI, Basel, Switzerland. This article is an open access article distributed under the terms and conditions of the Creative Commons Attribution (CC BY) license (https://creativecommons.org/licenses/by/4.0/).

1. Introduction

Noble metal nanoparticles have shown great potential for applications in various fields such as catalysis [1–3], sensing [4], fuel cells [5], and biomedicine [6,7] because of their high specific surface area, unique optoelectronic properties, and ease of synthesis [8]. Controlling the synthesis of nanoparticles, particularly colloidal dispersions for the preparation of noble metal nanoparticles, is essential for developing new custom catalysts for energy conversion and industrial and medical applications [9].

Noble metal colloids are usually prepared by chemical, physical, biosynthetic, and plasma methods. Traditional chemical methods use chemical reducing agents, such as sodium citrate [10], $NaBH_4$ [11], and ascorbic acid [12] mixed with noble metal salt precursors to synthesise noble metal colloids. The size and morphology of nanoparticles can be altered by changing the amount of organic solvent and reagent used in the preparation process. However, these chemical reduction agents are highly toxic and easily pollute the environment, which is not conducive to sustainable development and limits scale expansion [13]. Physical methods, such as laser ablative synthesis in a solution or ion beam sputtering, can also be used to prepare metal colloids; however, their equipment is expensive and energy-intensive [14]. The process of biosynthesis involves biological systems such as bacteria, fungi, and plant extracts, which are environmentally friendly and economical but also have drawbacks, including slow synthesis rates, limited material compatibility, pollution and purity, and variability between batches [15]. Compared with these traditional noble metal nanoparticle synthesis methods, the plasma synthesis process is simple and has high reactivity, which can quickly generate noble metal nanoparticles [16–18] and promote the phase transition and redispersion of metal nanoparticles in the later stage [19,20].

Many studies are dedicated to utilising plasma for the preparation of metal nanoparticles. One commonly used method is direct current glow discharge. For example, Dzimitrowicz et al. successfully synthesised fructose-stabilised, uniform, and monodisperse AgNPs through direct current glow discharge [21]. Li et al. successfully prepared nanoparticles using direct current glow discharge plasma, which was also able to rapidly and efficiently synthesise carbon quantum dots [22]. Although these methods have achieved success on a laboratory scale, scaling up for large-scale production still poses challenges. Another solution plasma produced by two electrodes immersed in liquid can also prepare nanoparticles, but it is also not suitable for amplification and will consume the electrode and even cause contamination of the final product of the electrode material [23]. Surface dielectric barrier discharge (DBD) is a typical non-equilibrium plasma at atmospheric pressure that can generate discharge at normal pressure and near room temperature without a vacuum. The high-energy electrons generated during discharge collide with the surrounding gas molecules, which can excite, dissociate, and ionise the gas molecules, producing numerous reactive radicals, excited atoms, molecules, and ions required for chemical reactions [24–26]. They are widely used in the preparation of highly dispersed metal nanoparticles because of the strong reducing power of the high-energy electrons (usually in inert gas plasmas such as Ar and He) or the hydrogen species in hydrogen-containing plasmas such as Ar/H_2 discharges [27,28]. The former reduces metal ions with positive standard redox potentials such as Au, Ag, Pt, Pd, Ir, and Rh, whereas the latter is more efficient than electron reduction. It not only reduces metal ions with positive standard redox potentials but also some metal ions with negative standard redox potentials, such as Co and Ni [29].

The synthesis of noble metal nanoparticles usually uses colloidal methods [30]. Therefore, the interaction between the plasma and liquid is significant, and there are many complex physical and chemical processes at the interface of the plasma and liquid contact [31]. When the plasma–liquid system contains an ethanol aqueous solution, ethanol and water molecules participate in the gas discharge plasma through different processes, namely, sputtering, electric-field-induced hydrated cation emission, and evaporation [32]. Isomeric radicals of C_2H_5O (CH_3CHOH, CH_2CH_2OH, and CH_3CH_2O), atomic hydrogen (H), and hydroxyl (OH) radicals can be generated by electron impact in the plasma [33]. Subsequently, the generated species are transported from the gaseous plasma to the liquid. In the positive half cycle of the discharge, the liquid acts as a cathode, and positive ions are accelerated across the cathode sheath to bombard the liquid surface, followed by evaporation, sputtering, and secondary electron emission from the liquid. In the negative half cycle, electrons are pulled out of the plasma to enter the liquid by the small electric field near the liquid anode. When these electrons enter the aqueous solution, they are hydrated to form the e^-_{aq} species (with a standard redox potential of -2.8 V) in the solution. In addition, photons emitted by excited particles in plasma may enter the liquid, triggering secondary processes, and neutral particles in the plasma are transported to the liquid surface by diffusion. Therefore, the synthesis process can be controlled in both the solution and plasma phases by controlling the solution and plasma parameters [34,35].

Generally, active species generated in plasma and solution systems can be divided into short- and long-lived species based on their lifetimes [36]. Short-lived active species include free electrons, hydrated electrons (e^-_{aq}), excited hydrogen atoms (H), hydroxide radicals (OH), negative hydrogen ions (H^-), and alcohol fragment radicals induced to dissociate from ethanol molecules by high-energy electrons or ultraviolet radiation, which disappear or rapidly decay after the plasma stops [37]. The long-lived active species is mainly H_2O_2, which can remain in liquids for a long time and can be detected within a few months of plasma irradiation [38]. Bjelajac et al. [39] synthesised AuNPs in a chloroauric acid solution with ethanol as a solvent using a DBD plasma torch at atmospheric pressure, and the results showed that plasma was conducive to the synthesis of more dispersed Au nanoparticles. Sauvageau et al. [40] synthesised platinum-group metal nanoparticles using a DBD hydrogen plasma and found that all three ions (Pt, Pd, and Rh) had high reduction rates.

In this study, atmospheric-pressure surface DBD cold plasma was developed and adopted to synthesise noble metal colloids by treating various metal precursor solutions. The results show that colloids of gold, platinum, and palladium can be successfully prepared by plasma in a few minutes. The synthesis process is fast, without the use of any other chemical reducing agent, and the prepared noble metal colloids have no significant change after 30 days of storage at room temperature. Moreover, in order to investigate the reduction mechanism of noble metal precursors by surface DBD plasma at atmospheric pressure, we synthesised noble metal colloids using both direct plasma treatment and plasma-activated ethanol aqueous solution treatment of the noble metal precursors, and a comparative analysis was then conducted. It was found that the difference in the metal colloids prepared by the two methods can be attributed to the distinct active species present in the solution, and the effect of these active species varies depending on the specific metal precursor employed.

2. Experimental Section

2.1. Materials

Gold trichloride ($AuCl_3 \cdot HCl \cdot 4H_2O$), chloroplatinic acid ($H_2PtCl_6 \cdot 6H_2O$), palladium chloride ($PdCl_2$), palladium nitrate ($Pd(NO_3)_2 \cdot 2H_2O$), silver nitrate ($AgNO_3$), and anhydrous ethanol (AR, $\geq$99%) used in this experiment were purchased from Kermel Chemical Reagent Co., Ltd. (Tianjin, China). Gold trichloride was dissolved in water before use to obtain an aqueous solution of chloroauric acid ($HAuCl_4$) with a concentration of 20 mM, and palladium chloride was dissolved in hydrochloric acid before use to obtain a solution of chloropalladic acid (H_2PdCl_4) with a concentration of 521 mM. Polyvinylpyrrolidone (PVP) (Mw 58000, K29-32) was purchased from Aladdin Chemical Reagent Co., Ltd. (Shanghai, China). High-purity argon (>99.999%) and hydrogen (>99.999%) were purchased from Zhonghao Guangming Chemical Research and Design Institute Co., Ltd. (Dalian, China).

2.2. Preparation of Noble Metal Colloids

2.2.1. Surface Dielectric Barrier Discharge (DBD) Reactor

The surface DBD reactor consists of a high-voltage electrode and a grounding electrode separated by a high-purity alumina dielectric layer (area: 9×5 cm^2; thickness: 1 mm), and the experimental setup is shown in Figure 1. Both electrodes are made of high-purity tungsten, the size of the grounding electrode is 1.7×0.5 cm^2, and the high-voltage electrode consists of nine comb-like tungsten wires connected at one end (tungsten width of 1 mm, wire spacing of 4 mm). The discharge voltage was measured by a high-voltage probe (Tektronix P6015A, Beaverton, OR, USA).

2.2.2. Direct Preparation of Au Colloid-P, Pt Colloid-P, and Pd Colloid-P with Plasma

First, 2 mL of an ethanol aqueous solution (50% water, 50% ethanol) containing 5% PVP was prepared, and then 25 μL of $HAuCl_4$ (20 mM) was added to it, and the precursor mixture solution was mixed thoroughly. The mixed solution was placed in a quartz reactor (3 cm diameter and 4 mm depth), and the height of the quartz reactor was adjusted such that the liquid level was 2 mm below the high-voltage electrode. The working gas was a mixture of Ar and H_2 with a total flow rate of 100 sccm and an Ar/H_2 molar ratio of 1:1. The rotational speed of the magnetic stirrer was set to 500 r·min^{-1}, the peak-to-peak sinusoidal applied voltage was 9.0 kV, the discharge frequency was 10.4 kHz, and the plasma treatment was carried out for 7 min. A wine-red solution was obtained after the plasma treatment. The solution was poured into a 5 mL measuring cylinder, and ethanol was added to compensate for the loss of liquid during the plasma treatment until the total volume of the mixed solution was 2 mL. After mixing, the solution was poured into a sample bottle, labelled as Au colloid-P, and stored in a refrigerator at 4 °C in the dark. The preparation processes for Pt colloid-P and Pd colloid-P were the same. H_2PtCl_6 and H_2PdCl_4 were used as noble metal precursors in the preparation process.

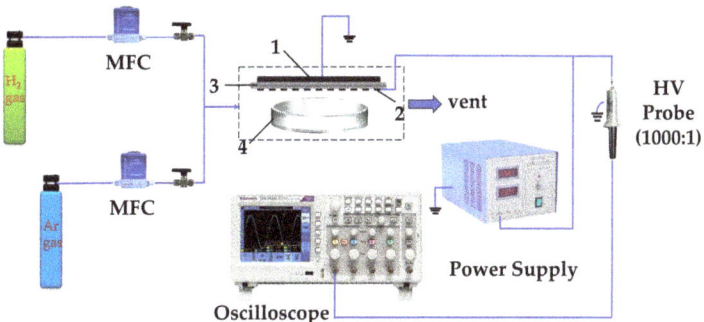

Figure 1. Diagram of the experimental setup for the preparation of noble metal colloids by surface dielectric barrier discharge: 1—grounding electrode, 2—high-voltage electrode, 3—dielectric layer, 4—noble metal precursor solution.

2.2.3. Preparation of Au Colloid-PA and Pd Colloid-PA with Plasma-Activated Solution

First, an ethanol aqueous solution containing 5% PVP (50% water, 50% ethanol, 2 mL) was prepared in a quartz reactor (diameter: 3 cm; depth: 4 mm), underwent plasma treatment (treatment conditions were consistent with Section 2.2.2), and then 25 μL of $HAuCl_4$ (20 mM) was added without plasma. The colour of the mixture quickly turned wine-red, indicating the presence of AuNPs. The obtained mixed solution was denoted as Au colloid-PA and stored in a refrigerator at 4 °C in the dark. The preparation process for Pd colloid-PA was the same. H_2PdCl_4 and $Pd(NO_3)_2$ were chosen as palladium precursors in the preparation process.

2.3. Characterisation

A UV-Vis spectrometer (Hitachi, U-3900, Tokyo, Japan) was used to record the absorption of the sample in the wavelength range of 200–800 nm. Sample analysis was performed using transmission electron microscopy (TEM) and high-resolution transmission electron microscopy (HRTEM) (JEOL JEM-2100F, Tokyo, Japan) at an acceleration voltage of 120 kV. More than 100 Au NPs, Pt NPs, and Pd NPs were selected from the corresponding TEM images of noble metal colloids, and the particle size and distribution of the three noble metal colloids were calculated.

3. Results and Discussion

3.1. Light-Absorbing Characteristics and Stability of Metal Colloids Prepared with Plasma

Using atmospheric-pressure surface DBD plasma at a discharge voltage of 9 kV and a discharge time of 7 min, solutions of chloroauric acid, chloroplatinic acid, and chloropalladic acid were treated separately to prepare gold colloids (Au colloid-P), platinum colloids (Pt colloid-P), and palladium colloids (Pd colloid-P), which were the most stable (Figures S1–S4). The UV–Vis absorption spectra and corresponding photographs of the three untreated metal precursor solutions and three freshly prepared metal colloids are shown in Figure 2. Figure 2a shows that the untreated chloroauric acid solution (0.25 mM) exhibited only weak absorption in the visible light region, and the sample colour appeared light yellow ((I) in Figure 2d). In comparison, the fresh Au colloid-P samples prepared by the plasma treatment showed obvious surface plasmon resonance (SPR) absorption peaks of gold at 500–550 nm, and the colour of the samples was burgundy ((II) in Figure 2d).

Meanwhile, the maximum SPR absorption peak of the Au colloid-P sample appeared at 521 nm, indicating that the particle size of the AuNPs was well controlled below 20 nm. As shown in Figure 2b, the untreated chloroplatinic acid solution (0.25 mM) exhibits significant absorption in the UV wavelength range of 200–300 nm, corresponding to the characteristic absorption band of a $PtCl_6^{2-}$ ion. The sample showed very weak absorption in the visible light region and appeared light orange ((IV) in Figure 2d). After plasma treatment, the Pt colloid-P sample was obtained, and the characteristic absorption band of an $PtCl_6^{2-}$ ion disappeared. The sample showed weak absorption in the visible light region, and the colour of the sample turned brown ((V) in Figure 2d), indicating that the plasma treatment can reduce $PtCl_6^{2-}$ ions to a Pt elemental substance. As shown in Figure 2c, the untreated chloropalladic acid solution (0.25 mM) exhibited weak absorption in the visible light region, and the sample appeared light brown ((VII) in Figure 2d). After plasma treatment, the Pd colloid-P sample was obtained, and its absorption intensity in the ultraviolet and visible light regions was slightly enhanced compared to the untreated H_2PdCl_4 solution, and the sample colour turned brown ((VIII) in Figure 2d), indicating that plasma treatment can reduce chloropalladic acid to Pd.

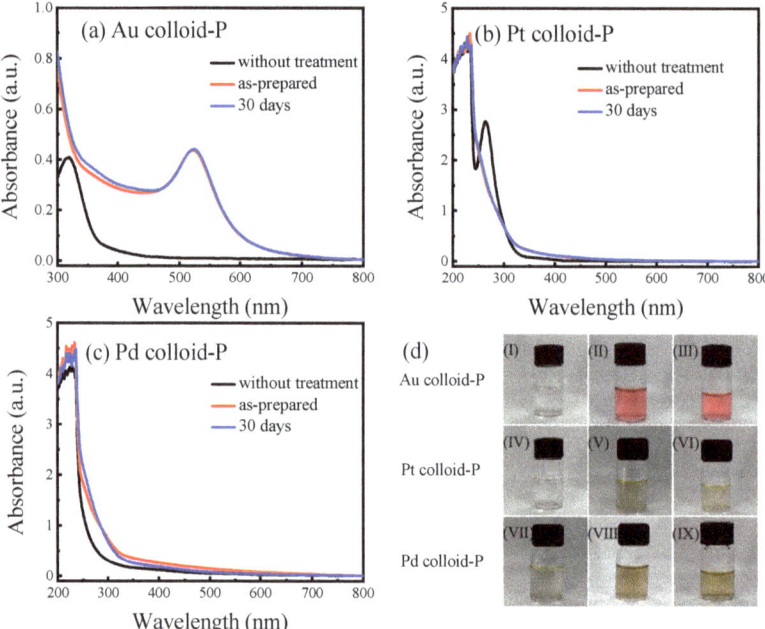

Figure 2. UV-Vis absorption spectra of (**a**) Au colloid-P, (**b**) Pt colloid-P, and (**c**) Pd colloid-P (using H_2PdCl_4 as Pd precursor) prepared by surface DBD cold plasma at atmospheric pressure and after 30 days of storage, and (**d**) corresponding photos before and after treatment and after 30 days of storage (gold colloids: (I)–(III); platinum colloids: (IV)–(VI); palladium colloids: (VII)–(IX)).

To study the stability of the plasma-prepared Au colloid-P, Pt colloid-P, and Pd colloid-P samples, the three samples were stored for 30 days and re-characterised by UV-Vis absorption spectroscopy, and the corresponding photographs were taken, as shown in Figure 1. Compared to the freshly prepared samples, after 30 d of storage, there was no significant change in the position and intensity of the absorption peaks in the UV–Vis absorption spectra of the three samples, and no precipitation was observed, indicating that the plasma-prepared Au colloid-P, Pt colloid-P, and Pd colloid-P samples had good stability.

3.2. Morphology and Particle Size of Metal Colloids Prepared with Plasma

Figure 3a–c show the transmission electron microscopy (TEM) and high-resolution transmission electron microscopy (HRTEM) images of the Au, Pt, and Pd colloid-P samples. From the figures, it can be observed that the metal nanoparticles in the three samples have good dispersibility and are in a spherical state, without obvious agglomeration. This is because PVP has a good stabilising effect on the metal nanoparticles in the system. Figure S5 shows that the prepared gold colloids are unstable when no PVP is present in the system. In addition, the HRTEM images of the three samples show clear lattice stripes, indicating that the metal nanoparticles in the prepared samples have a good degree of crystallisation. According to the HRTEM images, the lattice spacings of the Au colloid-P, Pt colloid-P, and Pd colloid-P samples were 0.210, 0.224, and 0.221 nm, corresponding to the Au (111), Pt (111), and Pd (111) crystal planes, respectively, all of which belong to the face-centred cubic structure (Fcc). Based on the TEM photographs of the samples, histograms of the particle size distribution of the metal nanoparticles of the Au colloid-P, Pt colloid-P, and Pd colloid-P samples were obtained by selecting more than 100 Au, Pt, and Pd nanoparticles, respectively, as shown in Figure 3d–f. In the Au colloid-P, Pt colloid-P, and Pd colloid-P samples, the average particle sizes of Au, Pt, and Pd are 11.0 ± 2.0 nm, 1.1 ± 0.1 nm, and 3.5 ± 0.3 nm, respectively. The larger size of the Au nanoparticles was mainly due to the influence of chloride ions in the precursor $HAuCl_4$ solution [41]. However, the particle sizes of Pt and Pd are very small, especially Pt, with a particle size of only 1.1 ± 0.1 nm, which is very suitable for use as metal catalysts [42,43].

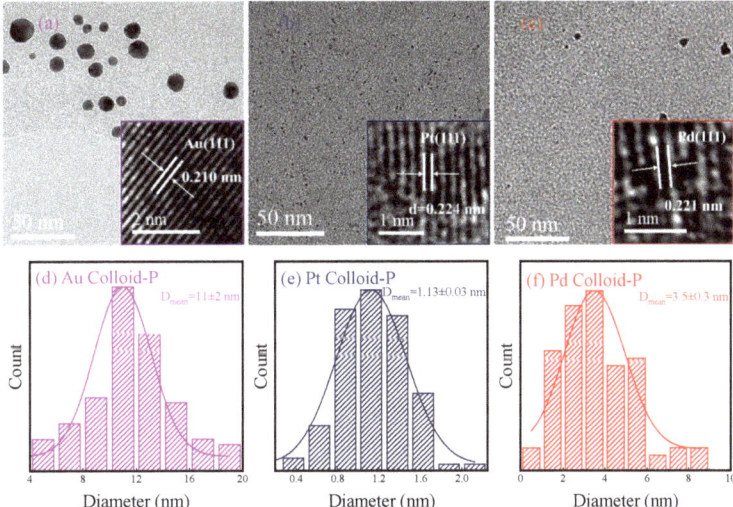

Figure 3. TEM photographs and HRTEM photographs (insets) of (**a**) Au colloid-P, (**b**) Pt colloid-P, and (**c**) Pd colloid-P and (**d**–**f**) histograms of the particle size distribution of the corresponding metal nanoparticles.

3.3. Light Absorption Properties and Morphology of Au Colloid-PA Prepared with Plasma-Activated Solution

To evaluate the light-absorbing characteristics and stability of the Au colloid-PA samples prepared from the plasma-activated solutions, UV-Vis absorption spectra were used to record the changes in their absorption spectra over a period of 15 d, as shown in Figure 4a. The figure shows that the SPR absorption peak of gold nanoparticles appeared at approximately 532 nm after adding $HAuCl_4$ to the plasma activation solution for only 10 min. With time, the position of the SPR absorption peak blue-shifted, and the intensity of the absorption peak gradually increased. Three days later, the SPR absorption peak of the Au colloid-PA sample blue-shifted to 528 nm, the intensity of the absorption peak no longer changed, and the sample appeared purplish-red (Figure 4a). Meanwhile, within 3–15 days, there was no significant change in the absorption peak intensity, position, or colour of the Au colloid-PA sample, indicating that the plasma activation solution could also reduce chloroauric acid to prepare metallic gold nanoparticles, and the stability of the prepared Au colloidal solution was good.

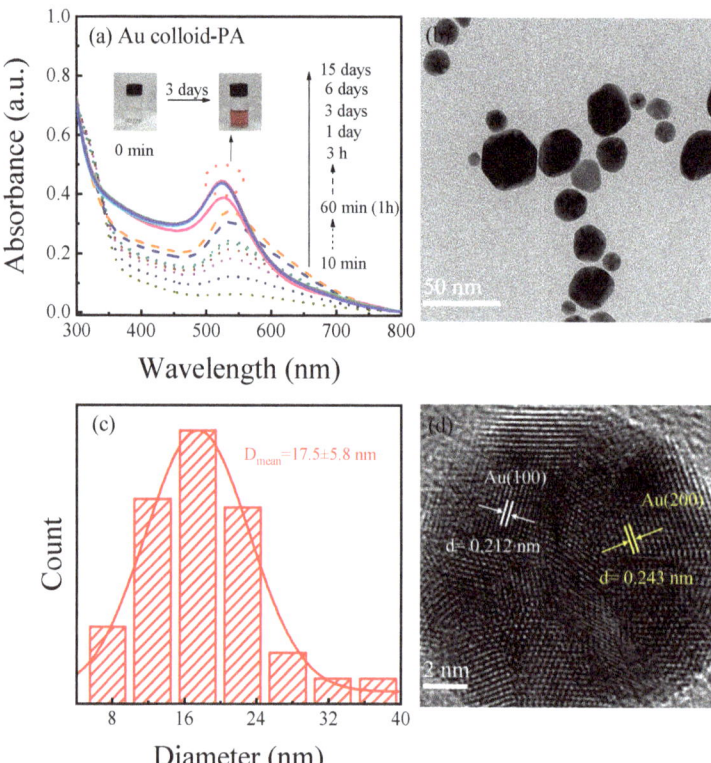

Figure 4. (**a**) UV-Vis absorption spectra of Au colloid-PA solution left from 10 min to 15 d, (**b**) TEM image of Au colloid-PA and corresponding (**c**) histogram of Au nanoparticle size distribution, (**d**) HRTEM image.

To compare the morphology and size of the gold nanoparticles in the plasma-activated solution-prepared Au colloid-PA and plasma-directly prepared Au colloid-P samples, the Au colloid-PA sample was characterised using TEM and HRTEM, and the results are shown in Figure 4b,d. The gold nanoparticles in the Au colloid-PA sample were well dispersed and did not undergo significant aggregation. This is because PVP plays a stabilising role in the colloidal solution of the system. Unlike the spherical gold nanoparticles in the Au colloid-P sample, the Au nanoparticles in the Au colloid-PA sample exhibited spherical, triangular, and hexagonal shapes. From Figure 4d, it can be observed that the lattice fringes in the HRTEM image of the Au colloid-PA sample are clear, with good crystallinity and two types of crystal planes: (111) and (200). By selecting more than 100 gold nanoparticles for the analysis, the average particle size of the gold nanoparticles in the Au colloid-PA sample was found to be 17.5 ± 5.8 nm (Figure 4c), which is approximately 6.5 nm larger than the average particle size of Au nanoparticles directly prepared by plasma in the Au colloid-P sample. This may be due to the electrostatic repulsion generated under direct plasma treatment, which hinders the aggregation and growth of AuNPs, whereas the long-lived active species in the plasma-activated solution slowly reduce HAuCl₄, resulting in weaker electrostatic repulsion and the generation of gold nanoparticles with different morphologies and larger particle sizes.

3.4. Mechanism of Plasma Preparation of Metal Colloids

Figure 5a–d show the UV–Vis absorption spectra and corresponding photographs of the untreated $HAuCl_4$, H_2PdCl_4, $AgNO_3$, and $Pd(NO_3)_2$ precursor solutions and the corresponding metal colloids prepared by direct plasma (DP) treatment and plasma activation (PA) treatment. It can be seen that the $HAuCl_4$, H_2PdCl_4, $AgNO_3$, and $Pd(NO_3)_2$ precursors can all be reduced to obtain the corresponding noble metal colloid by DP treatment. This is because during the DP treatment process, abundant short-lived reducing species can be continuously generated, such as free electrons, hydrated electrons (e^-_{aq}), excited hydrogen atoms (H), and reactive alcohol fragment radicals generated by high-energy electrons or the UV-induced dissociation of ethanol molecules, as well as the long-lived active species hydrogen peroxide. These reactive radicals have a lower redox potential and can effectively reduce the precursor solution ions to elemental nanoparticles.

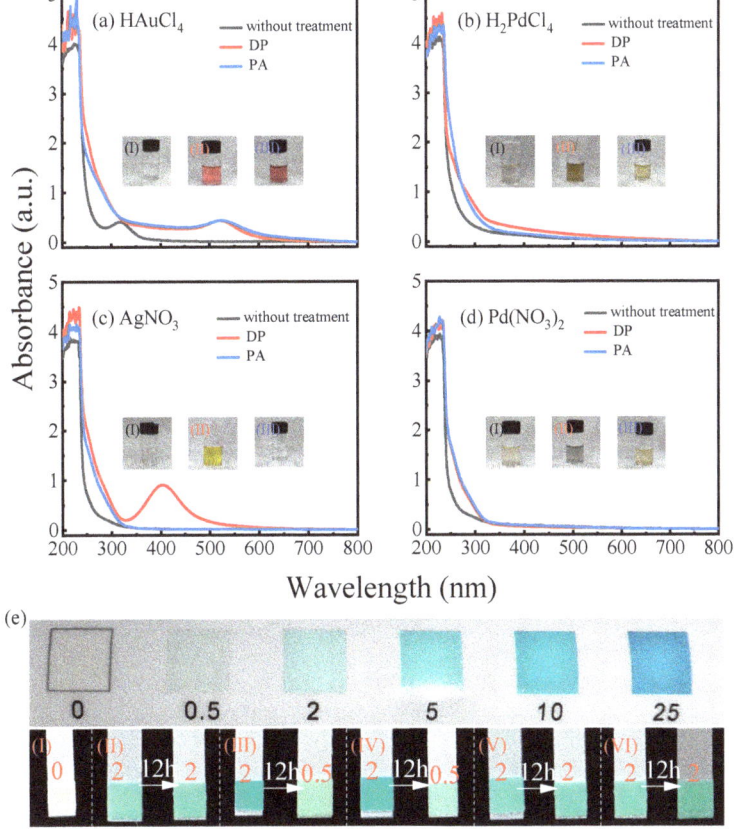

Figure 5. UV-Vis absorption spectra of (**a**) $HAuCl_4$, (**b**) H_2PdCl_4, (**c**) $AgNO_3$, and (**d**) $Pd(NO_3)_2$ under different treatments; (**e**) results of hydrogen peroxide detection strips: (I) untreated ethanol aqueous solution and (II) ethanol aqueous solution treated with plasma activation and left for 12 h. Results of adding (III) $HAuCl_4$, (IV) H_2PdCl_4, (V) $AgNO_3$, and (VI) $Pd(NO_3)_2$ to ethanol aqueous solution treated with plasma activation and left for 12 h. (H_2O_2 concentration unit: mg/L).

Nevertheless, only HAuCl$_4$ and H$_2$PdCl$_4$ precursors can be reduced to obtain gold and palladium colloids by PA treatment. Meanwhile, AgNO$_3$ and Pd(NO$_3$)$_2$ precursors cannot be reduced to obtain the corresponding metal colloids. This can be attributed to the effect of the residual long-lived species of H$_2$O$_2$, while the short-lived reducing species disappeared in the plasma-activated solution. HAuCl$_4$ can be reduced by the long-lived active species H$_2$O$_2$ via the following reaction [44]:

$$2AuCl_4^- + 3H_2O_2 \rightarrow 2Au + 3O_2 + 6H^+ + 8Cl^- \quad (1)$$

The reduction mechanism of H$_2$PdCl$_4$ is similar to that of HAuCl$_4$. And the long-lived species of hydrogen peroxide in solution may be produced by OH radicals generated in the gas-phase plasma dissolved in the liquid, as shown in Equations (2)–(6) [33,45].

$$H_2O_g \xrightarrow{plasma} OH_g + H_g \quad (2)$$

$$C_2H_5OH_g \xrightarrow{plasma} C_2H_{5g} + OH_g \quad (3)$$

$$CH_3CHOH \xrightarrow{plasma} C_2H_{4g} + OH_g \quad (4)$$

$$OH_g \longrightarrow OH_{aq} \quad (5)$$

$$OH_{aq} + OH_{aq} \longrightarrow H_2O_{2aq} \quad (6)$$

Studies have shown that both Ag$^+$ reduction and oxidation processes occur during AgNP synthesis [46,47]. The e$^-$$_{aq}$, H, and alcohol fragment radicals are the main reducing agents for Ag$^+$ to AgNPs, whereas H$_2$O$_2$ generated by the plasma–liquid interaction oxidises AgNPs to Ag$^+$, as shown in Equation (7).

$$Ag^+ \xleftrightarrow{e_{aq}^-/H_2O_2} Ag \quad (7)$$

Because e$^-$$_{aq}$, H, and reactive alcohol fragment radicals are short-lived species that rapidly decay or disappear after plasma quenching, the long-lived species H$_2$O$_2$, in an acidic environment, exerts etching effects on Ag, preventing the formation of AgNPs even in the presence of ethanol in the solution [48]. Hence, AgNO$_3$ treated with plasma activation (PA) cannot obtain silver colloids. The inability to obtain palladium colloids by adding Pd(NO$_3$)$_2$ to plasma-activated ethanol aqueous solution may be attributed to the strong oxidising effect on PdNPs by H$_2$O$_2$ under acidic conditions combined with the existence of NO$_3^-$.

To verify the presence of H$_2$O$_2$ in the plasma activation solution and whether H$_2$O$_2$ species react during the synthesis process, we used H$_2$O$_2$ detection strips to test the solutions before and after plasma activation; the results are shown in Figure 5e. The untreated ethanol aqueous solution could not change the colour of the H$_2$O$_2$ detection strips ((I) of Figure 5e), whereas the plasma-activated ethanol aqueous solution turned the H$_2$O$_2$ detection strips into light green, and there were no significant changes after 12 h ((II) of Figure 5e), indicating that the long-lived species H$_2$O$_2$ was formed in the solution during the plasma treatment. When HAuCl$_4$ and H$_2$PdCl$_4$ were added to the activated solution, the colour of the hydrogen peroxide detection strips slightly deepened ((III) and (IV) of Figure 5e). This may be because the colour of the solution changed from colourless to light yellow and light brown, which affected the colour of the detection strips but did not hinder the determination of H$_2$O$_2$ consumption. The results showed that the H$_2$O$_2$ detection strips in an ethanol aqueous solution containing HAuCl$_4$ and H$_2$PdCl$_4$ faded significantly after 12 h compared to the initial solution (as shown in (III) and (IV) of Figure 5e), indicating that H_2O_2 played the role of a reducing agent and was consumed during the reduction of $HAuCl_4$ and H_2PdCl_4. However, there was no significant change in the H_2O_2 detection strips measured 12 h before and after the addition of $AgNO_3$ and $Pd(NO_3)_2$ to the ethanol aqueous solution system (as shown in (V) and (VI) of Figure 5e), confirming that $AgNO_3$ and $Pd(NO_3)_2$ in Figure 5c,d cannot be reduced.

4. Conclusions

In this study, we demonstrated a simple, fast, and environmentally friendly method for synthesising noble metal colloids through the interactions between atmospheric-pressure cold plasma and liquids. The experimental results indicated that atmospheric-pressure argon–hydrogen surface DBD plasma can be used to successfully prepare gold, platinum, and palladium colloids within a few minutes. The synthesis process was fast and did not require the use of any other chemical reducing agents. The prepared gold, platinum, and palladium colloids remained stable for 30 d without significant changes. The Au, Pt, and Pd in noble metal colloids exhibited uniform spherical shapes, with average particle sizes of 11.0 ± 2.0 nm, 1.1 ± 0.1 nm, and 3.5 ± 0.3 nm, respectively. In addition, in order to investigate the reduction mechanism of noble metal precursors by surface DBD plasma, we also synthesised noble metal colloids using a plasma-activated ethanol aqueous solution treatment of the noble metal precursors. The comparative results showed that the direct plasma treatment of the ethanol aqueous solutions continuously generated active species such as e^-_{aq}, H, reactive alcohol fragment radicals, and hydrogen peroxide, thereby enabling the rapid reduction of noble metal precursor solutions to obtain gold, silver, platinum, and palladium colloids. The short-lived active species generated by the plasma activation treatment of the ethanol aqueous solution quickly disappeared, and the long-lived species H_2O_2 in the activated ethanol aqueous solution oxidised AgNPs and PdNPs, making it impossible to reduce $AgNO_3$ and $Pd(NO_3)_2$. The H_2O_2 test strip quickly faded after adding $HAuCl_4$ and H_2PdCl_4 to the plasma-activated ethanol aqueous solution, and the activated ethanol aqueous solution was able to reduce $AuCl_4^-$ and $PdCl_4^{2-}$ to obtain gold and palladium nanoparticles, indicating that H_2O_2 acted as a reducing agent.

Supplementary Materials: The following supporting information can be downloaded at https://www.mdpi.com/article/10.3390/ma17050987/s1: Figure S1: UV-Vis absorption spectra of $HAuCl_4$ precursor solution reduced with different contents of ethanol under the same plasma treatment condition; Figure S2: Time-dependent UV-Vis absorption spectra of the reduced $HAuCl_4$ precursor solution at different discharge times with a discharge voltage of 6 kV and 5% PVP; Figure S3: Time-dependent UV-Vis absorption spectra of the reduced $HAuCl_4$ precursor solution with different discharge voltages at a discharge time of 7 min and 5% PVP; Figure S4: Time-dependent UV-Vis absorption spectra of $HAuCl_4$ precursor solutions reduced with different contents of PVP at a discharge voltage of 6 kV and a discharge time of 7 min; Figure S5: UV-Vis absorption spectra of gold colloid solution without PVP protection over time.

Author Contributions: Y.P.: Conceptualisation, formal analysis, investigation, methodology, data curation, writing—original draft, software, and writing—review and editing. H.L.: Writing—review and editing, supervision, funding acquisition, and project administration. Y.H.: Writing—review and editing. X.Z.: Writing—review and editing. L.D.: Writing—review and editing, funding acquisition, supervision, and project administration. All authors have read and agreed to the published version of the manuscript.

Funding: This work was supported by the National Natural Science Foundation of China (Grant No. 12005031, 52077024, 21773020), the Xingliao Talents Program (Grant No. 2022RJ16, XLYC2203147), the Yunnan Police College (Grant No. YJKF003), and the Interdisciplinary project of Dalian University (Grant No. DLUXK-2023-QN-001).

Data Availability Statement: Data are contained within the article and supplementary materials.

Conflicts of Interest: The authors declare no conflicts of interest.

References

1. Khan, Z.; Al-Thabaiti, S.A.; Rafiquee, M.Z.A. Cu-based tri-metallic nanoparticles with noble metals (Ag, Pd, and Ir) and their catalytic activities for hydrogen generation. *Int. J. Hydrogen Energy* **2021**, *46*, 39754–39767. [CrossRef]
2. Kim, H.; Yoo, T.Y.; Bootharaju, M.S.; Kim, J.H.; Chung, D.Y.; Hyeon, T. Noble Metal-Based Multimetallic Nanoparticles for Electrocatalytic Applications. *Adv. Sci.* **2021**, *9*, 2104054. [CrossRef]
3. Quinson, J. Colloidal surfactant-free syntheses of precious metal nanoparticles for electrocatalysis. *Curr. Opin. Electrochem.* **2022**, *34*, 100977. [CrossRef]
4. Longato, A.; Vanzan, M.; Colusso, E.; Corni, S.; Martucci, A. Enhancing Tungsten Oxide Gasochromism with Noble Metal Nanoparticles: The Importance of the Interface. *Small* **2022**, *19*, 2205522. [CrossRef]
5. Konwar, D.; Basumatary, P.; Lee, U.; Yoon, Y.S. P-doped SnFe nanocubes decorated with PdFe alloy nanoparticles for ethanol fuel cells. *J. Mater. Chem. A* **2021**, *9*, 10685–10694. [CrossRef]
6. Azharuddin, M.; Zhu, G.H.; Das, D.; Ozgur, E.; Uzun, L.; Turner, A.P.F.; Patra, H.K. A repertoire of biomedical applications of noble metal nanoparticles. *Chem. Commun.* **2019**, *55*, 6964–6996. [CrossRef]
7. Rai, M.; Ingle, A.P.; Gupta, I.; Brandelli, A. Bioactivity of noble metal nanoparticles decorated with biopolymers and their application in drug delivery. *Int. J. Pharm.* **2015**, *496*, 159–172. [CrossRef]
8. de Oliveira, P.F.M.; Torresi, R.M.; Emmerling, F.; Camargo, P.H.C. Challenges and opportunities in the bottom-up mechanochemical synthesis of noble metal nanoparticles. *J. Mater. Chem. A* **2020**, *8*, 16114–16141. [CrossRef]
9. Quinson, J.; Bucher, J.; Simonsen, S.B.; Kuhn, L.T.; Kunz, S.; Arenz, M. Monovalent Alkali Cations: Simple and Eco-Friendly Stabilizers for Surfactant-Free Precious Metal Nanoparticle Colloids. *ACS Sustain. Chem. Eng.* **2019**, *7*, 13680–13686. [CrossRef]
10. Sivaraman, S.K.; Kumar, S.; Santhanam, V. Monodisperse sub-10nm gold nanoparticles by reversing the order of addition in Turkevich method—The role of chloroauric acid. *J. Colloid Interface Sci.* **2011**, *361*, 543–547. [CrossRef]
11. Iqbal, M.; Usanase, G.; Oulmi, K.; Aberkane, F.; Bendaikha, T.; Fessi, H.; Zine, N.; Agusti, G.; Errachid, E.-S.; Elaissari, A. Preparation of gold nanoparticles and determination of their particles size via different methods. *Mater. Res. Bull.* **2016**, *79*, 97–104. [CrossRef]
12. Britto Hurtado, R.; Cortez-Valadez, M.; Aragon-Guajardo, J.R.; Cruz-Rivera, J.J.; Martínez-Suárez, F.; Flores-Acosta, M. One-step synthesis of reduced graphene oxide/gold nanoparticles under ambient conditions. *Arab. J. Chem.* **2020**, *13*, 1633–1640. [CrossRef]
13. Hossain, M.M.; Robinson Junior, N.A.; Mok, Y.S.; Wu, S. Investigation of silver nanoparticle synthesis with various nonthermal plasma reactor configurations. *Arab. J. Chem.* **2023**, *16*, 105174. [CrossRef]
14. Seitkalieva, M.M.; Samoylenko, D.E.; Lotsman, K.A.; Rodygin, K.S.; Ananikov, V.P. Metal nanoparticles in ionic liquids: Synthesis and catalytic applications. *Coord. Chem. Rev.* **2021**, *445*, 213982. [CrossRef]
15. Bhattacharya, T.; Das, D.; Borges e Soares, G.A.; Chakrabarti, P.; Ai, Z.; Chopra, H.; Hasan, M.A.; Cavalu, S. Novel Green Approaches for the Preparation of Gold Nanoparticles and Their Promising Potential in Oncology. *Processes* **2022**, *10*, 426. [CrossRef]
16. Zhang, T.; Ouyang, B.; Zhang, X.; Xia, G.; Wang, N.; Ou, H.; Ma, L.; Mao, P.; Ostrikov, K.; Di, L.; et al. Plasma-enabled synthesis of Pd/GO rich in oxygen-containing groups and defects for highly efficient 4-nitrophenol reduction. *Appl. Surf. Sci.* **2022**, *597*, 153727. [CrossRef]
17. Hua, Y.; Zhang, J.; Zhang, T.; Zhu, A.; Xia, G.; Zhang, X.; Di, L. Plasma synthesis of graphite oxide supported PdNi catalysts with enhanced catalytic activity and stability for 4-nitrophenol reduction. *Catal. Today* **2023**, *418*, 114069. [CrossRef]
18. Hua, Y.; Zhao, L.; Zhao, Q.; Xia, G.; Zhang, X.; Di, L. Cold Plasma for Preparation of Pd/graphene Catalysts toward 4-nitrophenol Reduction: Insight into Plasma Treatment. *Mod. Low Temp. Plasma* **2023**, *1*, 7. [CrossRef]
19. Zhang, J.; Hua, Y.; Li, H.; Zhang, X.; Shi, C.; Li, Y.; Di, L.; Wang, Z. Phase reconstruction of Co_3O_4 with enriched oxygen vacancies induced by cold plasma for boosting methanol-to-formate electro-oxidation. *Chem. Eng. J.* **2023**, *478*, 147288. [CrossRef]
20. Di, L.; Fu, Z.; Dong, M.; Zhu, A.; Xia, G.; Zhang, X. Cold plasma-prepared Ru-based catalysts for boosting plasma-catalytic CO_2 methanation. *Chem. Eng. Sci.* **2023**, *280*, 119056. [CrossRef]
21. Dzimitrowicz, A.; Motyka-Pomagruk, A.; Cyganowski, P.; Babinska, W.; Terefinko, D.; Jamroz, P.; Lojkowska, E.; Pohl, P.; Sledz, W. Antibacterial Activity of Fructose-Stabilized Silver Nanoparticles Produced by Direct Current Atmospheric Pressure Glow Discharge towards Quarantine Pests. *Nanomaterials* **2018**, *8*, 751. [CrossRef]
22. Li, Y.; Zhong, X.; Rider, A.E.; Furman, S.A.; Ostrikov, K. Fast, energy-efficient synthesis of luminescent carbon quantum dots. *Green Chem.* **2014**, *16*, 2566–2570. [CrossRef]
23. Burakov, V.; Kiris, V.; Nedelko, M.; Tarasenka, N.; Nevar, A.; Tarasenko, N. Plasmas in and in contact with liquid for synthesis and surface engineering of carbon and silicon nanoparticles. *J. Phys. D Appl. Phys.* **2018**, *51*, 484001. [CrossRef]
24. Dvořák, P.; Talába, M.; Obrusník, A.; Kratzer, J.; Dědina, J. Concentration of atomic hydrogen in a dielectric barrier discharge measured by two-photon absorption fluorescence. *Plasma Sources Sci. Technol.* **2017**, *26*, 085002. [CrossRef]
25. Mouele, E.S.M.; Tijani, J.O.; Badmus, K.O.; Pereao, O.; Babajide, O.; Fatoba, O.O.; Zhang, C.; Shao, T.; Sosnin, E.; Tarasenko, V.; et al. A critical review on ozone and co-species, generation and reaction mechanisms in plasma induced by dielectric barrier discharge technologies for wastewater remediation. *J. Environ. Chem. Eng.* **2021**, *9*, 105758. [CrossRef]
26. Vanraes, P.; Bogaerts, A. The essential role of the plasma sheath in plasma–liquid interaction and its applications—A perspective. *J. Appl. Phys.* **2021**, *129*, 220901. [CrossRef]

27. Ramos, S.V.; Cisquini, P.; Nascimento, R.C., Jr.; Franco, A.R., Jr.; Vieira, E.A. Morphological changes and kinetic assessment of Cu$_2$O powder reduction by non-thermal hydrogen plasma. *J. Mater. Res. Technol.* **2021**, *11*, 328–341. [CrossRef]
28. Morales-Lara, F.; Abdelkader-Fernández, V.K.; Melguizo, M.; Turco, A.; Mazzotta, E.; Domingo-García, M.; López-Garzón, F.J.; Pérez-Mendoza, M. Ultra-small metal nanoparticles supported on carbon nanotubes through surface chelation and hydrogen plasma reduction for methanol electro-oxidation. *J. Mater. Chem. A* **2019**, *7*, 24502–24514. [CrossRef]
29. Sabat, K.C. Production of Nickel by Cold Hydrogen Plasma: Role of Active Oxygen. *Plasma Chem. Plasma Process.* **2022**, *42*, 833–853. [CrossRef]
30. Hühn, J.; Carrillo-Carrion, C.; Soliman, M.G.; Pfeiffer, C.; Valdeperez, D.; Masood, A.; Chakraborty, I.; Zhu, L.; Gallego, M.; Yue, Z.; et al. Correction to Selected Standard Protocols for the Synthesis, Phase Transfer, and Characterization of Inorganic Colloidal Nanoparticles. *Chem. Mater.* **2021**, *33*, 4830. [CrossRef]
31. Bruggeman, P.J.; Kushner, M.J.; Locke, B.R.; Gardeniers, J.G.E.; Graham, W.G.; Graves, D.B.; Hofman-Caris, R.C.H.M.; Maric, D.; Reid, J.P.; Ceriani, E.; et al. Plasma–liquid interactions: A review and roadmap. *Plasma Sources Sci. Technol.* **2016**, *25*, 053002. [CrossRef]
32. Ulejczyk, B.; Nogal, Ł.; Młotek, M.; Krawczyk, K. Hydrogen production from ethanol using dielectric barrier discharge. *Energy* **2019**, *174*, 261–268. [CrossRef]
33. Levko, D.; Shchedrin, A.; Chernyak, V.; Olszewski, S.; Nedybaliuk, O. Plasma kinetics in ethanol/water/air mixture in a 'tornado'-type electrical discharge. *J. Phys. D Appl. Phys.* **2011**, *44*, 145206. [CrossRef]
34. De Vos, C.; Baneton, J.; Witzke, M.; Dille, J.; Godet, S.; Gordon, M.J.; Sankaran, R.M.; Reniers, F. A comparative study of the reduction of silver and gold salts in water by a cathodic microplasma electrode. *J. Phys. D Appl. Phys.* **2017**, *50*, 105206. [CrossRef]
35. Adamovich, I.; Agarwal, S.; Ahedo, E.; Alves, L.L.; Baalrud, S.; Babaeva, N.; Bogaerts, A.; Bourdon, A.; Bruggeman, P.J.; Canal, C.; et al. The 2022 Plasma Roadmap: Low temperature plasma science and technology. *J. Phys. D Appl. Phys.* **2022**, *55*, 373001. [CrossRef]
36. Privat-Maldonado, A.; Gorbanev, Y.; Dewilde, S.; Smits, E.; Bogaerts, A. Reduction of Human Glioblastoma Spheroids Using Cold Atmospheric Plasma: The Combined Effect of Short- and Long-Lived Reactive Species. *Cancers* **2018**, *10*, 394. [CrossRef]
37. Xu, C.; Chaudhuri, S.; Held, J.; Andaraarachchi, H.P.; Schatz, G.C.; Kortshagen, U.R. Silver Nanoparticle Synthesis in Glycerol by Low-Pressure Plasma-Driven Electrolysis: The Roles of Free Electrons and Photons. *J. Phys. Chem. Lett.* **2023**, *14*, 9960–9968. [CrossRef]
38. He, X.; Lin, J.; He, B.; Xu, L.; Li, J.; Chen, Q.; Yue, G.; Xiong, Q.; Liu, Q.H. The formation pathways of aqueous hydrogen peroxide in a plasma-liquid system with liquid as the cathode. *Plasma Sources Sci. Technol.* **2018**, *27*, 085010. [CrossRef]
39. Bjelajac, A.; Phillipe, A.-M.; Guillot, J.; Fleming, Y.; Chemin, J.-B.; Choquet, P.; Bulou, S. Gold nanoparticles synthesis and immobilization by atmospheric pressure DBD plasma torch method. *Nanoscale Adv.* **2023**, *5*, 2573–2582. [CrossRef]
40. Sauvageau, J.F.; Turgeon, S.; Chevallier, P.; Fortin, M.A. Colloidal Suspensions of Platinum Group Metal Nanoparticles (Pt, Pd, Rh) Synthesized by Dielectric Barrier Discharge Plasma (DBD). *Part. Part. Syst. Charact.* **2018**, *35*, 1700365. [CrossRef]
41. Zhao, L.; Jiang, D.; Cai, Y.; Ji, X.; Xie, R.; Yang, W. Tuning the size of gold nanoparticles in the citrate reduction by chloride ions. *Nanoscale* **2012**, *4*, 5071–5076. [CrossRef]
42. Darwish, M.; Mafla-Gonzalez, C.; Kolenovic, B.; Deremer, A.; Centeno, D.; Liu, T.; Kim, D.-Y.; Cattabiani, T.; Drwiega, T.J.; Kumar, I.; et al. Rapid synthesis of metal nanoparticles using low-temperature, low-pressure argon plasma chemistry and self-assembly. *Green Chem.* **2022**, *24*, 8142–8153. [CrossRef]
43. Quinson, J.; Neumann, S.; Wannmacher, T.; Kacenauskaite, L.; Inaba, M.; Bucher, J.; Bizzotto, F.; Simonsen, S.B.; Theil Kuhn, L.; Bujak, D.; et al. Colloids for Catalysts: A Concept for the Preparation of Superior Catalysts of Industrial Relevance. *Angew. Chem. Int. Ed.* **2018**, *57*, 12338–12341. [CrossRef]
44. Liu, Z.; Chen, Q.; Liu, Q.; Ostrikov, K. Visualization of gold nanoparticles formation in DC plasma-liquid systems. *Plasma Sci. Technol.* **2021**, *23*, 075504. [CrossRef]
45. Chen, Q.; Li, J.; Chen, Q.; Ostrikov, K. Recent advances towards aqueous hydrogen peroxide formation in a direct current plasma–liquid system. *High Volt.* **2022**, *7*, 405–419. [CrossRef]
46. Wu, H.; Liu, Z.; Xu, L.; Wang, X.; Chen, Q.; Ostrikov, K. The Ag+ Reduction Process in a Plasma Electrochemical System Tuned by the pH Value. *J. Electrochem. Soc.* **2021**, *168*, 123508. [CrossRef]
47. Gong, X.; Ma, Y.; Lin, J.; He, X.; Long, Z.; Chen, Q.; Liu, H. Tuning the Formation Process of Silver Nanoparticles in a Plasma Electrochemical System by Additives. *J. Electrochem. Soc.* **2018**, *165*, E540–E545. [CrossRef]
48. Lu, P.; Kim, D.-W.; Park, D.-W. Simple reactor for the synthesis of silver nanoparticles with the assistance of ethanol by gas–liquid discharge plasma. *Plasma Sci. Technol.* **2019**, *21*, 044005. [CrossRef]

Disclaimer/Publisher's Note: The statements, opinions and data contained in all publications are solely those of the individual author(s) and contributor(s) and not of MDPI and/or the editor(s). MDPI and/or the editor(s) disclaim responsibility for any injury to people or property resulting from any ideas, methods, instructions or products referred to in the content.

Article

Characterization of Bulgarian Copper Mine Tailing as a Precursor for Obtaining Geopolymers

Darya Ilieva [1], Lyudmila Angelova [1], Temenuzhka Radoykova [1], Andriana Surleva [1,*], Georgi Chernev [2], Petrica Vizureanu [3,4,*], Dumitru Doru Burduhos-Nergis [3] and Andrei Victor Sandu [3]

[1] Faculty of Chemical Technologies, University of Chemical Technology and Metallurgy, 8 "St. Kl. Ohridski" Blvd., 1756 Sofia, Bulgaria; daryailieva@abv.bg (D.I.); lyudmila@uctm.edu (L.A.); nusha_v@uctm.edu (T.R.)
[2] Faculty of Metallurgy and Materials Science, University of Chemical Technology and Metallurgy, 8 "St. Kl. Ohridski" Blvd., 1756 Sofia, Bulgaria; g.chernev@uctm.edu
[3] Faculty of Materials Science and Engineering, "Gheorghe Asachi" Technical University of Iasi, 67 Prof. D. Mangeron Blvd., 700050 Iasi, Romania; doru.burduhos@tuiasi.ro (D.D.B.-N.); sav@tuiasi.ro (A.V.S.)
[4] Technical Sciences Academy of Romania, 26 Dacia Blvd., 030167 Bucharest, Romania
* Correspondence: surleva@uctm.edu (A.S.); peviz2002@yahoo.com (P.V.)

Abstract: Valorization of high-volume mine tailings could be achieved by the development of new geopolymers with a low CO_2 footprint. Materials rich in aluminum and silicon with appropriate solubility in an alkaline medium can be used to obtain a geopolymer. This paper presents a study of copper mine tailings from Bulgaria as precursors for geopolymers. Particle size distribution, chemical and mineralogical composition, as well as alkaline reactivity, acidity and electroconductivity of aqueous slurry are studied. The heavy metal content and their mobility are studied by leaching tests. Sequential extraction was applied to determine the geochemical phase distribution of heavy metals. The studied samples were characterized by high alkalinity, which could favor the geopolymerization process. The water-soluble sulphates were less than 4%. The Si/Al ratio in mine tailing was found to be 3. The alkaline reactivity depended more so on the time of extraction than on the concentration of NaOH solution. The main part of the heavy metals was found in the residual fraction; hence, in high alkaline medium during the geopolymerization process, they will stay fixed. Thus, the obtained geopolymers could be expected to exert low environmental impact. The presented results revealed that studied copper mine tailing is a suitable precursor for geopolymerization.

Keywords: mine tailing; geopolymers; alkaline reactivity; characterization; heavy metals; leaching tests; valorization; raw materials

1. Introduction

Transformation of industrial wastes into a resource is one of the pillars of the circular economy. The need of waste valorization is widely recognized, and a large amount of effort is focused on the research and development of industrial processing activities for reusing or recycling industrial wastes. The mine activities and metal extraction processes generate an enormous amount of tailings usually stored in tailing dumps [1,2]. The disadvantages of this practice are well recognized, and the industry is eager to find alternatives to landfill practice [2,3].

Geopolymer technology offers a valuable approach to reuse mine tailings as well as other types of industrial wastes by applying low CO_2 footprint processes. An advantage of the geopolymerization process is that the new composite materials could be obtained by fine tuning the precursor mixture containing low and high reactive silica and alumina to obtain a material with desired characteristics [4,5]. The blending of raw material mixtures allowed in some cases for the skipping of the activation step of low reactive precursors, thus ensuring eco-friendly process [4]. Coal combustion byproducts, blast furnace slag, fly ash, glass fibers, metakaolin, etc. were studied as raw materials with high reactive Al and

Si [4,5]. The thorough characterization of mine tailing is required in order to finely adjust the technology to the particular characteristics of raw materials. Due to the specific chemical composition and characteristics of mine tailings from different sources, a deep study is needed for each tailing source. The characterization strategy should include not only the required characteristics for the obtaining technology, but also the assessment of some parameters, taking into account the application of the final product as well as the generated flows into the environment during its production, usage and post-application fate.

Over 55 billion cubic meters of mine tailings (MTs) are stored globally, with a 23% increase by 2025 [3]. Many storage facilities are vulnerable, with over 100 large dam collapses since 1960 causing death and environmental consequences [6]. Utilizing mine tailings as a replacement can conserve natural resources for 4 to 5 years, reducing the use of virgin raw materials in concrete production. Zhang et al. [7] explored the use of mine tailings as precursors for geopolymers, partially replacing fly ash (FA) with molybdenum tailings for cost-effective waste valorization. However, a higher percentage of FA substitution with MT leads to mechanical performance loss due to an increased macropore volume fraction. Qing et al. [8] produced geopolymer concrete with a compressive strength of 47.6 MPa using the alkali-hydrothermal activation of quartz powder at 300 °C. Although it meets 42.5 cement standards, high temperatures may hinder its sustainability. Orozco et al. [9] developed sustainable bricks by activating gold mine tailings with NaOH or $(CaOH)_2$ and curing at 80 °C. Opiso et al. [10] found that adding 10% palm oil fuel ash to gold mine tailings-based geopolymer bricks improved their mechanical qualities and allowed for room-temperature curing, resulting in cost and CO_2 emission savings. However, creating geopolymers with suitable mechanical characteristics using MTs as raw materials is challenging due to the large number of nonreacting phases, particularly quartz. Vizureanu et al. [6] also concluded that the properties of the geopolymers that use mine tailings as pre-cursors strongly depend on the characteristics of the raw materials. Next to the characteristics of the mine tailings, the activator parameters (type and concentration) and the curing conditions (temperature and time) will also affect the performance of the geopolymers. Krishna et al. [5] found low reactivity of metallurgical tumbling in mine tailings for a geopolymer manufacture. They suggested a customized processing approach for each location and available MT, considering factors like raw material properties, activation parameters and end product type.

Geopolymers based on copper mine tailings with and without blending with additional sources of aluminosilicates have been proposed for potential use as construction materials (bricks, pavements, road construction, etc.) or as an approach for hazardous element encapsulation [11–18]. As copper tailings showed low reactivity [12,17], preliminary activation was proposed, such as mechanical activation [16] or curing at elevated temperatures [11,13,16]. Castillo et al. [12] reported that heating at 90 °C promoted the dissolution of aluminosilicates in an alkaline medium and favored geopolymer hardening processes based on copper tailing. In contrast, Tian et al. [13] found that moderate heating (80 °C) promoted the dissolution of aluminosilicates, whereas higher temperatures negatively influenced the microstructure of the obtained geopolymers. The optimal curing temperature should be established empirically for the given raw materials and should depend on their reactivity and alkali concentration [11]. Geopolymer materials appropriate for low-strength applications were developed based on copper tailing and alkali activated fly ash or low-calcium slag as additional sources of reactive aluminosilicates [14,16]. High-calcium additives were reported to result in better characteristics of the obtained geopolymers [17]. A geopolymer concrete based on copper tailing and blast furnace slag was reported to show promising behavior in marine-related environments [18].

Mine tailing dumps in Bulgaria are well studied from a environmental or technological point of view. Some results on mine tailing valorization have been recently discussed in detail [19–21]. However, the mine tailing potential for application as raw materials for geopolymer obtaining is scarcely reported. The main challenges for the incorporation of mine tailings into geopolymer manufacturing are related to heavy metal contaminations and leaching. Wang et al. [22] reviewed the literature on geopolymers containing heavy metals and found that the process produces safe and long-lasting products through physical encapsulation, covalent bonding, ion exchange, and compound creation mechanisms. However, experimental validation is necessary due to the particularities of each mine tailing and geopolymer manufacturing method, especially when blended systems are researched.

The aim of the present study is characterization of copper mine tailing as a raw material for obtaining geopolymers with a low CO_2 footprint. To achieve this goal, a specific algorithm for the chemical characterization of the item was developed to test not only the chemical composition parameters, but also the behavior of potentially hazardous components and their mobility. From a geopolymer technology point of view, the following parameters were studied: particle size distribution, chemical and mineralogical composition, characteristics of aqueous leachate, and alkaline reactivity. From an environmental point of view, the following characteristics were studied: heavy metal fractionalization, the composition of water leachate, and the mobility of hazardous components. The results revealed that the copper mine tailing from the Assarel Concentrator plant in Bulgaria has a great potential for valorization by using it as a precursor for obtaining geopolymers. This study presents for the first time a detailed characterization of copper mine tailing from Assarel, Bulgaria as a raw material for geopolymers obtained by alkali activation and low CO_2 footprint technology.

2. Materials and Methods

2.1. Raw Materials Analysis

Copper mine tailings have been collected from the Assarel Concentrator plant in Bulgaria and comprehensively evaluated to establish its suitability as a precursor for geopolymers. The fly ash used in the study was from the burning of sub-bituminous coal in thermal power plants (TPP "Bobov dol", Bulgaria) and could be classified as class C (Table 1).

Table 1. The oxide chemical composition of the raw materials.

Oxide	SiO_2	Fe_xO_y	Al_2O_3	CaO	SO_3	K_2O	TiO_2	Na_2O	BaO	SrO	ZrO_2	MgO	P_2O_5	Oth.
FA wt.%	41.20	21.80	15.90	12.30	3.50	2.70	1.20	0.67	0.34	0.18	0.06	-	-	<0.05
MT wt.%	65.50	2.67	19.20	1.33	3.13	4.70	0.32	0.97	-	0.03	0.01	1.90	0.19	<0.05

2.1.1. Particle Size Distribution, pH, EC and Eh

Twenty grams of mine tailing were transferred in a baker, and 20 mL of dist. H_2O were added. The mixture was homogenized and left to settle for 10 min or overnight. The pH was measured by a combined glass electrode and Hanna HI5522-02 multimeter. The redox potential and electroconductivity of suspension were measured using a combined platinum electrode with Ag/AgCl ref. (HI3230B, Hanna Instruments, Woonsocket, RI, USA) and conductometric cell (HI-763100, Hanna Instruments), respectively. The measured values of redox potential were adjusted to corresponding values vs. SHE by adding the standard potential of Ag/AgCl reference electrode (+224 mV S.H.E.). The particle size distribution was determined by sieving and determining the amount of material retained on a series of sieves with different sized apertures.

2.1.2. Alkaline Reactivity

An accurately weighed sample of one gram was transferred into a plastic beaker. Twenty mL of 3.0; 6.5 or 10 M NaOH were added to each sample. The samples were agitated on reciprocal shaker at 100 min^{-1} for 10 min; 30 min; 60 min; 4 h; 24 h; 48 h; and 72 h. After the specified time, the samples were centrifuged for 5 min at 6000 rpm. A volume of 15.0 mL of the obtained supernatant was filtered and transferred into a 50.00 mL volumetric flask containing 10 mL of distilled water and 15 mL of conc. HNO_3. The obtained solutions were diluted to volume by distill. H_2O and sent to ICP-OES measurement of concentration of dissolved Al, Si and Ca. The concentrations of the solutions measured by ICP-OES were used to calculate the content of alkali dissolved Al, Si and Ca in mg/kg solid sample, taking into account the dilution factor.

2.1.3. Total Heavy Metals Content of Copper Mine Tailing

The total heavy metal content was determined by ICP-OES after open acid digestion. A sample of 0.2–0.3 g was digested in aqua regia (HNO_3 and HCl 1:3 v/v) (15 mL) and boiled for 25 min in a beaker covered with a watch glass. A total volume of 15 mL of the acid mixture was added in portions of 5 mL. The extract was filtered and diluted up to 50.00 mL with distilled water. The resulted solution was analyzed by ICP-OES.

2.1.4. Water Leaching Characteristics

The leaching characteristics of the studied copper mine tailing were determined by a protocol based on EN 12457-2:2002 [23]. Ten grams air dried sample was mixed with 100 mL dist. water in a plastic beaker. The samples were agitated on a reciprocal shaker for 24 h at room temperature, 25 ± 2 °C. The dissolution process was followed by measuring pH and EC. In 24 h, the approximately stable results for pH and EC of supernatant were obtained; hence, the dissolution process was accepted to be completed. The samples were filtered to separate supernatant from solid residue. The obtained solutions were sent to ICP-OES for determination of elements and for UV-Vis for determination of nitrate, phosphate and sulfate ions. The obtained data for the concentration of dissolved ions in mg/L were calculated on a solid sample base and presented as mg/kg dry solid sample. Thus, the part of elements/ions of the sample soluble in water at pH of the sample for 24 h were estimated.

2.1.5. Sequential Extraction of Heavy Metals

A modified four-step sequential extraction protocol based on the method proposed by the European Community Bureau Reference (BCR)—now Standard, Measurements and Testing Programme and modified by Gitari et al. [24]—was applied in the present study.

- Extraction step 1—determination of water/acid soluble and exchangeable fraction: 40 mL of 0.11 M HOAc was added to one g of accurately weighed dry sample (±0.0001 g). The mixture was shaken for 16 h on a reciprocal shaker (30 rpm) at 25 °C. The supernatant was filtered and collected in a dry vessel.
- Extraction step 2—determination of reducible or bound to Fe oxides fraction 2: 40 mL of 0.5 M $NH_2OH.HCl$ (adjusted to pH = 1.5 with HNO_3) was added to the residue from step 1 and shaken for 16 h at 25 °C. The supernatant was filtered and collected in a dry vessel.
- Extraction step 3—determination of oxidizable or bound to sulphides and organic phases fraction: 10 mL of 8.8 M H_2O_2 was added to the residue from step 2 and digested for 1 h at 25 °C, and then for 1 h at 85 °C in a water bath with a second volume of H_2O_2. The solution was evaporated to a few milliliters. A total of 50 mL of 1 M NH_4OAc (adjusted to pH = 2.0 with HNO_3) was added to the residue. The suspension was shaken for 16 h at 25 °C. The supernatant was filtered and collected in a dry vessel.
- Extraction step 4—determination of residual or bound to silicates fraction: The final residue obtained from step 3 was dried at 120 °C for two hours. A total of 0.5 g of the dry residue was subjected to open vessel aqua regia digestion for 30 min without boiling. The solution was filtered and diluted to the final volume of 50 mL in a volumetric flask.

The obtained solutions were analyzed by ICP-OES for determination of heavy metals content.

2.1.6. ICP-OES Analysis of Leachates

ICP-OES method for determination of heavy metals and metalloids in leachates from mine tailing was previously validated [25]. An ICP-OES spectrometer (Prodigy High Dispersion ICP-OES, Teledyne Leeman Labs, Hudson, NH, USA) with a dual-view torch, cyclonic spray chamber, and concentric nebulizer was used. The wavelengths free from spectral interferences at the studied concentration range were chosen [26]. Emission intensity of the analytes was a mean of at least three replicates. The analytical function was calibrated in two concentration ranges (5–100 µg/L or 1–10 mg/L). The calibration solutions were prepared by appropriate dilution of appropriate reference materials. Calibration curves and backgrounds correction were obtained by ICP-OES building software Salsa, Prodigy, HD ICP OES 2009, USA.

2.2. Geopolymers Obtaining and Characterization

Fly ash and mine tailing blended systems have been designed to analyze the influence of the Bulgarian copper mine tailings on the mechanical properties of geopolymers.

2.2.1. Geopolymers Design and Preparation

The chemical composition of both raw materials has been analyzed by X-ray fluorescence (XRF) using an XRF S8 Tiger (Bruker GmbH, Karlsruhe, Germany) in order to establish a suitable activation method. The mineralogical composition of raw materials was determined by XRD analysis using X-ray diffractometer Empyrean, Panalytical with CuK radiation and building software. The step interval, integration time and angle interval used were $0.0530°$; 53.8 s; and 5–80 2θ, respectively. Also, to assure experiment repeatability, the collected wastes have been dried until a constant weight (as described in [27]) was attained. As can be seen, both types of raw materials have Si and Al oxides in significant amounts. Also, the XRF analysis reveals a high content of Fe in the FA, which could negatively influence the mechanical performance of the geopolymers [28]. However, according to previous studies [29], it seems that in the case of fly ash-based geopolymers, the iron content has no influence. The FA also contains CaO, which could contribute to early strength development, along with decreasing setting time and curing temperature [30].

As activators, commercially available sodium silicate solution (S.C. KYNITA S.R.L., Valcea, Romania) and sodium hydroxide flakes (98% purity) from the same supplier were chosen. Prior to mixing with the Na_2SiO_3 solution, the NaOH flakes were dissolving in tap water at the desired concentration. The sodium silicate solution had a density of 1.52 g/cm^3 and a pH of 11.5. Also, according to its quality certificate, the solution contains sodium silicate min. 44.8%, min. 31.10% SiO_2, min. 13.70% Na_2O and additives. Considering the chemical composition of the sodium silicate, the $SiO_2:Na_2O$ ratio was determined as 2.27; therefore, 5.5 g of NaOH is necessary for each 100 g of sodium silicate in order to obtain the best alkaline activator [31]. Further, depending on the desired Na:Al ratio, the solid component is calculated based on the Na_2O from the obtained activator and the Al_2O_3 from the aluminosilicate precursor [31].

To establish the influence of the MT addition on the mechanical properties of geopolymers, (i) three mixtures comprising different percentages of fly ash and mine tailings designated as B1 = 100% FA, B2 = 75% FA + 25% MT, and B3 = 50% FA + 50 MT (Table 2), (ii) three different liquid-to-solid ratios (0.70, 0.75, and 0.80), and (iii) three different Na:Al ratios (0.5, 0.75, and 0.1) were used. The Na:Al ratio was calculated considering the value of Al from the XRF analysis and the value of Na from the composition of the activator. The design of experiments with three factors and three levels is shown in Table 3. Considering the L9 orthogonal matrix Taguchi method, nine different mixtures were required to establish the influence of all factors involved. The mixture design is presented in Table 4.

Table 2. The mix design of solid components for geopolymers.

Mixture Code	Composition of Mixtures, wt.%	
	FA	MT
B1	100	0
B2	75	25
B3	50	50

Table 3. Experimental factors and levels.

Factor	Level 1	Level 2	Level 3
A. Binder content	B1	B2	B3
B. Na:Al ratio	0.5	0.75	1
C. Liquid-to-solid ratio	0.7	0.75	0.8

Table 4. Mixing combinations of geopolymer samples.

Sample Code	S1	S2	S3	S4	S5	S6	S7	S8	S9
Factor A	B1	B1	B1	B2	B2	B2	B3	B3	B3
Factor B	0.5	0.75	1	0.5	0.75	1	0.5	0.75	1
Factor C	0.7	0.75	0.8	0.75	0.8	0.7	0.8	0.7	0.75

The procedure for obtaining the geopolymers consisted of the following steps: The raw materials (FA and MT) were dried, weighted, and mixed for 3 min (until a homogeneous mixture was obtained). Before mixing, the FA was dried, and only the particles lower than 100 μm in diameter were kept. During the process, sodium silicate and sodium hydroxide solution were mixed according to the composition. Afterwards, the liquid component (activator) was added over the solid component (raw materials) and mixed for 3 min to obtain a homogeneous mixture. The mixing of the components was conducted using a planetary mixer with variable speed, according to the EN 196-1:2016 standard [32]. A schematic representation of geopolymers obtained is presented in Figure 1. After mixing the liquid and the solid component, the obtained binder was poured into molds with sample dimensions of 40 mm × 40 mm × 160 mm, and their vibration was applied in order to obtain a uniform mixture with reduced air bubble content. The filled molds were then cured at room temperature (22 ± 2 °C) up to the testing age. For the first 24 h, the samples were covered with a plastic sheet in order to reduce the evaporation rate of the liquid, then kept in open air.

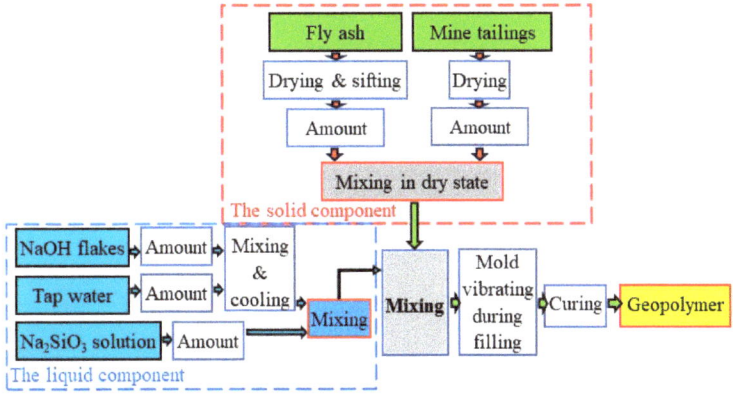

Figure 1. Schematic representation of the technological flow of geopolymers obtaining.

The most energy-efficient raw materials and processing procedures were examined to ensure that the developed technology is eco-friendly and has a low CO_2 footprint. Moreover, in industrial use, both the drying and sifting stages can be removed since the water content can be evaluated through a moisture content, and then the amount already available in the raw materials can be reduced from the activator, while some impurities of large dimensions can exist in the raw materials will not significantly influence the properties of large, real-size, products.

2.2.2. Geopolymers Analysis Methods

Both the compressive and flexural strengths of the developed materials have been tested at 14 and 28 days, respectively. The sample size and testing conditions have been followed according to SR EN 196-1:2016 requirements [32]. Accordingly, the flexural strength tests have been conducted on six specimens for each mixture (three have been tested at 14 days and three at 28 days) measuring 40 mm × 40 mm × 160 mm. The samples that resulted from flexural strength tests (two specimens split from the 40 mm × 40 mm × 160 mm specimen) have been used further to evaluate the compressive strength. As per standard requirements, a loading rate of 50 ± 10 N/s was chosen.

The microstructural analysis of the obtained geopolymers was conducted on the fracture surface after the mechanical tests using a scanning electron microscope (SEM) type FEI Quanta FEG 450 (FEI Company, Washington, DC, USA).

3. Results

3.1. Mine Tailing Caracterization

3.1.1. Particle Size Distribution

The particle size distribution of copper mine tailing was studied as received without additional grinding. The particles were finer than 1250 µm. The main part of the particles (89%) was below 315 µm; almost half of the total amount of particles (46%) was below 200 µm. The fraction of particles with a size favorable for geopolymerization (<100 µm) was 17%. The finest fraction (between 90 and 63 µm) was 3%. The average values of particle size were 320, 210 and 90 µm for d_{90}, d_{50} and d_{10}, respectively (Figure 2). The particle size distribution curve was very steep, indicating a uniformly graded material [33]. Generally, the particle size distribution of the studied copper-mine tailing (90% below 0.3 µm) showed a fine sized material suitable for the geopolymerization process, according to the criteria proposed by Aseniero et al. [34]. However, the finer particle size was proven to increase the reactivity and to improves the geopolymers characteristics [35]. In this study, a relatively low percentage of the particles of mine tailing (17%) was below 100 µm, and the dimension was characterized with higher reactivity; thus, additional activation or blending with reach of aluminosilicates raw materials with fine particles was needed to obtain geopolymers with desired characteristics.

3.1.2. Physicochemical Characteristics: pH, EC and Eh

The pH of the copper tailing at 1:1 solid-to-liquid ratio and 10 min settling time was 9.01, the electrical conductivity was 1796 µS/cm, and the redox potential was 122.4 mV (vs. Ag/AgCl). The alkaline pH was mainly due to the abundance of aluminosilicate minerals, such as clinochlore and muscovite found by XRD study (Figure 3). The redox potential of the aqueous supernatant was E = 366 mV (vs. SHE), indicating oxidative conditions in the slurry [36].

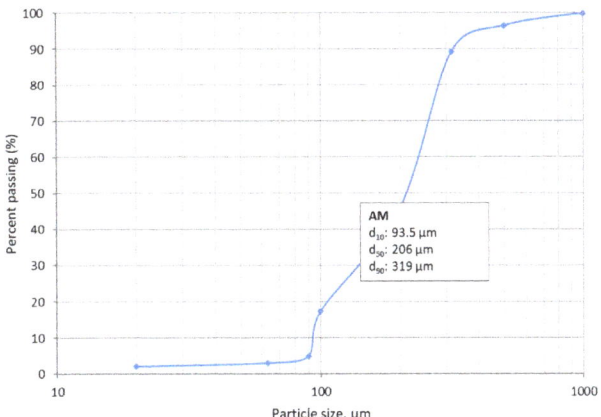

Figure 2. Particle size distribution of copper mine tailing.

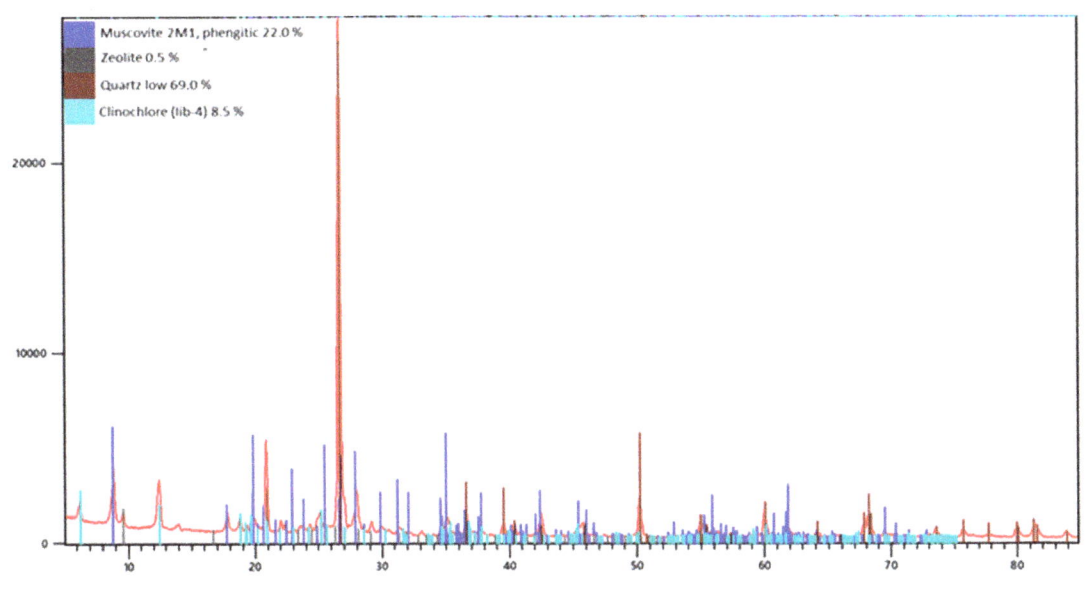

Figure 3. X-ray powder diffractogram of copper tailings sample. The main crystalline phases are presented in the Table 5.

Table 5. The main crystalline phases in the studied copper mine tailing by XRD.

Reference	Crystalline Phase	Composition
98-008-3849	Quartz low	SiO_2
98-008-4262	Clinochlore (IIb-4)	$Mg_5Al(AlSi_3O_{10})(OH)_8$
98-017-0492	Zeolite	SiO_2
98-007-7495	Muscovite 2M1, phengitic	$K(Al,Mg)_2(OH)_2(Si,Al)_4O_{10}$

3.1.3. Chemical and Mineralogical Composition

The chemical composition of copper mine tailing obtained by XRF is presented in Table 1. The results showed that the copper mine tailing was reached in SiO_2 (66%) and

Al_2O_3 (19%). The high content of Si and the significant content of Al indicated the presence of aluminosilicate minerals. The total content of alumina and silica was above 70%, which favored the geopolymerization process [37]. The ratio between Si/Al was calculated to be equal to 3. Moreover, the studied copper mine tailing contained a relatively low quantity of SO_3 (3%), which was found to enhance the stability of the obtained geopolymer products [38–40]. Besides the Si and Al bearing minerals, elements such as Fe, Mg and Ca, can react and participate in the geopolymer matrix by forming other solid phases, which can affect the final properties of the structure [41]. In the studied sample, Fe_2O_3, MgO and CaO were between 2 and 3%. A relatively low content of Ca-bearing minerals (3% CaO) will affect the workability of the geopolymer paste, and additional sources of Ca should be added to the mixture. Between 1 and 4% of MgO, CaO, and K_2O was found in the tailing. The presence of easily soluble basic components corresponded to high pH values measured in the water/tailing mixture (see Section 3.1.2). Although the sample contained Fe and S (around 3% as corresponding oxides), the high pH of the water leachate indicated that Fe was not in the form of Fe sulphide minerals [24].

The hazardous elements (Pb, Cu, Mn, Fe, As, Zn) were found at the levels of impurities (<1%) or traces (<0.1%). The concentrations of Pb, Cu and Mn were typical for industrial terrains.

An important prerequisite for raw materials is a higher content of Al_2O_3 and SiO_2, preferably in amorphous reactive form [12,37,42]. The mineralogical composition of the studied copper mine tailing is presented on Figure 3 and Table 5. The primary minerals are quartz (SiO_2) 69.0%, clinochlore (($Mg, Fe^{2+})_5Al_2Si_3O_{10}(OH)_8$) 8.5%, zeolite ($SiO_2$) 0.5% and muscovite ($KAl_2(AlSi_3O_{10})$) 22.0%. Alumosilicate minerals presented in the sample and the zeolite phase could be expected to favor the geopolymerization process, resulting in geopolymers with a high compressive strength [12,42,43].

3.1.4. Alkaline Reactivity

The alkaline reactivity of mine tailing was studied at different concentrations of alkali (3.0; 6.5 and 10 M NaOH). The level of dissolution of Ca, Al and Si in alkaline media was followed at different time intervals, starting from 10 min up to 72 h by the ICP-OES determination of the concentration of the studied components in the obtained solutions. The alkaline reactivity of studied mine tailing was estimated as mg dissolved component per kg solid sample. The results are presented in Figure 4a–c. The concentration ratio of dissolved Al to Si as a function of molarity of NaOH and time of leaching is presented on the Figure 4d.

Alkali dissolution of Al from mine tailing followed the pattern presented on Figure 4a. The concentration of dissolved Al constantly increased with time at each of studied concentration of NaOH. Increasing the alkali concentration up to 10 M solution in the first 48 h resulted in a decrease in the level of dissolved Al. At 72 h of contact between the solid and the liquid, the following results were obtained: 340 mg/kg in 3 M NaOH, 460 mg/kg in 6.5 M NaOH and 540 mg/kg in 10 M NaOH. The detailed results are presented in Table 3. At shorter periods (10 min to 4 h), the influence of alkali concentration was less pronounced: 93 mg/kg in 3 M; 125 mg/kg in 6.5 M and 190 mg/kg in 10 M NaOH.

Alkali dissolution of Si from the mine tailing followed the same trend as Al (Figure 4b). The concentration of dissolved Si constantly increased with time at each of the studied concentrations of NaOH. Up to 4 h of contact, the increased concentration of NaOH exerted a slight effect on the Si dissolution (Figure 4b inset): 306 mg/kg in 3 M NaOH; 403 mg/kg in 6.5 M NaOH and 330 mg/kg in 10 M NaOH. At the reaction times from 10 to 60 min, the quantity of dissolved Si slightly increased at each of the studied concentrations of alkali. The positive influence of the concentration of NaOH could be seen at these reaction times: highest dissolved Si was observed in 10 M NaOH. As can be seen from Figure 4b (inset), in the first 60 min, increasing the concentration of alkali resulted in a higher slope in the curve, which correspond to the increased rate of Si dissolution during the first contact. Moreover, in 10 M NaOH, the quantity of dissolved Si was approximately constant until 60 min:

270 mg/kg, indicating that the most reactive Si was dissolved in less than 10 min. More noticeable increases in the rate of Si dissolution in each of the studied alkali concentrations were observed at prolonged reaction times > 1 h.

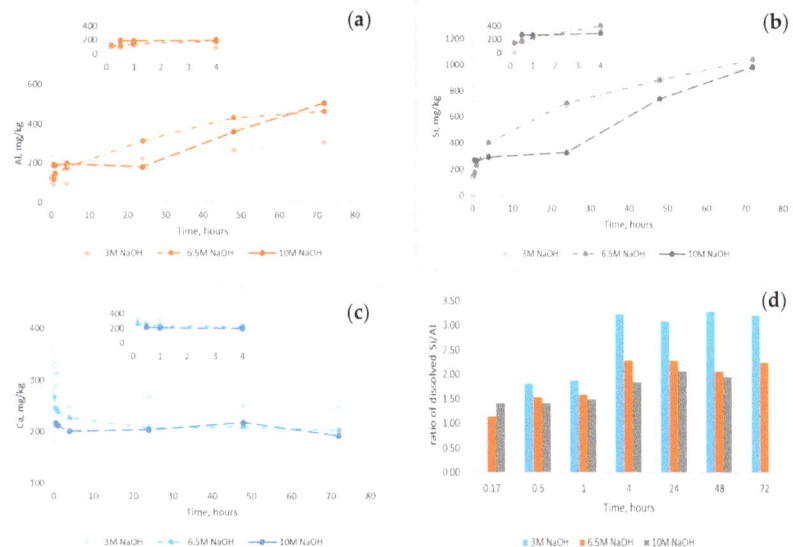

Figure 4. Alkali-dissolved Ca, Al and Si of the copper mine tailing as a function of time at different concentrations of NaOH: (**a**) Ca; (**b**) Al; (**c**) Si and (**d**) ratio of alkali-dissolved Si/Al as a function of alkali concentration and time of leaching. The alkaline reactivity is presented as mg dissolved component per kg solid sample. In-sets present the time intervals up to 4 h.

As can be seen from Figure 4c, the concentration of dissolved Ca was higher in 3 M NaOH and decreased with increasing alkali concentration. The highest concentration of dissolved Ca was observed at the beginning of the contact between alkali and mine tailing, reaching a steady value after 4 h. At 10 min, the concentration of Ca varied from 330 mg/kg in 3 M NaOH to 210 mg/kg in 10 M NaOH. After 4 h until the end of the studied period, the concentration of dissolved Ca reached a steady value depending on the alkali concentration: 260 mg/kg in 3 M to 200 mg/kg in 6.5 and 10 M. After 4 h of contact between mine tailing and 6.5 and 10 M NaOH solution, the concentration of alkali did not influence the concentration of dissolved Ca. In contrast, 3 M NaOH still dissolve a higher amount of Ca.

Figure 4d presents the influence of alkali concentration and reaction time on the dissolved Si/Al ratio. As can be seen, increasing the alkali concentration resulted in a lower Si/Al ratio. This could be explained by the higher rate of Al dissolution in more concentrated NaOH solutions compared to Si. Hence, the alkali concentration exerted more pronounced effects on the dissolution of Al than on the dissolution of Si. The trend was observed in the whole studied period from 10 min to 72 h. It should be noted that the highest difference in Si/Al ratio was observed when increasing concentrations from 3 to 6.5 M. Higher concentrations of alkali showed slightly leveling effect.

The results presented in the Table 6 demonstrated the alkaline solubility of the main geopolymer forming components in the studied copper mine tailing. The concentration of dissolved Al in mg per kg dry solid sample was taken as a base for ratio calculations. The concentration ratio of dissolved Ca and Si followed the same pattern: the increased concentration of NaOH resulted in lower Ca to Al and Si to Al ratio. The same trend in Ca to Al ratio is observed regarding the contact time. The findings could be explained by the increased concentration of Al with increasing the concentration of alkali and the contact time. In contrast, the Si to Al ratio increased with shorter reaction times, reaching state value after 4 h. A possible explanation could be found in the faster rate of dissolution of Si than of Al in the studied sample in the beginning of contact between alkali and solid material; after 4 h, the rate of dissolution of both components is almost equal. At shorter reaction times (<60 min), the Al/Si ratio is not influenced by the concentration of NaOH and is found to be approximatively 1:2 in all studied concentrations. More pronounced effects of alkali were observed at longer contact times (>60 min). In 3 M NaOH, the Al/Si ration was 1:3 and lowered to 1:2 in 10 M alkali.

Table 6. Ratio of dissolved Ca/Al/Si in copper mine tailing as a function of alkali concentration and reaction time.

Concentration of NaOH, M	Reaction Time, h	Dissolved Ca/Al/Si Ratio [1]
3	10′	3.6/1/0
	30′	3.2/1/1.8
	60′	2.5/1/2
	4 h	2.6/1/3.2
	24 h	1.2/1/3.1
	48 h	0.9/1/3.3
	72 h	0.8/1/3.2
6.5	10′	2.1/1/1.1
	30′	2.1/1/1.5
	60′	1.6/1/1.6
	4 h	1.3/1/2.3
	24 h	0.7/1/2.3
	48 h	0.5/1/2.1
	72 h	0.4/1/2.2
10	10′	1.1/1/1.4
	30′	1/1/1.5
	60′	1.1/1/1.8
	4 h	0.6/1/2.1
	24 h	0.4/1/1.9
	48 h	0.4/1/1.9
	72 h	1.1/1/1.4

[1] Concentration of dissolved Al in mg/kg dry material was taken as 1 for calculation of the ratio.

3.1.5. Heavy Metals Content and Mobility in Aqueous Medium

Table 7 presents the chemical composition of copper mine tailings determined by ICP-OES after aqua regia digestion and the composition of the generated aqueous leachate. The obtained results allowed for the environmental impact of mine tailing to be estimated and give some indication about the chemical composition of generated plume after interaction between the tailing and rainwater [24]. The generated leachates from the copper mine tailings were alkaline with pH 9.53 and were stable in the studied period of 24 h. The electrical conductivity of leachate was 750 µS/cm after 2 h and reached 850 µS/cm after 24 h.

The concentrations of Cd and Co in the generated leachate were below <0.005 mg/L, and Cr and As were below 0.010 mg/L. The highest concentrations in leachate were observed for (in decreased order) Fe > Ca > Mg > Al. Additionally, the concentrations were calculated in a mg/kg air dried sample to present the concentrations of ions in solid-sample dissoluble in water at sample pH and accordingly were capable of moving into the aqueous solution. The results showed that less than 0.5 mg/kg of Cd, Co, Cr and As in copper mine tailing were in water soluble form; the main part of these components should be in non-labile geochemical fractions. Fe, Al, Ca and Mg showed the highest water-soluble concentrations: 2462, 694, 389 and 699 mg/kg, respectively, and govern the electrical conductivity of the leachate. The concentration in leachate of the studied anions decreased in the following order $SO_4^{2-} > NO_3^- > PO_4^{3-}$.

Table 7. Chemical composition of copper mine tailing and generated aqueous leachate.

Component	Concentration in Mine Tailing, mg/kg	Concentration in Leachate, mg/L	MPL [1], mg/L	Component	Concentration in Mine Tailing, mg/kg	Concentration, mg/L	MPL [1], mg/L
Ca	9491	123	150	Cd	<0.5	<0.010	0.005
Mg	699	2.93	80	Ni	6.700	<0.010	0.02
Na	278	9.28	200	Co	<0.5	<0.005	
Al	694	0.27	0.2	Pb	traces	traces	0.010
Fe	18,629	<0.005	0.2	NO_3^-		1.0	0.05
Cu	375	<0.005	0.002	SO_4^{2-}		370	250
Zn	72.5	<0.010	4	PO_4^{3-}		<0.5	0.5
Mn	186	<0.010	0.50	Cl^-			250
As	<2	<0.010	0.010	pH		9.53	
S	3734	198		EC, μS/cm		750	
Cr	<0.5	<0.010	0.05				

[1] Maximum permissible level according to the Bulgarian regulation for the quality of drinking water [44].

3.1.6. Heavy Metals Distribution in Mine Tailing Fractions

Heavy metal distribution in geochemical fractions of copper mine tailing was studied by a BCR sequential extraction procedure, including the additional step of the analysis of the residual fraction. The heavy metal fractionalization was strongly correlated with their availability and possible toxicity to the environment. The sum of the heavy metals in exchangeable fraction (F1), iron and manganese oxide fraction (F2) and organic matter and reducible fraction (F3) presented labile species, which were susceptible to dissolution and transferred into liquid phase. The residual fraction (F4) presented the non-labile phase. The following metals were studied: Fe, Cu, Zn, Mn, Cd, Ni, Pb and As. The results showed that Cd, Ni and As were below method LOQ (<0.5 mg/kg) in the studied copper tailing sample. The total content of heavy metals showed that Fe is the most abandoned element, followed by (in decreasing order) Cu, Mn, Zn and Pb. The sum of F1, F2 and F3 was considered as a mobile part of the heavy metals. The following decreasing trend in concentrations was observed: Fe > Cu > Mn > Pb > Zn. Considering the first three fractions, it could be seen that all of the studied metals showed the highest concentration in liquid phase in the F3: organic matter and reducible fraction. The residual fraction was considered to contain the non-labile forms of the heavy metals. In this fraction the elements were arranged as follows: Fe > Cu > Mn > Zn > Pb. The results showed that the highest content of the studied hazardous elements was found in the residual fraction, indicating the limited mobility and bioavailablity of hazardous components of the copper mine tailing.

The partitional pattern of Fe showed that the main part of Fe in the copper tailing (approximatively 50%) was in a mobile form (sum F1 + F2 + F3). The highest concentration was observed in Fraction 3: organic matter and reducible fraction.

The concentration of dissolved Cu in sequential fractions decreased in the following order: F3 > F1 > F2. A total of 77% of the total Cu was found in the mobile fractions, and 23% was found in the residual fraction. The main part of the Cu species was extracted in the F3 organic matter and reducible fraction. The minor part of Cu was bound to Fe- and Mn oxides.

The main part of Zn was found in the residual fraction, indicating its low lability and corresponding bioavailability. Around 36% of the total Zn in the copper mine tailing was extracted in the sequential extraction steps. The highest quantity of Zn was bound to organic matter and the reducible fraction. The quantity of Zn in exchangeable fraction and Zn bound to Fe and Mn oxide were approximatively equal.

The main part (60%) of Mn in the copper mine tailing was found in the non-libile form in the residual fraction. The highest concentration of dissolved Mn was observed in the F3 fraction, and the lowest was in the F2 fraction. Around 12% of the total Mn was observed in the F1 exchangeable fraction, 6% in F2 and 23% in F3. The results indicated the low mobility of Mn in the studied sample.

The concentration of dissolved Pb followed the pattern F4 > F3 > F2 > F1. The main part (60%) of Pb in the sample was bound into the mobile fractions. However, it could be observed that less than 10% of the available Pb was found in the easily exchangeable fraction.

3.2. Geopolymers Analysis

3.2.1. Mechanical Properties

Compressive strength. Figure 5 shows the evolution of compressive strength from 14 to 28 days for the obtained geopolymers. As can be seen, the mixture specific to S1 did not perform too well at this test, being one of the samples with almost the same lost value at 28 days. In this case, S6 exhibited the optimum composition despite the aging time, while S1 and S3 showed a slight decrease after 28 days of curing. The increase in compressive strength was in the range of 22 to 54%; the highest evolution was presented by S9, while the lowest was shown by S8. Overall, it can be observed that the compressive strength depends on the raw material mixture and the Na:Al ratio; higher ratios result in better compressive strength. A replacement of FA with MT seems to promote the best compressive strength, despite the Na:Al ratio. A 25 wt.% replacement of FA with MT showed the optimum replacement, yet at 50 wt.% of MT, the compressive strength of the blended geopolymer is superior to that based on FA at almost all testing ages and Na:Al ratios. Moreover, all mixtures showed a low deviation value, which confirms that the obtained samples had a homogenous structure and composition.

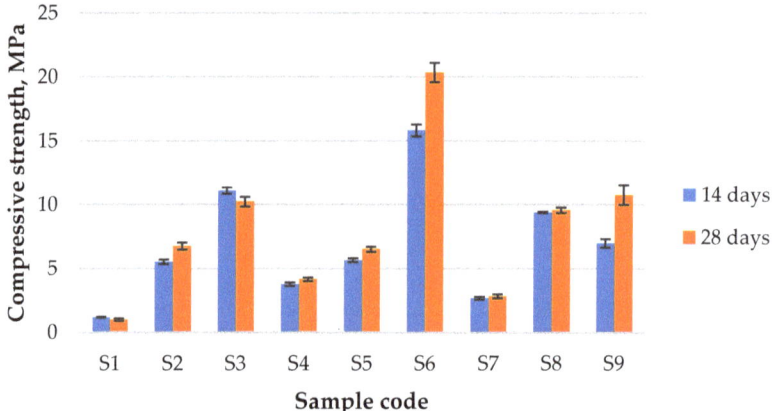

Figure 5. The compressive strength of the obtained geopolymers.

To better understand the influence of the obtained parameters on the mechanical properties of the obtained geopolymers, 3D plots were drawn. Figure 6 shows the influence of MT content and the liquid-to-solid ratio on the compressive strength. As can be seen, higher liquid content will result in poor mechanical properties. The optimum mixture is the one with 25 wt.% MT and a 0.7 liquid-to-solid ratio. This behavior could be related to activator loss during curing. The OH^- from the activator, especially from the NaOH solution, is responsible for the leaching process of Si^{4+} and Al^{3+} ions; therefore, a loss of activator will also result in a loss of the necessary ions to assure the dissolution of the aluminosilicate material [45]. Moreover, during this drying stage, the water evacuation will also promote crack formation [46]. The samples with only FA as a precursor showed an increase in compressive strength with the increase in liquid-to-solid ratio, probably due to the nature of the FA particles sponge-like structure, which will absorb the activator and release it much slower compared to the compact particles from MT [47].

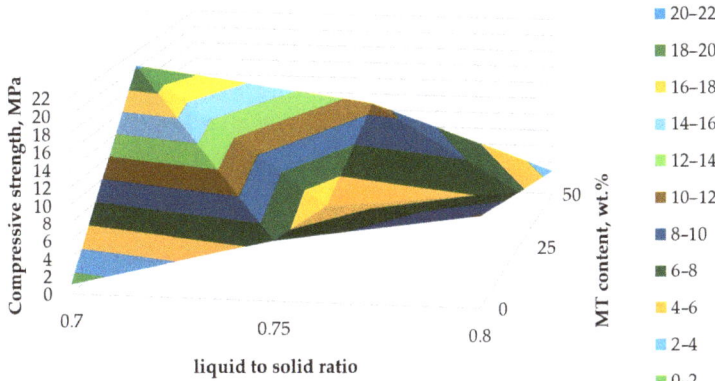

Figure 6. Influence of liquid-to-solid ratio and MT content on compressive strength at 28 days.

Considering the dependence between the amount of MT and the Na:Al ratio (Figure 7), it can be observed that the 25 wt.% replacement of FA with MT will not produce the best compressive strength, despite the activating conditions. For example, at a Na:Al ratio of 0.75, the mixture with 50 wt.% MT performed better than all other geopolymers. Even though the increase in Na:Al ratio resulted in better mechanical properties for all mixtures, only the FA-based geopolymer seems to present a continuous trend from a Na:Al of 0.5 to a Na:Al of 1. In the case of the other mixtures, the Na:Al increase seems to have a negative effect at 0.75 for the geopolymer with 25 wt.% MT or at 1 for the geopolymer with 50 wt.% MT.

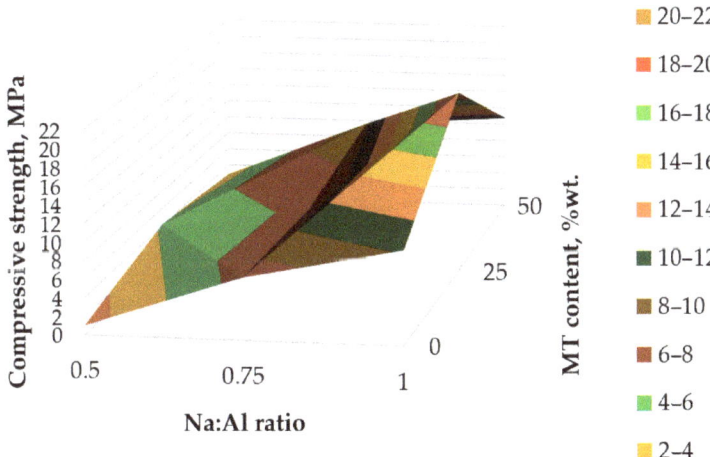

Figure 7. Influence of Na:Al ratio and MT content on compressive strength at 28 days.

The data used to create the 3D graphs from Figures 6 and 7 are presented in Supplementary File S1.

Flexural strength. The flexural strength of the obtained geopolymers decreased in most of the cases from 14 days to 28 days (Figure 8), except for samples S5, S6 and S8, where a significant improvement in flexural strength can be observed. For example, from 14 to 28 days, the value of S8 and S6 increased almost two times, and the lowest increase was that of S5, which is approximately 20% higher. As observed, all mixture with FA as raw materials showed a decrease in flexural strength, probably due to the high Ca content that will react faster than Al, while at latter stages, the Si will start reacting, resulting in a much more brittle structure [48]. On the contrary, most of the samples that have MT content showed an increase in flexural strength over time, probably due to a lower concentration of Ca from the system. Considering the Na:Al ratio, it can be seen that only the mixtures with a Na:Al of 0.5 showed a decrease in flexural strength over time. At 14 days, the optimum composition was that of S2, while at 28 days, S8 showed the highest value.

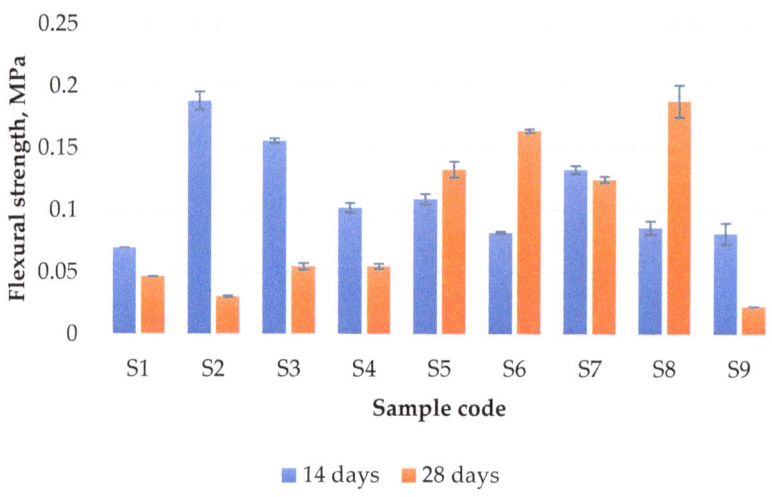

Figure 8. The flexural strength of the obtained geopolymers.

3.2.2. Microstructural Analysis of the Obtained Geopolymers

To emphasize the link between the morphological characteristics of the created materials and their mechanical performances, the combination with the lowest compressive strength was compared to the mixture with the highest value for each mixture of raw materials. As shown in Figure 9, the microstructure of the S1 geopolymer (Figure 9a) contains multiple unreacted coal ash particles, particularly those with large diameters. Moreover, there are large areas that were not dissected, which act as defects, resulting in a poor structure in terms of mechanical properties. Compared to the S1 sample, the sample with a Na:Al ratio of 1 and FA as raw material, i.e., S3, showed a much more compact matrix with better dissolution; however, the matrix also exhibits multiple pores and cracks distributed all over the analyzed surface (Figure 9b). When mine tailings are introduced in the mixture, the morphology of the structure significantly changes; the matrix shows very few cracks, but there are still some areas where unreacted FA particles can be observed (Figure 9c). The samples with the highest compressive strength, i.e., S6, show a substantially more compact matrix than all other mixtures (Figure 9d). Some unreacted FA particles can be observed, but most of them are embedded in the compact matrix. By increasing the MT content to 50 wt.%, the dissolution of the raw materials seems to decrease (Figure 9e), while a significant increase in the number of cracks occurs (Figure 9f).

As can be seen in Figure 10, at higher resolution, the differences between the best (sample S6) and the worst (sample S1) samples in terms of compressive strength are even more visible. Compared to S6 (Figure 10a), the microstructure of S1 (Figure 10b) shows activation only in limited areas (the smooth area from the image), while high areas show morphology specific to the nonreacted particles. However, in the case of S6, most particles seem to be activated, while those that remain unreacted are embedded in the matrix.

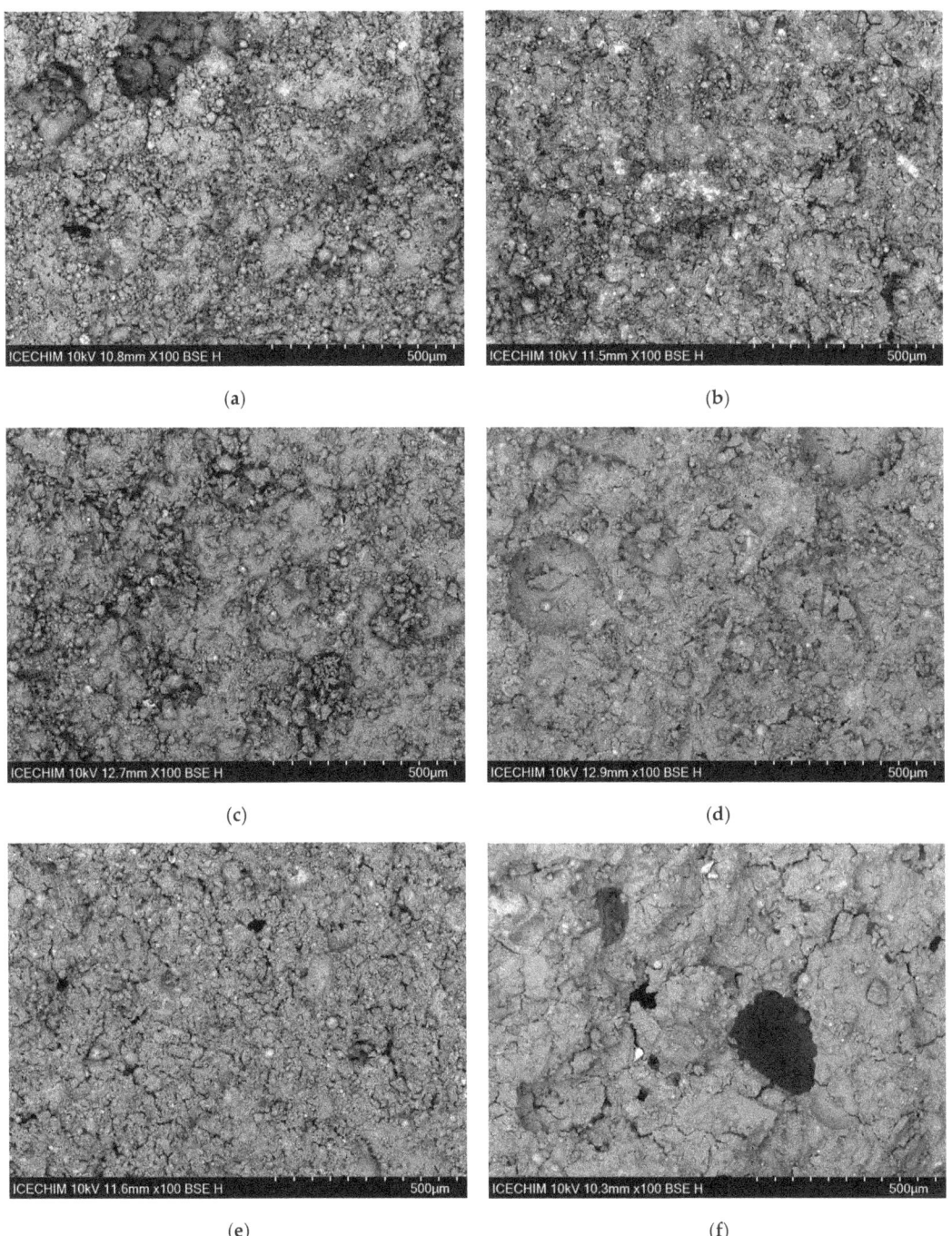

Figure 9. Microstructural analysis of the obtained geopolymers: (**a**) sample S1; (**b**) sample S3; (**c**) Sample S4; (**d**) sample S6; (**e**) sample S7; (**f**) sample S9.

(a) (b)

Figure 10. The microstructure of the obtained geopolymers at 1k× magnification: (**a**) sample S1; (**b**) sample S6.

Considering the composition of the compared mixture, it can be stated that higher Na:Al ratios will increase the dissolution of the aluminosilicate source and increase the compactness of the matrix, promoting better mechanical properties.

4. Discussion

A detailed characterization of the Bulgarian copper mine tailing from Assarel-Medet as a precursor for geopolymer obtaining was reported for the first time. The chemical and mineralogical composition, leaching behavior, alkaline reactivity amd heavy metal distribution in the geochemical phases of mine tailing were studied. The conclusions drawn were focused on its utilization as a raw material for obtaining geopolymers.

The chemical composition of the studied copper mine tailings showed their perspective to be reused as raw materials for obtaining geopolymers (Table 1). The high content in Si and significant content in Al-bearing minerals are favorable for the geopolymerization process [37]. The ratio of Si/Al in the copper mine tailing was in the optimal ratio of 1–3 [12,24,35,49,50] thus making the studied tailing a promising precursor for geopolymers. However, an additional source of Al could be added to enhance the geopolymerization process and improve the characteristics of the obtained products [24,51]. In a recent review, Lazorenko et al. summarized the chemical composition of various mine tailings used for geopolymers. The SiO_2 content in copper mine tailing from different sources varied between 28 and 65%, and for Al_2O_3 from 4 to 14% [52]. Comparing the composition of the studied material, it could be concluded that it is a suitable precursor for geopolymerization, having the benefit of higher Al content.

The study of mineralogical composition proved the presence of alumosilicate minerals (muscovite, clinochlore), as well as SiO_2 in the zeolite and quartz phases. The presence of alumosilicates in the tailings could be expected to exert a neutralizing or pH buffering effect on generated leachates [24], which will mitigate the environmental effect during rainfall. The study of the alkaline reactivity of copper mine tailing (Figure 4) revealed that the alkali-leachable Si и Al were between 0.2 and 0.5% from the total content of SiO_2 and Al_2O_3 in the raw material. Hence, the low extent of dissolution of crystalline phases in the studied conditions 3, 6.5 and 10 M NaOH at different times up to 72 h was observed. The results were in support of the findings of Cristello et al., who demonstrated that crystalline phases in copper tailing were inert [14]. The ratio of reactive Si and Al in copper tailing (Table 6) was also studied, as it was known to influence the formation of geopolymer networks [14,15,17]. However, a convincing answer for which ratio was optimal for designing a geopolymer product with the intended characteristics and application could not be found. The reason was that additional sources of reactive Si were added to the precursor mixture during the obtaining of geopolymer material: (i) sodium silicate and (ii) fly ash. Thus, the properties of the obtained materials depended on the complex combinations of parameters that govern the alkali activation process [5,15,53]. It was observed that increasing the molarity of NaOH in activator solution resulted in a higher strength of the obtained geopolymer materials. The results were in line with previous studies [12,14,15]. It could be explained only partially by the increased dissolution rate of aluminosilicates, and accordingly, the concentrations of reactive Si and Al, as fly ash addition, could positively influence the development of the geopolymer network [5,14].

Additionally, the studied mine tailing contained fine particles; hence, the mine tailing possessed a potential to be an appropriate raw material for geopolymerization.

The results from sequential extraction demonstrated the low potential environmental risk, as it was found that the main part of hazardous heavy metals was immobilized in the non-labile phase. It could be concluded that the copper mine tailing is a promising precursor for obtaining building products based on the geopolymerization process. Additionally, due to the encapsulation of heavy metals and other environmentally contaminants in the geopolymer matrix, the mobility of hazardous compounds could be expected to be lower in the final geopolymer product [40].

The chemical composition, the behavior in alkaline media, the leaching characteristics of the studied copper mine tailing, as well as the fractionalization of the potential environmental contaminants demonstrated its high potential for the valorization of the tailings in the form of geopolymer products for building engineering.

5. Conclusions

Copper mine tailings from the Assarel Concentrator Plant (Bulgaria) can be successfully used as precursors for geopolymers. The replacement of 25 wt.% of fly ash with mine tailings promoted better compressive strength. By increasing the replacement of FA with MT to 50 wt.%, a significant decrease in compressive strength can be observed, especially for the mixture with a Na to Al ratio of 1.

From the three mixtures of raw materials, three different liquid-to-solid ratios, and three different Na-to-Al ratios, the system with 75 wt.% FA, 25 wt.% MT, a liquid-to-solid ratio of 0.7, and a Na-to-Al ratio of 1 exhibits the highest compressive strength at 28 days of curing. From a flexural strength point of view, the system with 50 wt.% MT, a liquid-to-solid ratio of 0.7, and a Na-to-Al ratio of 0.75 was optimum, but at a very low difference from the system that also exhibited the highest compressive strength. Therefore, from all nine mixtures, it can be considered that the one that had the highest compressive strength is optimal from a mechanical properties point of view.

The microstructural analysis showed a clear relationship between the homogeneity of the matrix and the mechanical performance of the mixture. Accordingly, it was observed that compact matrixes with fewer cracks and unreacted particles would perform better.

Supplementary Materials: The supporting information can be downloaded at: https://www.mdpi.com/article/10.3390/ma17030542/s1, Table S1 (Table for Figure 6): Influence of liquid-to-solid ratio and MT content on compressive strength; Table S2 (Table for Figure 7): Influence of Na:Al ratio and MT content on compressive strength at 28 days.

Author Contributions: Conceptualization, A.S., A.V.S. and D.I.; methodology, A.V.S. and D.I.; investigation, L.A., T.R., D.I., G.C and D.D.B.-N.; resources, P.V., T.R. and G.C.; writing—original draft preparation, L.A., T.R., D.I. and D.D.B.-N.; writing—review and editing, A.S. and P.V.; visualization, L.A. and A.V.S.; supervision, project administration and funding acquisition, A.S. and P.V. All authors have read and agreed to the published version of the manuscript.

Funding: This research was funded by BNSF, grant number KP-06-DO02/5 "RecMine—Environmental footprint reduction through eco-friendly technologies of mine tailings recycling" in the frame of ERA-MIN3 program. Also, this work was supported by a grant of the Ministry of Research, Innovation and

Digitization, Romania CNCS/CCCDI -UEFISCDI, project number COFUND-ERAMIN-3-RecMine, contract no. 307/2022, within PNCDI III.

Institutional Review Board Statement: Not applicable.

Informed Consent Statement: Not applicable.

Data Availability Statement: Data are contained within the article and supplementary materials.

Acknowledgments: The support of Assarel Medet Ltd. (Panagyurishte, Bulgaria) by providing copper mine tailing samples is highly acknowledged.

Conflicts of Interest: The authors declare no conflicts of interest. The funders had no role in the design of the study; in the collection, analyses, or interpretation of data; in the writing of the manuscript; or in the decision to publish the results.

References

1. Kossoff, D.; Dubbin, W.E.; Alfredsson, M.; Edwards, S.J.; Macklin, M.G.; Hudson-Edwards, K.A. Mine tailings dams: Characteristics, failure, environmental impacts, and remediation. *Appl. Geochem.* **2014**, *51*, 229–245. [CrossRef]
2. Cacciuttolo, C.; Atencio, E. An Alternative Technology to Obtain Dewatered Mine Tailings: Safe and Control Environmental Management of Filtered and Thickened Copper Mine Tailings in Chile. *Minerals* **2022**, *12*, 1334. [CrossRef]
3. GRID-Arendal Global Tailings Dam Portal | GRID-Arendal. Available online: https://www.grida.no/publications/472 (accessed on 13 December 2020).
4. Xiao, R.; Huang, B.; Zhou, H.; Ma, Y.; Jiang, X. A state-of-the-art review of crushed urban waste glass used in OPC and AAMs (geopolymer): Progress and challenges. *Clean. Mater.* **2022**, *4*, 100083. [CrossRef]
5. Krishna, R.S.; Shaikh, F.; Mishra, J.; Lazorenko, G.; Kasprzhitskii, A. Mine Tailings-Based Geopolymers: Properties, Applications and Industrial Prospects. *Ceram. Int.* **2021**, *47*, 17826–17843. [CrossRef]
6. Vizureanu, P.; Nergis, D.D.B.; Sandu, A.V.; Nergis, D.P.B.; Baltatu, M.S. The Physical and Mechanical Char-acteristics of Geopolymers Using Mine Tailings as Precursors. In *Advances in Geopolymer-Zeolite Composites*; Vizureanu, P., Krivenko, P., Eds.; IntechOpen: Rijeka, Croatia, 2021.
7. Zhang, J.; Fu, Y.; Wang, A.; Dong, B. Research on the Mechanical Properties and Microstructure of Fly Ash-Based Geopolymers Modified by Molybdenum Tailings. *Constr. Build. Mater.* **2023**, *385*, 131530. [CrossRef]
8. Qing, L.; Chuanming, L.; Huili, S.; Junxiang, W.; Xianjun, L. Use of Activated Quartz Powder as an Alkaline Source for Producing One-Part Ca-Rich Slag Based Cementitious Materials: Activation Mechanism, Strength, and Hydration Reaction. *J. Build. Eng.* **2023**, *64*, 105586. [CrossRef]
9. Orozco, C.R.; Castro, K.D.L.T.; De Boda, M.M.T. Valorization of Waste Mill Tailings from Small-Scale Mining through Geopolymerization: Strength, Durability, and Heavy Metal Leaching Potential. *J. Soils Sediments* **2023**, *23*, 1985–1997. [CrossRef]
10. Opiso, E.M.; Tabelin, C.B.; Maestre, C.V.; Aseniero, J.P.J.; Arima, T.; Villacorte-Tabelin, M. Utilization of Palm Oil Fuel Ash (POFA) as an Admixture for the Synthesis of a Gold Mine Tailings-Based Geopolymer Compo-site. *Minerals* **2023**, *13*, 232. [CrossRef]
11. Ahmari, S.; Zhang, L. Production of eco-friendly bricks from copper mine tailings through geopolymerization. *Constr. Build. Mater.* **2012**, *29*, 323–331. [CrossRef]
12. Castillo, H.; Droguett, T.; Vesely, M.; Garrido, P.; Palma, S. Simple Compressive Strength Results of Sodium-Hydroxide- and Sodium-Silicate-Activated Copper Flotation Tailing Geopolymers. *Appl. Sci.* **2022**, *12*, 5876. [CrossRef]
13. Tian, X.; Xu, W.; Song, S.; Rao, F.; Xia, L. Effects of curing temperature on the compressive strength and microstructure of copper tailing-based geopolymers. *Chemosphere* **2020**, *253*, 126754. [CrossRef] [PubMed]
14. Cristelo, N.; Coelho, J.; Oliveira, M.; Consoli, N.C.; Palomo, A.; Fernández-Jiménez, A. Recycling and Application of Mine Tailings in Alkali-Activated Cements and Mortars—Strength Development and Environmental Assessment. *Appl. Sci.* **2020**, *10*, 2084. [CrossRef]
15. Yu, L.; Zhang, Z.; Huang, X.; Jiao, B.; Li, D. Enhancement Experiment on Cementitious Activity of Copper-Mine Tailings in a Geopolymer System. *Fibers* **2017**, *5*, 47. [CrossRef]
16. Manjarrez, L.; Nikvar-Hassani Shadnia, R.; Zhang, L. Experimental Study of Geopolymer Binder Synthesized with Copper Mine Tailings and Low-Calcium Copper Slag. *J. Civil. Eng.* **2019**, *31*, 04019156. [CrossRef]
17. He, X.; Yuhua, Z.; Qaidi, S.; Isleem, H.F.; Zaid, O.; Althoey, F.; Ahmad, J. Mine tailings-based geopolymers: A comprehensive review. *Ceram. Int.* **2022**, *48*, 24192–24212. [CrossRef]
18. Li, J.; Yang, L.; Rao, F.; Tian, X. Gel Evolution of Copper Tailing-Based Green Geopolymers in Marine Related Environments. *Materials* **2022**, *15*, 4599. [CrossRef] [PubMed]
19. Demeusy, B.; Madanski, D.; Bouzahzah, H.; Gaydardzhiev, S. Mineralogical study of electrum grain size, shape and mineral chemistry in process streams from the Krumovgrad mine, Bulgaria. *Miner. Eng.* **2023**, *198*, 108080. [CrossRef]
20. Banov, M.; Bech, J.; Ivanov, P.; Tsolova, V.; Zhyianski, M.; Blagiev, M. Resolving of environmental problems caused by the processing of copper ore. *Bulg. J. Agric. Sci.* **2014**, *20*, 581–589.

21. Mitov, I.; Stoilova, A.; Yordanov, B.; Krastev, D. Technological research on converting iron ore tailings into a marketable product. *J. S. Afr. Inst. Min. Metall.* **2021**, *121*, 181–186. [CrossRef]
22. Wang, S.; Liu, B.; Zhang, Q.; Wen, Q.; Lu, X.; Xiao, K.; Ekberg, C.; Zhang, S. Application of Geopolymers for Treatment of Industrial Solid Waste Containing Heavy Metals: State-of-the-Art Review. *J. Clean. Prod.* **2023**, *390*, 136053. [CrossRef]
23. *EN 12457-2:2002*; Characterisation of Waste—Leaching—Compliance Test for Leaching of Granular Waste Materials and Sludges—Part 2: One Stage Batch Test at a Liquid to Solid Ratio of 10 l/kg for Materials with Particle Size below 4 mm (without or with Size Reduction). iTec: Newark, DE, USA, 2002.
24. Gitari, M.W.; Akinyemi, S.A.; Thobakgale, R.; Ngoejana, P.C.; Ramugondo, L.; Matidza, M.; Mhlongo, S.E.; Dacosta, F.A.; Nemapate, N. Physicochemical and mineralogical characterization of Musina mine copper and New Union gold mine tailings: Implications for fabrication of beneficial geopolymeric construction materials. *J. Afr. Earth Sci.* **2018**, *137*, 218–228. [CrossRef]
25. Ilieva, D.; Surleva, A.; Murariu, M.; Drochioiu, G.; Al Bakri Abdullah, M.M. Evaluation of ICP-OES Method for Heavy Metal and Metalloids Determination in Sterile Dump Material. *Solid State Phenom.* **2018**, *273*, 159–166. [CrossRef]
26. Ilieva, D.; Argirova, M.; Angelova, L.; Gradinaru, R.V.; Drochioiu, G.; Surleva, A. Application of chemical and biological tests for estimation of current state of a tailing dump and surrounding soil from the region of Tarnița, Suceava, Romania. *Environ. Sci. Pollut. Res.* **2020**, *27*, 1386–1396. [CrossRef] [PubMed]
27. Burduhos-Nergis, D.D.; Vizureanu, P.; Sandu, A.V.; Burduhos-Nergis, D.P.; Bejinariu, C. XRD and TG-DTA Study of New Phosphate-Based Geopolymers with Coal Ash or Metakaolin as Aluminosilicate Source and Mine Tailings Addition. *Materials* **2022**, *15*, 202. [CrossRef] [PubMed]
28. Ke, Y.; Liang, S.; Hou, H.; Hu, Y.; Li, X.; Chen, Y.; Li, X.; Cao, L.; Yuan, S.; Xiao, K.; et al. A zero-waste strategy to synthesize geopolymer from iron-recovered Bayer red mud combined with fly ash: Roles of Fe, Al and Si. *Constr. Build. Mater.* **2022**, *322*, 126176. [CrossRef]
29. Kumar, S.; Djobo, J.N.Y.; Kumar, A.; Kumar, S. Geopolymerization behavior of fine iron-rich fraction of brown fly ash. *J. Build. Eng.* **2016**, *8*, 172–178.
30. Castillo, H.; Collado, H.; Droguett, T.; Sánchez, S.; Vesely, M.; Garrido, P.; Palma, S. Factors Affecting the Compressive Strength of Geopolymers: A Review. *Minerals* **2021**, *11*, 1317. [CrossRef]
31. Davidovits, J. *Geopolymer Chemistry and Applications*, 5th ed.; Institut Geopolymere: Saint-Quentin, France, 2008; Volume 680.
32. *EN 196-1:2016*; Methods of Testing Cement Determination of Strength. BIS: Basel, Switzerland, 2016.
33. Adajar, M.A.; Beltran, H.E.; Calicdan, C.A.; Duran, T.R.; Ramos, C.D.; Galupino, J. Assessment of Gold Mine Tailings as Based Geopolymer Binder in Concrete. In Proceedings of the DLSU Research Congress 2021 de la Salle University, Manila, Philippines, 7–9 July 2021.
34. Aseniero, J.P.; Opiso, A.M.; Banda, M.H.; Tabelin, C. Potential utilization of artisanal gold-mine tailings as geopolymeric source material: Preliminary investigation. *SN Appl. Sci.* **2019**, *1*, 35. [CrossRef]
35. Sata, V.; Sathonsaowaphak, A.; Chindaprasirt, P. Resistance of lignite bottom ash geopolymer mortar to sulfate and sulfuric acid attack. *Cem. Concr. Compos.* **2012**, *34*, 700–708. [CrossRef]
36. Queiroz, H.M.; Ruiz, F.; Deng, Y.; Souza, J.V.S.; Ferreira, A.D.; Otero, X.L.; Camêlo, D.L.; Bernardino, A.F.; Ferreira, T.O. Mine tailings in a redox-active environment: Iron geochemistry and potential environmental consequences. *Sci. Total Environ.* **2022**, *807*, 151050. [CrossRef]
37. Al Bakri Abdullah, M.M.; Ming, L.Y.; Yong, H.C.; Tahir, M.F.M. Clay-Based Materials in Geopolymer Technology. In *Cement Based Materials*; Intech Open: London, UK, 2018; Chapter 14 [CrossRef]
38. Chen, X.; Zhang, J.; Lu, M.; Chen, B.; Gao, S.; Bai, J.; Zhang, H.; Yang, Y. Study on the effect of calcium and sulfur content on the properties of fly ash based geopolymer. *Constr. Build. Mater.* **2022**, *314*, 125650. [CrossRef]
39. Jun, Y.; Oh, J.E. Use of Gypsum as a Preventive Measure for Strength Deterioration during Curing in Class F Fly Ash Geopolymer System. *Materials* **2015**, *8*, 3053–3067. [CrossRef]
40. Ofer-Rozovsky, E.; Bar-Nes, G.; Katz, A.; Haddad, M.A. Alkali Activation of Fly Ash in the Presence of Sodium Nitrate. *Waste Biomass Valorization* **2021**, *13*, 2425–2446. [CrossRef]
41. Provis, J.; Van Deventer, J.S.J. (Eds.) *Alkali Activated Materials: State-of-the-Art Report, RILEM TC 224-AAM*; Springer Science & Business Media: Berlin/Heidelberg, Germany, 2014. [CrossRef]
42. Farhan, K.Z.; Johari, M.A.M.; Demirboğa, R. Assessment of important parameters involved in the synthesis of geopolymer composites: A review. *Constr. Build. Mater.* **2020**, *264*, 120276. [CrossRef]
43. Xu, H.; Van Deventer, J.S.J. Factors Affecting the Geopolymerization of Alkali-Feldspars. *Miner. Metall. Process.* **2002**, *19*, 209–214. [CrossRef]
44. Bulgarian National Legislation. *Regulation Nr 9. the Quality of Water Intended for Drinking and Household Purposes*; Council Directive 98/83/EC on the Quality of Water Intended for Human Consumption Health; Bulgarian Ministry of Environment and Water, Bulgarian Ministry of Health, Ministry of Regional Development and Public Works: Sofia, Bulgaria, 2018.
45. Zuhua, Z.; Xiao, Y.; Huajun, Z.; Yue, C. Role of Water in the Synthesis of Calcined Kaolin-Based Geopolymer. *Appl. Clay Sci.* **2009**, *43*, 218–223. [CrossRef]
46. Yang, X.; Zhang, Y.; Li, Z.; Wang, M. Research on Compressive and Flexural Properties of Coal Gangue-Slag Geopolymer under Wetting-Drying Cycles and Analysis of Micro-Mechanism. *Polymers* **2021**, *13*, 4160. [CrossRef]

47. Luna-Galiano, Y.; Leiva, C.; Arenas, C.; Fernández-Pereira, C. Fly Ash Based Geopolymeric Foams Using Silica Fume as Pore Generation Agent. Physical, Mechanical and Acoustic Properties. *J. Non-Cryst. Solids* **2018**, *500*, 196–204. [CrossRef]
48. Yaghoubi, M.; Arulrajah, A.; Disfani, M.M.; Horpibulsuk, S.; Bo, M.W.; Darmawan, S. Effects of Industrial By-Product Based Geopolymers on the Strength Development of a Soft Soil. *Soils Found.* **2018**, *58*, 716–728. [CrossRef]
49. Abdila, S.R.; Abdullah, M.M.A.B.; Ahmad, R.; Burduhos Nergis, D.D.; Rahim, S.Z.A.; Omar, M.F.; Sandu, A.V.; Vizureanu, P.; Syafwandi. Potential of Soil Stabilization Using Ground Granulated Blast Furnace Slag (GGBFS) and Fly Ash via Geopolymerization Method: A Review. *Materials* **2022**, *15*, 375. [CrossRef]
50. Zhang, L.; Saeed, A.; Zhang, J. Synthesis and characterization of fly ash modified mine tailings—Based geopolymers. *Constr. Build. Mater.* **2011**, *25*, 3773–3781. [CrossRef]
51. Vizureanu, P.; Burduhos Nergis, D.D. *Green Materials Obtained by Geopolymerization for a Sustainable Future*; Materials Research Foundations; Materials Research Forum LLC: Millersville, PA, USA, 2020; Volume 90, ISBN 978-1-64490-112-0.
52. Lazorenko, G.; Kasprzhitskii, A.; Shaikh, F.; Krishna, R.S.; Mishra, J. Utilization potential of mine tailings in geopolymers: Physicochemical and environmental aspects. *Proc. Saf. Environ. Prot.* **2021**, *147*, 559–577. [CrossRef]
53. Le, V.Q.; Do, Q.M.; Hoang, M.D.; Nguyen, H.T. The role of active silica and alumina in geopolymerization. *Vietnam J. Sci. Technol. Eng.* **2018**, *60*, 16–23. [CrossRef] [PubMed]

Disclaimer/Publisher's Note: The statements, opinions and data contained in all publications are solely those of the individual author(s) and contributor(s) and not of MDPI and/or the editor(s). MDPI and/or the editor(s) disclaim responsibility for any injury to people or property resulting from any ideas, methods, instructions or products referred to in the content.

Article

Plasmonic Nanocomposites of ZnO-Ag Produced by Laser Ablation and Their Photocatalytic Destruction of Rhodamine, Tetracycline and Phenol

Elena D. Fakhrutdinova [1], Anastasia V. Volokitina [1,2], Sergei A. Kulinich [2,*], Daria A. Goncharova [1], Tamara S. Kharlamova [3] and Valery A. Svetlichnyi [1]

[1] Laboratory of Advanced Materials and Technology, Tomsk State University, 634050 Tomsk, Russia; fakhrutdinovaed@gmail.com (E.D.F.)
[2] Research Institute of Science and Technology, Tokai University, Hiratsuka, Kanagawa 259-1292, Japan
[3] Laboratory of Catalytic Research, Tomsk State University, 634050 Tomsk, Russia; kharlamova83@gmail.com
* Correspondence: skulinich@tokai-u.jp

Abstract: Hydrosphere pollution by organic pollutants of different nature (persistent dyes, phenols, herbicides, antibiotics, etc.) is one of the urgent ecological problems facing humankind these days. The task of water purification from such pollutants can be effectively solved with the help of modern photocatalytic technologies. This article is devoted to the study of photocatalytic properties of composite catalysts based on ZnO modified with plasmonic Ag nanoparticles. All materials were obtained by laser synthesis in liquid and differed by their silver content and preparation conditions, such as additional laser irradiation and/or annealing of produced powders. The prepared ZnO-Ag powders were investigated by electron microscopy, X-ray diffraction and UV-Vis spectroscopy. Photocatalytic tests were carried out with well-known test molecules in water (persistent dye rhodamine B, phenol and common antibiotic tetracycline) using LED light sources with wavelengths of 375 and 410 nm. The introduction of small concentrations (up to 1%) of plasmonic Ag nanoparticles is shown to increase the efficiency of the ZnO photocatalyst by expanding its spectral range. Both the preparation conditions and material composition were optimized to obtain composite photocatalysts with the highest efficiency. Finally, the operation mechanisms of the material with different distribution of silver are discussed.

Keywords: ZnO-Ag nanoparticles; plasmonic nanoparticles; pulsed laser ablation; photocatalysis; organic pollutants

1. Introduction

Environmental problems associated with environmental pollution continue to remain not just a threat to the modern sustainable development of human civilization, but also to life on earth as a whole. The problem of polluting aquatic environments with various toxicants is still acute, and only a small part of the population has sufficient access to clean water resources [1]. One of the leaders in technogenic water pollution are complex organic compounds (OCs), textile and other dyes, antibiotics (especially those used in animal husbandry), pesticides and bacteriological pollutants [2–7]. One of the effective and environmentally friendly methods for purifying water from toxic organics is photocatalysis [8]. At this stage, the design, development and production of promising nanomaterials (photocatalysts) capable of decomposing various organic pollutants under the influence of light becomes a very important task.

Among the first materials that showed good efficiency in decomposing OC were wide-gap oxide semiconductors TiO_2 and ZnO with E_g~3.0–3.4 eV [9,10], which are still the reference materials for photocatalysis. In addition to their efficiency, they are easily accessible and have low toxicity. Thus, a large number of methods for their synthesis have been developed so far [11–14]. At the same time, ZnO has a number of advantages over TiO_2 specifically in the decomposition of OC, since during the operation of the latter, instead of effective decomposition to simple products, photosorption and incomplete decomposition of persistent dyes often occur. For example, during photocatalysis of rhodamine B, the most effective process is not the complete destruction of the aromatic structure, but only N-diethelation [15,16]. The disadvantages inherent in wide-gap monophasic oxides associated with inactivity in the visible region of the spectrum and insufficient charge separation are leveled due to the creation of defects (doping and self-modoping) [17–19], or heterostructures [20]. Another effective approach to improve photocatalytic properties is the modification of the surface of semiconductor catalysts, such as ZnO, with noble metals having surface plasmon resonance (SPR) in the visible region of the spectrum (Ag, Au) [21]. The addition of a noble metal with a high electron yield work provides a negative shift of the Fermi level and the formation of the Schottky barrier, which significantly affects the separation of photogenerated charge carriers, whereas the excitation of the composite catalyst into the SPR band significantly increases the formation rate of photoinduced charge carriers [22–24]. In many cases, modification of ZnO NPs with metals such as Ag and Au was reported to significantly increase their photocatalytic activity in water purification from organic pollutants [25–27]. In the case of addition of Ag NPs, the increase in antibacterial activity of such nanocomposites was also reported [28]. It should be noted that when designing such composites with noble metals for practical application, it is important that the maximum photocatalytic and antibacterial effect is achieved at a relatively small level of additives.

Pulsed laser ablation (PLA) is rightfully considered one of the promising methods for creating modern nanomaterials for photocatalysis [29] and biomedical applications [30]. This method is cost-efficient, environmentally friendly, does not need complex precursors and is simple enough to produce complex structures, including nanocomposites [31,32]. The PLA method is excellent for producing ZnO, as well as increasing its photocatalytic properties, through modification with metals with surface plasmon resonance (SPR) to increase the spectral range of operation to the visible region [33]. Due to the suitable thermophysical characteristics of zinc, the power density of laser radiation and the productivity of ZnO preparation by PLA metal target in air are much higher than the productivity of obtaining TiO_2 from Ti target [34,35].

There are various options for producing nanoparticles (NPs) in the ZnO-Ag system using PLA: (i) using metallic silver and (ii) using precursors (usually $AgNO_3$). For instance, ablation of a silver target was carried out in a colloidal solution of commercially available ZnO [36], while others produced ZnO via PLA method too [37,38]. This approach was supplemented by irradiation of the resulting ZnO-Ag colloid with UV laser [39]. The second approach to the preparation of such metal oxide nanostructures involves the reduction of $AgNO_3$ in the presence of ZnO NPs under the action of various radiation sources. Jung et al. carried out photoreduction of $AgNO_3$ on the surface of ZnO NPs obtained by PLA under irradiation with light from a powerful Xe lamp [40]. In contrast, Whang and coworkers reduced $AgNO_3$ by pulsed laser radiation of the second harmonic of a Nd:YAG laser on the surface of commercial ZnO NPs [41], while in [33] both the ZnO generation and its decoration via reduction were carried out by laser irradiation.

In this work, we used the metallic silver approach to prepare ZnO-Ag composite nanostructures. Both components of the composite particle were first obtained by laser ablation using nanosecond IR Nd:YAG lasers (1064 nm). The as-generated colloids were then mixed and thoroughly homogenized by means of ultrasound, after which several series of ZnO-Ag composites were prepared by further laser processing, post-preparation annealing or a combination thereof. In this work, for the first time, we used additional processing of colloidal mixtures based on Ag and ZnO NPs by focused laser radiation. This led to the formation of laser-induced plasma in the focal region of the lens, which provided conditions for more efficient interaction of the colloidal components and the formation of the ZnO-Ag interface with improved photocatalytic properties. The prepared composite catalysts were tested for the degradation of the persistent model dye Rhodamine B (Rh B), the widespread antibiotic tetracycline (TC) and the persistent organic pollutant phenol (Phen), all under low-power (0.3 W) irradiation of an LED source. The obtained results were analyzed from the viewpoint of morphology and composition, as well as surface properties of the newly reported catalysts.

2. Results and Discussion

2.1. Structure and Morphology of NPs

The results of sample phase composition obtained by means of XRD are presented in Table 1 and Figure S1, while the morphology of prepared NPs is shown in Figure 1.

Table 1. Characteristics of samples.

Sample	Sample Composition		CSR *, nm	Ag Content, wt.% **	S_{BET} m^2/g	Band Gap *	
	Phase	%				Tauc, eV	DASF, eV
Non-modified samples (as-prepared)							
ZnO	ZnO	90	37	–	36	3.29	3.31
	$Zn_2(CO_3)_2(OH)_6$	10					
ZnO-400	ZnO	100	43	–	21	3.25	3.27
Non-modified samples + ALT							
ZnO_hν	ZnO	95	62	–	23	3.11	3.28
	$Zn_2(CO_3)_2(OH)_6$	5					
ZnO-hν-400	ZnO	100	62	–	13	3.14	3.25
Ag-modified samples							
ZnO-01Ag	ZnO	90	36	0.13	40	3.30	3.31
	$Zn_2(CO_3)_2(OH)_6$	10					
ZnO-025Ag	ZnO	94	36	0.28	40	3.29	3.32
	$Zn_2(CO_3)_2(OH)_6$	6					
ZnO-05Ag	ZnO	93	37	0.49	40	3.29	3.32
	$Zn_2(CO_3)_2(OH)_6$	6					
	Ag	~1					
ZnO-1Ag	ZnO	93	37	0.98	40	3.30	3.32
	$Zn_2(CO_3)_2(OH)_6$	6					
	Ag	1					
Ag modification + annealing							
ZnO-01Ag-400	ZnO	100	40	0.13	18	3.25	3.26
ZnO-025Ag-400	ZnO	100	38	0.28	19	3.25	3.26
ZnO-05Ag-400	ZnO	99	35	0.49	21	3.24	3.26
	Ag	~1					
ZnO-1Ag-400	ZnO	99	35	0.98	26	3.25	3.27
	Ag	1					

Table 1. Cont.

Sample	Sample Composition		CSR *, nm	Ag Content, wt.% **	S_{BET} m^2/g	Band Gap *	
	Phase	%				Tauc, eV	DASF, eV
Ag modification + ALT							
ZnO-01Ag-hv	ZnO	95	40	0.13	39	3.10	3.29
	$Zn_2(CO_3)_2(OH)_6$	5					
ZnO-025Ag-hv	ZnO	95	38	0.28	38	3.10	3.29
	$Zn_2(CO_3)_2(OH)_6$	5					
ZnO-05Ag-hv	ZnO	93	39	0.49	35	3.10	3.31
	$Zn_2(CO_3)_2(OH)_6$	7					
ZnO-1Ag-hv	ZnO	93	40	0.98	36	3.06	3.31
	$Zn_2(CO_3)_2(OH)_6$	7					
Ag modification + ALT+ annealing							
ZnO-01Ag-hv-400	ZnO	100	41	0.13	27	3.12	3.25
ZnO-025Ag-hv-400	ZnO	100	42	0.28	29	3.13	3.27
ZnO-05Ag-hv-400	ZnO	100	40	0.49	26	3.11	3.27
ZnO-1Ag-hv-400	ZnO	100	38	0.98	26	3.10	3.28

*: for phase ZnO, **: according to XRF data.

The initial ZnO sample (obtained by PLA of metallic Zn in water) is seen to have two phases: the dominant ZnO phase with the wurtzite structure (PDF Card # 04-008-8198) and an admixture of monoclinic phase of zinc hydroxycarbonate (PDF Card # 04-013-7572). Drying the as-produced colloidal solution in the air was found to lead to the formation of the so-called corrosion products of metallic Zn, i.e., zinc hydrocarbonate. This is known to occur through the interaction of zinc, oxygen and water, with the formation of zinc oxide and hydroxide, which react with dissolved carbonate species (CO_3^{2-}, HCO_3^-) with the subsequent formation of hydrogen carbonates [42]. The formation of $Zn_2(CO_3)_2(OH)_6$ during PLA in water was previously reported [43]. Additional laser treatment (ALT) is seen in Table 1 to lead to a decrease in the content of the hydroxycarbonate phase, as well as an increase in the crystallite size and a decrease in the specific surface area (see the adsorption-desorption isotherms of samples in Supplementary Materials Figure S2). Thermal treatment at 400 °C was found to lead to the decomposition of hydroxycarbonate (decomposition temperature of $Zn_2(CO_3)_2(OH)_6$ is about 260 °C, as seen in Supplementary Materials Figure S3) and the formation of 100% wurtzite phase, as well as a decrease in the surface area of NPs. This is similar to findings previously reported in [44]. According to XRF analysis, the Ag content in the samples was close to the claimed content and was the same for all four series of composite samples as they were prepared based on the same ZnO-XAg series.

Addition of Ag was found to have almost no effect on the phase composition of the initially generated particles. At high Ag content ≤0.5 wt%, a reflex in the region of 37.5° 2θ appears in XRD patterns which belongs to metallic Ag cubic syngony (PDF Card # 04-003-1472, Figure S1a). At the same time, silver was found to prevent particle enlargement during heat treatment, as the crystallite size of ZnO remains ~36–37 nm, while their specific surface area even increases slightly to 40 m^2/g. ALT of mixed colloids obtained by PLA of Zn and Ag was also observed to prevent NP aggregation, while XRD patterns of samples with 0.5 and 1 wt% showed no peaks of metallic Ag. This implies a greater dispersion of Ag on the surface of ALT-processed ZnO (Figure S1) in comparison with their non-treated counterparts. Further heat treatment of this series of samples with low Ag addition did not lead to the appearance of metallic Ag peaks (Figure S1d).

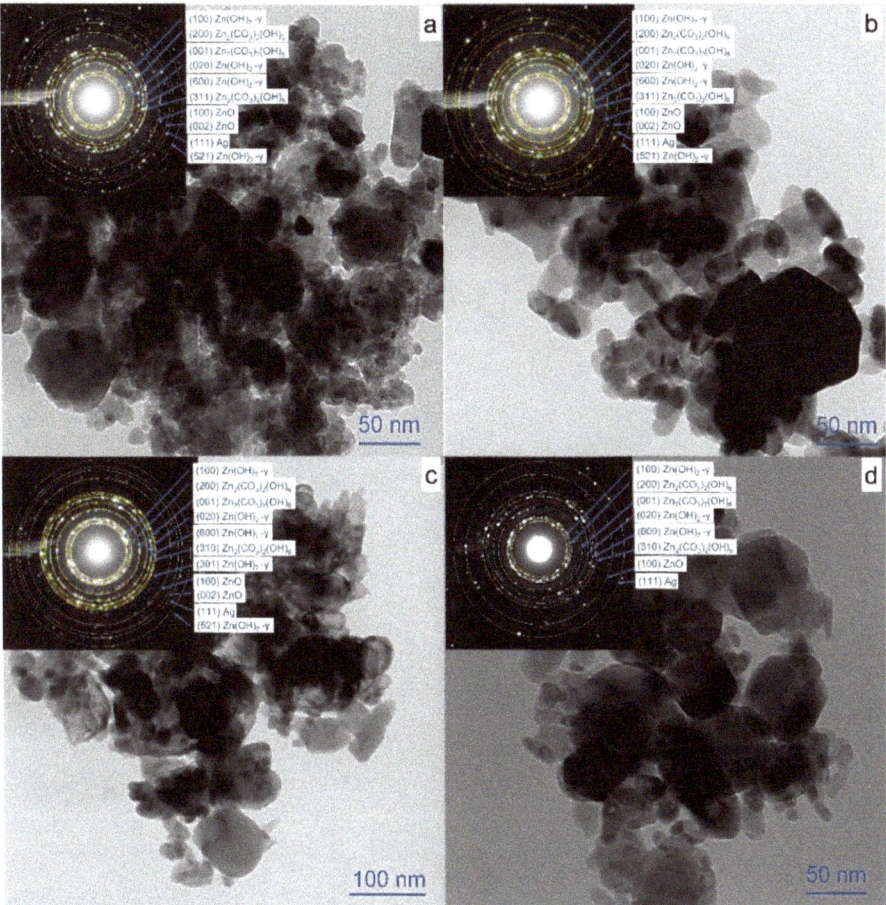

Figure 1. TEM images of samples (**a**) ZnO-1Ag, (**b**) ZnO-1Ag-400, (**c**) ZnO-1Ag-hν and (**d**) ZnO-1Ag-hν-400. Insets present corresponding SAED patterns.

The size and shape of the obtained NPs were examined by transmission electron microscopy (TEM). Figure 1a–d presents micrographs of composite particles with Ag content equal to 1 wt.%. The powders are seen to consist of agglomerated particles of irregular shape. In addition to the main fine fraction with the size of 5–30 nm, some larger particles as big as 100 nm are present (Figure 1a). After heat treatment at 400 °C, the smallest particles were found to grow bigger (Figure 1b) due to decomposition of hydroxycarbonates. ALT of the colloids is seen in Figure 1c, leading to some amorphization of the surface and formation of fused particles of irregular shape. Heat treatment of such samples at 400 °C did not result in a significant enlargement of particles, with their average size being mainly unchanged (Figure 1d).

Selected area electron diffraction (SAED) was used to clarify the structure of the obtained materials (Figure 1). For non-annealed samples, both SAED patterns and XRD analysis revealed phases of wurtzite ZnO (crystallographic plane (111)), zinc hydroxycarbonate $Zn_2(CO_3)_2(OH)_6$ (planes (200), (001) and (310)), and cubic metallic Ag (plane (111)). Also, the phase of unstable gamma zinc hydroxide—$Zn(OH)_2$ (crystallographic planes (100), (020), (600), (521), PDF Card # 00-020-1437)—was found in non-annealed samples. SAED analysis of annealed samples (insets in Figure 1b,d) also show the presence of zinc hydroxycarbonate and zinc hydroxide, both phases not being detectable by XRD. This is probably owing to the formation of $-CO_3^{2-}$ and OH^- groups in the near-surface and surface layers during storage in air. The more blurred rings with a low content of even reflections in SAED pattern observed in Figure 1c (inset) also confirm some degree of amorphization of NPs caused by ALT, which is in good agreement with from TEM images.

To analyze the distribution of silver in NPs, Figure 2a–d shows EDS mapping spectra of the elements Zn, O and Ag. For sample ZnO-1Ag, it is clear that silver is present throughout the sample surface, not only in the form of homogeneous small clusters, but also as relatively larger agglomerates (Figure 2a). After ALT processing, Ag is seen in Figure 2c to be more dispersed. Finally, heat treatment is seen in Figure 2b,d to lead to some enlargement of individual particles, while small clusters still remained present throughout the particle surface.

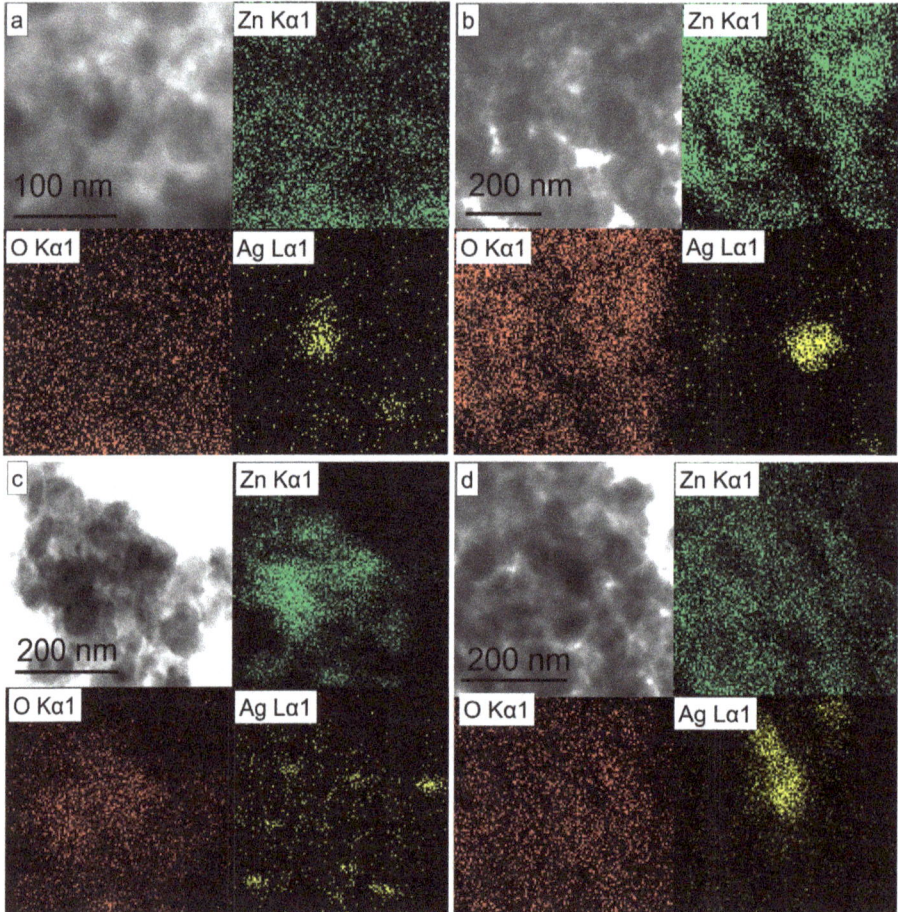

Figure 2. TEM image and EDS mapping of Zn, AgO and O in NPs: (**a**) Ag ZnO-1Ag, (**b**) ZnO-1Ag-400, (**c**) ZnO-1Ag-hv and (**d**) ZnO-1Ag–hv-400.

2.2. Electrokinetic Properties

Figure 3 and Table 2 present the values of zeta-potential of particles dispersed in media with different pH. The zeta-potential of all analyzed powders is seen to be positive (from +16.5 to +28.5 mV), with the pH of their dispersions being around 7.5. Thus, based on the obtained results, it can be concluded that the surface state of the analyzed ZnO samples is similar and is largely determined by the Zn–OH/Zn–OH^{2+} and Zn-O-/Zn-OH equilibria on the particle surface, which is also in agreement with the SAED and SEM data. ALT of colloids does not appear to lead to significant changes in the electrokinetic properties of the particle surface (Figure 3a). Doping of NPs with silver was found to result in a small (about −0.5) shift of the isoelectric point (IEP) to the region of lower pH values. This indicates that the surface state of ZnO NPs does not change significantly upon its modification with silver (Figure 3b).

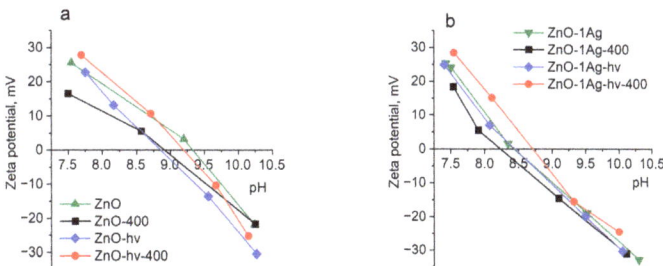

Figure 3. Zeta potential dependence on pH measured for sample (**a**) ZnO and composite sample (**b**) ZnO-Ag.

Table 2. Electrokinetic properties of samples.

Sample	pH$_o$	ζ_o, mV	pH$_{IEP}$
ZnO	7.6	25.5	9.3
ZnO-400	7.5	16.5	8.9
ZnO-hν	7.7	22.8	8.8
ZnO-hν-400	7.7	27.8	9.2
ZnO-1Ag	7.4	25.3	8.4
ZnO-1Ag-400	7.5	18.4	8.2
ZnO-1Ag-hν	7.4	25.0	8.4
ZnO-1Ag-hν-400	7.6	28.5	8.4

2.3. Optical Properties of NPs

UV-Vis spectra of powders investigated by diffuse reflectance spectroscopy (DRS) are presented in Figure 4, while the results of the band gap energy (E_g) estimation are summarized in Table 1. Since the introduction of silver strongly affects the edge of the absorption band, E_g values were calculated by two methods: the classical Tauc method (Figure S4) and the derivation of absorption spectrum fitting (DASF) method (Figure S5) [45].

As seen in Figure 4a, the absorption band of the initial samples ZnO and ZnO-hν is in the region of 380 nm, corresponding to the band gap value of zinc oxide. The long-wavelength edge of the absorption band is diffused and extends into the visible region of the spectrum. This is owing to the presence of defects of different nature, most of which are related to oxygen vacancies and to interstitial zinc both in its ground and ionic states [46,47]. The absorption of zinc hydroxycarbonate is known to lie far in the UV range (E_g = 5.5 eV) and does not affect the absorption band edge of ZnO [48]. ALT of the colloid was found to have no significant effect on its absorption spectrum. Annealing at 400 °C of undoped samples leads to a slight long-wavelength shift, which can probably be explained by particle enlargement and the increase in their crystallite size (Table 1). There is no decrease in absorption in the visible region associated with the defective structure, as defectivity is preserved during annealing.

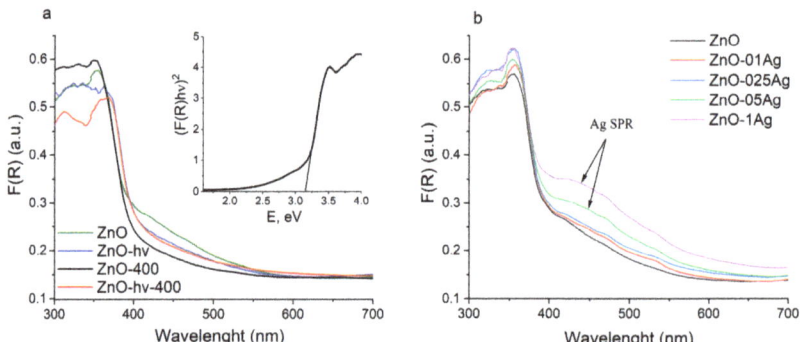

Figure 4. Diffuse reflectance spectra for NPs without Ag (**a**) and with Ag (**b**). Inset in panel (**a**) gives an example of E_g estimation.

Figure 4b presents the absorption spectra of doped materials that were not subjected to additional laser treatment (spectra of the other samples are exhibited in Figure S4). When the amount of added Ag was 0.5 and 1 wt%, a characteristic shoulder in the region of 420–480 nm appears in the spectra, which is associated with the surface plasmon resonance (SPR) of silver. The SPR band is known to be sensitive to the size and shape of metallic NPs, as well as to the refractive index of the medium in which they are dispersed [49,50]. For small spherical NPs of metallic Ag, the peak of the SPR band is in the region of 390–420 nm, while for the ZnO-Ag composites, the Ag bands are strongly broadened and shifted toward higher wavelengths. This is believed to be due to the distribution of silver on the surface of zinc oxide and to the strong interfacial electronic interaction between Ag clusters and ZnO particles [51,52].

2.4. Photocatalytic Properties of NPs

2.4.1. Photocatalytic Decomposition of Rh B

The photocatalytic activity of samples was studied on the model Rh B dye irradiated by LEDs with wavelengths of 375 nm (soft UV-A) and 410 nm (visible region). Decomposition of Rh B was not observed under irradiation without photocatalysts (Figure S6a). Figure 5a shows how the absorption spectra of Rh B changed over time in presence of sample ZnO-025Ag irradiated with LED with λ = 375 nm. Under UV excitation, dye decomposition occurred with a slight shift of its main absorption peak (553 nm) to the short-wave region of the spectrum. This is related to the N-diethylation of Rh B and the formation of intermediate Rhodamine 110 [32,53]. Upon further irradiation, Rhodamine 110 is also effectively decomposed. The decrease in absorption in the entire visible range of the spectrum and discoloration of solutions indicates the destruction of aromatic rings of the Rh B structure. Kinetic decomposition curves for different series of samples are given in Figure 5b–e, and the rate constants are listed in Table S1. The initial powders of ZnO and ZnO-hν demonstrated relatively low photocatalytic activity (Figure 5b,d), which was due to the presence of zinc hydroxycarbonates on the surface of their particles. Annealing at 400 °C reduced the content of this phase, leading to an increase in the rate of Rh B decomposition. As a result, for both samples ZnO-400 and ZnO-hν-400, the long-wavelength band of Rh B completely disappeared after 8 h and the solution became discolored (Figure 5c,d).

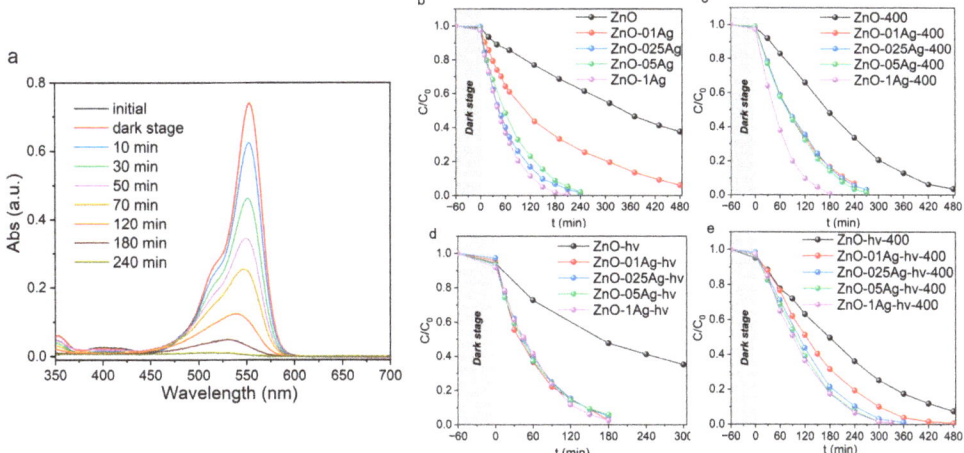

Figure 5. Absorption spectra of Rh B during decomposition in presence of sample (**a**) ZnO-025Ag. Decomposition kinetics curves of Rh B for series (**b**) ZnO-Ag, (**c**) ZnO-Ag-400, (**d**) ZnO-Ag-hν and (**e**) ZnO-Ag-hν. Irradiation: LED with λ = 375 nm.

The addition of Ag was found to lead to increased photocatalytic activity. For the initial series of samples without annealing and without additional laser treatment, the decomposition rate increases with increasing silver content in composite NPs (Figure 5b). At contents of 0.25–1 wt.%, complete Rh B decomposition was observed after 3–4 h. Additional annealing at 400 °C resulted in a decrease in the decomposition rate (Figure 5c). At Ag content of 0.1–0.5 wt%, complete decomposition was achieved in 4.5 h, while at an Ag content of 1 wt%, it was achieved in 3 h. For the series of samples with ALT, the decomposition rate is seen in Figure 5d to be almost independent of the Ag content, which is probably due to the increased dispersibility of Ag. All samples of this series were observed to degrade the dye within 3 h of irradiation. Annealing of such ALT-processed samples at 400 °C also reduced their photocatalytic efficiency (Figure 5e), which is probably due to the enlargement of Ag clusters dispersed on the surface of ZnO NPs.

Figure 6a shows how the absorption spectra of Rh B (irradiated with LED with λ = 410 nm) changed over time in the presence of sample ZnO-1Ag-hν. The use of longer-wavelength blue radiation led to slower photodegradation of Rh B when compared with UV exposure and no complete decolorization of the dye was observed even after 8 h. Figure 6b shows kinetic dependences for a number of samples with different silver content obtained with ALT but without annealing. This series showed the best activity under visible light irradiation. The introduction of Ag significantly increased the photocatalytic activity of its NPs, similar to the case of UV light irradiation (Figure 5d).

In addition to efficiency, photostability is also an important characteristic of the catalyst (Figure 7). Cyclic stability curves of NPs were measured under UV exposure at λ = 375 nm without removing the catalysts from the reactor. After each Rh B degradation cycle, 10–25 μL of concentrated dye solution was added to the reactor to restore the initial concentration (control was performed by the optical density of the solution). Figure 7a shows the cyclic stability curves for sample ZnO. Here, each cycle lasted 8 h, but no complete decomposition of the dye occurred (Figure 5b). The efficiency of the catalyst is maintained for three cycles, after which it begins to decrease. At the same time, the ZnO-Ag composite catalysts completely decolorized Rh B solutions. Sample ZnO-1Ag is seen in Figure 7b not to lose its efficiency for four cycles. The efficiency slightly decreased after cycles five and six (by 3–5%,) dropping by ~40% after cycle seven. Sample ZnO-1Ag-hν, which was ALT-processed, is seen in Figure 7c not to lose its efficiency during five cycles, after which it began to degrade rapidly. The best stability was shown by annealed samples. Figure 7d shows the performance for sample ZnO-1Ag-400, which demonstrated the best activity (Figure 5c). This photocatalyst is seen to work stably during all seven cycles and completely decolorizes Rh B. Its ALT-processed counterpart, sample ZnO-1Ag-hv-400, showed similar results.

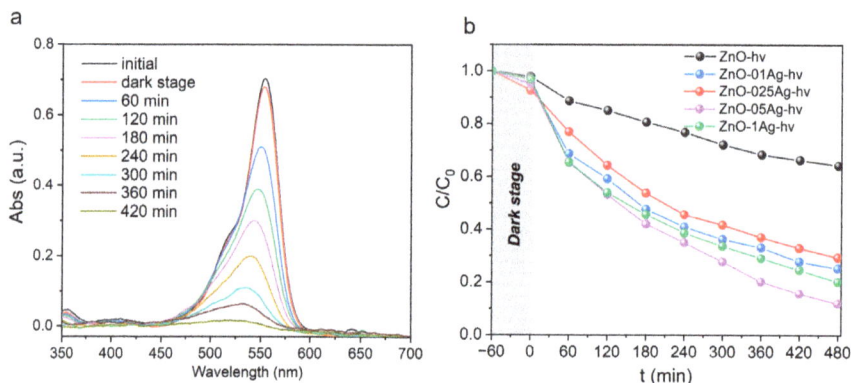

Figure 6. Absorption spectra of Rh B during decomposition in presence of sample (**a**) ZnO-1Ag-hv. Decomposition kinetic curves of Rh B for the (**b**) ZnO-Ag-hv series. Irradiation: LED with λ = 410 nm.

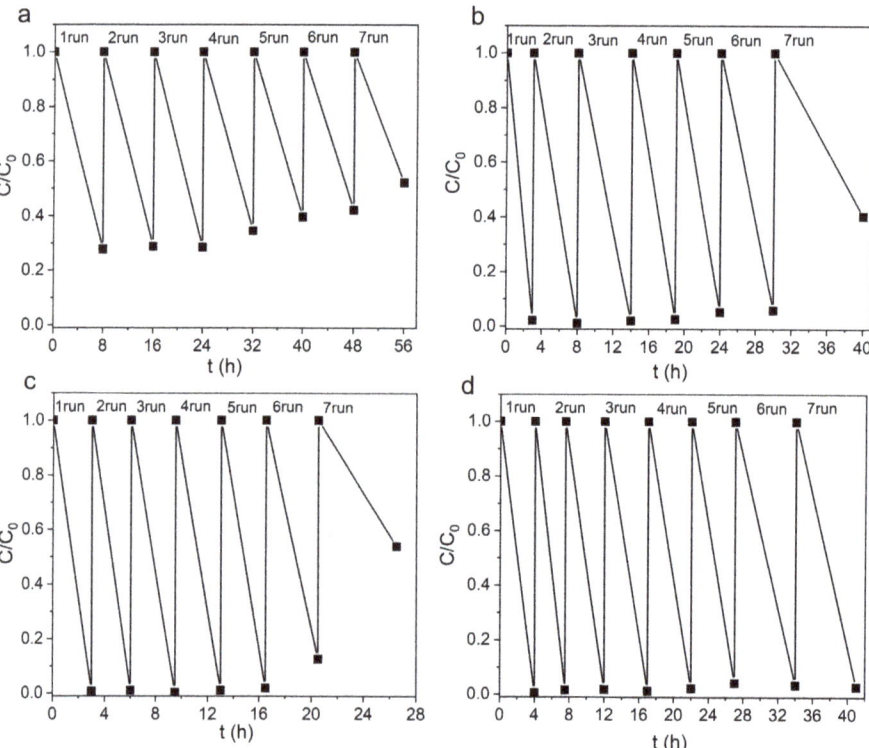

Figure 7. Cyclic stability curves of samples (**a**) ZnO, (**b**) ZnO-1Ag, (**c**) ZnO-1Ag-hv and (**d**) ZnO-1Ag-400 during decomposition of Rh B. Irradiation: LED with λ = 375 nm.

The influence of silver on the mechanism of ZnO-Ag photocatalyst operation can be presented in the form of the scheme of photogenerated charge transfers shown in Figure 8. Several processes of photogenerated electron transfer are possible at the interface between ZnO and metallic Ag. At UV irradiation λ = 375 nm, electrons are transferred from the conduction zone of ZnO to Ag clusters with a formation of the Schottky barrier, which is characteristic of most noble metals with high yield work [54,55]. Since PLA-prepared ZnO NPs have many defects of different natures (for example, oxygen vacancies and interstitial zinc atoms in their ground and ionic states [46]), such defects can act as either electron acceptors or hole traps. Therefore, the longer-wavelength irradiation (λ = 410 nm) can result in the transition of charge carriers from defective levels of ZnO to the levels of metallic Ag [56]. At excitation in the surface plasmon resonance (SPR) band, it is also possible to excite electrons of silver particles [57]. Such electrons can also participate in the generation of active particles involved in redox processes or migrate to the conduction zone of ZnO NPs. Electron transfer from the Fermi state (E_f) of Ag NPs to defect levels of ZnO near the conduction zone is also possible. All the above-described processes with the involvement of silver NPs contribute to the enhanced photocatalytic efficiency of the composite nanostructures.

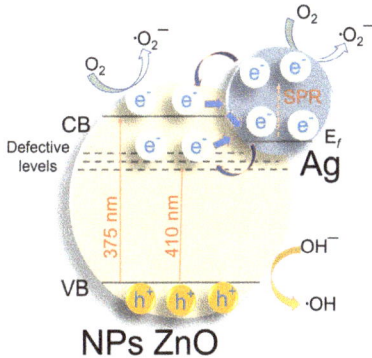

Figure 8. Schematic diagram of energy states and separation of electrons and holes in composite ZnO-Ag photocatalyst.

2.4.2. Photocatalytic Decomposition of Tetracycline

The photocatalytic activity of ZnO-Ag NPs towards the decomposition of the antibiotic tetracycline (TC) was studied under LED irradiation with λ = 375 nm. TC is known to have two characteristic absorption peaks in the UV range located at 275 and 357 nm (Figure 9a). The shorter-wave peak at 275 nm is associated with the structure of aromatic ring A including enolic hydroxyl, amide and ketone groups, while its longer-wave peak at 357 nm belongs to the structure consisting of aromatic rings B, C and D [58]. Under irradiation without a catalyst, TC is stable and only ~6% of its molecules decompose within 8 h (Figure S4b).

Figure 9a shows the spectra of TC decomposition in the presence of sample ZnO-05Ag. Even at the dark stage of the experiment, when absorption-desorption equilibrium is established, a shift of the long-wavelength absorption band of TC from 357 nm to 375 nm is observed. This shift is due to the interaction of surface oxygen vacancies of ZnO NPs with numerous OH groups of TC molecules when they absorb on the catalyst surface. This interaction was previously reported to increase the degree of π-conjugation in the system, leading to a red shift of the absorption peak [59]. The photodegradation of TC in the presence of the catalyst results in a drop in optical density in both the 357 nm and 257 nm regions due to the decomposition of all aromatic rings of the molecule which are commonly designated as A, B, C and D (see Figure 9a). The obtained kinetic curves for all series of samples are presented in Figure 9b–e, and their corresponding rate constants are given in Table S2. In the presence of non-decorated ZnO NPs, complete degradation of TC was observed within 3 h. When composite ZnO-Ag NPs were used, the degradation rate increased along the silver content, so that in the presence of sample ZnO-1Ag, TC degraded after just 90 min (Figure 9b). Both ALT processing and annealing of ZnO-Ag powders at 400 °C were found to have practically no effect on their efficiency, as the average time of TC degradation was in the range 90–120 min.

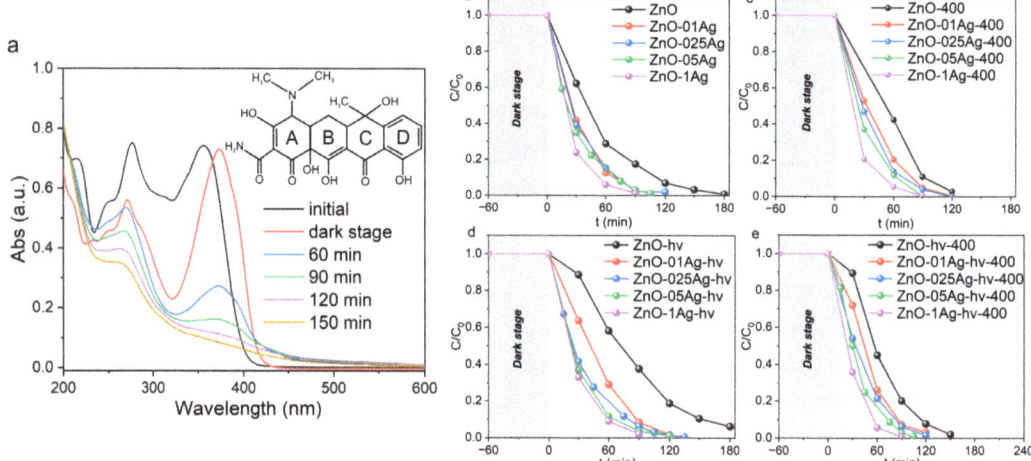

Figure 9. Absorption spectra of TC during decomposition in presence of sample (**a**) ZnO-05Ag. Decomposition kinetics curves of TC for series (**b**) ZnO-Ag, (**c**) ZnO-Ag-400, (**d**) ZnO-Ag-hν and (**e**) ZnO-Ag-hν. Irradiation: LED with λ = 375 nm.

2.4.3. Photocatalytic Decomposition of Phenol

Phenol is a rather stable molecule that absorbs in the UV range of the spectrum shorter than 300 nm, which is why it is not decomposed when irradiated with LED radiation with λ = 375 nm without a catalyst (Figure 4c). Its photocatalytic decomposition in the presence of catalysts occurs with the formation of a number of intermediate products. Hence, after the photocatalytic reaction, its absorption spectra are seen in Figure 10a to show additional absorption bands at 292 and 246 nm that belong to the decomposition products hydroquinone and p-benzoquinone, respectively [32,60].

Since the absorption spectra of both the decomposition products and phenol itself overlap, this prevents the determination of phenol's concentration. Therefore, the decomposition of Phen was determined from its photoluminescence spectra (inset, Figure 10a). After as long as 8 h of irradiation in the presence of catalysts, no complete decomposition of phenol was observed. The non-decorated sample ZnO and sample ZnO-hν showed the smallest photocatalytic efficiency, with decomposition efficiency in their presence being only 8–10% (Figure 10b,d). After annealing the catalysts, their efficiency was found to increase. As seen in Figure 10c,e, after 8 h of irradiation, 22 and 60% of Phen molecules were decomposed by samples ZnO-400 and ZnO-hν-400, respectively. The addition of silver to ZnO NPs increased the rate of Phen degradation for all series of catalysts. The highest photocatalytic activity was demonstrated by the ZnO-Ag-hν-400 series, which decomposed up to 70% of the ecotoxicant molecules after 8 h of irradiation (Figure 10e).

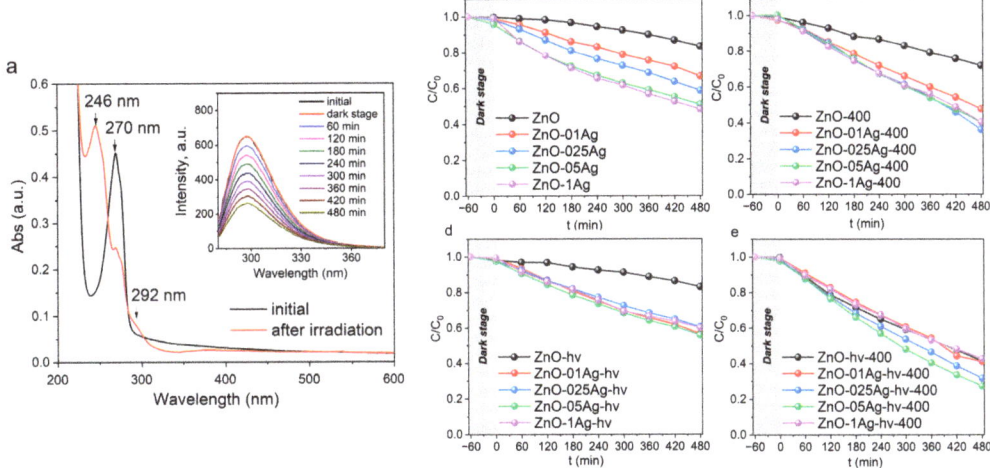

Figure 10. Absorption (and fluorescence, inset) spectra of Phen during decomposition in presence of sample (**a**) ZnO-1Ag-hv-400. Decomposition kinetics curves of Phen for series (**b**) ZnO-Ag, (**c**) ZnO-Ag-400, (**d**) ZnO-Ag-hv, and (**e**) ZnO-Ag-hv-400. Irradiation: LED with λ = 375 nm.

Table 3 compared the photocatalytic performance of our samples with that of ZnO-Ag nanocomposites previously reported by others and prepared by both laser-based and wet-chemistry-based methods.

Table 3. Photocatalytic properties of ZnO-Ag nanocomposites.

Synthesis Conditions	Parameters of Photocatalytic Experiment		Reaction Rate Constant for Best Sample	Refs.
	Pollutant, Concentration/ Catalyst Loading	Light Source, Power		
Laser approach synthesis				
PLA Zn plate in H$_2$O + PLA Ag plate in ZnO colloid (Nd:YAG laser, 1064 nm, 160 µs, 100 mJ)	Rhodamine 6G, ~10^{-5} M/ 2 mL of colloidal NPs	UV lamps (Sankyo Denki, Japan), 8 W, peak at 352 nm	0.0167 min^{-1} ZnO-Ag3 (Ag 23.4%)	[37,38]
PLA Zn plate in H$_2$O (Nd:YAG laser (1064 nm, 7 ns, 90 mJ), calcined at 500 °C + photoreduction of Ag from AgNO$_3$	Lindane (C$_6$H$_6$Cl$_6$) 5 × 10^{-5} M/0.5 g/L NPs	UV–vis xenon lamp, 200 W	0.0352 min^{-1} ZnO/Ag (Ag~3%)	[40]
PLA Zn plate in isopropanol (Nd:YAG laser (532 nm, 7 ns, 25 mJ) + laser photoreduction of Ag from AgNO$_3$ + calcined at 500 °C	Methylene blue (MB) 5 µg/0.15 g NPs	UV-Vis high-pressure sodium lamp	0.00547 min^{-1} 2 wt%Ag/ZnO (at pH 11)	[41]
PLA Zn plate in H$_2$O + PLA Ag plate in ZnO colloid (800 nm, 90 fs, 3.5 mW, 1 kHz)	MB 10 mg/L/ 0.33 g/L NPs	250 W metal halide lamp (GE ARC250/T/H/ 960E40)	0.0419 min^{-1} 6 wt% Ag/ZnO at pH 10)	[49]
PLA Ag-coated ZnO target in H$_2$O (Nd:YAG laser (1064 nm, 5 ns, 300 mJ)	MB 2.7 × 10^{-5} M/ ~0.1 g/l NPs	UV-vis Hg lamp, VIS 2.11 klx, UVA 0.2 mW/cm^2, UVB 0.02 mW/cm^2, UVC 0.08 mW/cm^2	0.0233 min^{-1} Zn_1000Ag (0.32%)	[61]
PLA Zn plate in H$_2$O + PLA Zn plate in H$_2$O + ALT of mixed solution (1064 nm, 7 ns, 150 mJ)	Rh B 5 × 10^{-6} M Phen 5 × 10^{-5} M TC 5 × 10^{-5} M/ 0.5 g/L NPs	LED 375 nm, 50 mW	Rh B, 0.0209 min^{-1} ZnO-1Ag-hv Phen, 0.0019 min^{-1} ZnO-1Ag-hv TC, 0.0589 min^{-1} ZnO-1Ag-hv	This work

Table 3. Cont.

Synthesis Conditions	Parameters of Photocatalytic Experiment		Reaction Rate Constant for Best Sample	Refs.
	Pollutant, Concentration/ Catalyst Loading	Light Source, Power		
Other ways of synthesis				
Ultrasonic microwave-assisted method for ZnO + thermal reduction of Ag from AgNO$_3$	Rh B, methylene orange 1×10^{-5} M/ 0.5 g/L NPs	500 W Xe lamp with a 400 nm cut-off filter	0.0431 min^{-1} ZnO/Ag (Ag~10%)	[62]
Microwave-assisted one-pot method of Ag/ZnO synthesis with thermal reduction of Ag from AgNO$_3$	Rh B 2.1 $\times 10^{-5}$ M/ 1.5 g/L NPs	Xe lamp 300 W and AM 1.5 filter were used as the simulated solar light.	0.1732 min^{-1} Ag:ZnO (8:92)	[27]
Hydrothermal method for ZnO and further loaded via precipitation with Ag (photo deposition from AgNO$_3$) and CDots	TC 6.8 $\times 10^{-5}$ M/ 1 g/L NPs	UV-vis xenon lamp, 150 W (300–780 nm)	0.03389 min^{-1} Ag/ZnO 0.0489 min^{-1} CDots/Ag/ZnO	[63]

The presented results indicate that the nanocomposites prepared in this study exhibited high photocatalytic activity when using relatively low-power radiation sources (LED, 50 mW) and a low loading of Ag (0.25–0.5 wt.%).

3. Research Methods and Material Preparation

3.1. Obtaining Materials Using PLA

Pulsed laser ablation of metal Zn (99.9% purity) Ag (99.99%) plates was carried out using the fundamental harmonic radiation of a Nd:YAG laser (LS2131M-20 model from LOTIS TII, Minsk, Belarus) with the following parameters: wavelength λ = 1064 nm, pulse duration 7 ns, frequency 20 Hz and pulse energy 150 mJ. At the beginning, two colloids were prepared separately by ablating the metal Zn target in 80 mL of distilled water for 30 min and Ag target in 80 mL of water for 1–5 min. The concentration of particles in the prepared dispersions was determined from the loss of target mass after ablation. The mass concentration of generated NPs (by metal mass) in colloids was ~300 mg/L (for Zn) and 10–30 mg/L (for Ag). Then the colloids were mixed in such proportions that the Ag content in the samples was 0.1, 0.25, 0.5 and 1 wt.% with respect to ZnO.

A part of mixed colloids was sonicated for 15 min and then dried in the air at ~60 °C to a powder state. Below, this series of samples is denoted as ZnO-XAg where X is the mass fraction of Ag. The sample without the addition of silver is designated as ZnO. Another part of mixed colloids was additionally irradiated with the same focused pulsed laser radiation as during their preparation. Such an additional laser treatment (ALT) of colloids was carried out for 1.5 h with constant stirring using a magnetic stirrer, after which the processed dispersions were also dried to a powder state. The use of a focused laser beam during ALT provided plasma locally generated inside the processed colloidal mixture [64], which stimulated the efficient formation of composite particles. This series of samples was denoted ZnO-XAg-hv where X is the mass fraction of introduced Ag. Similar to its non-irradiated counterpart, the sample without silver was denoted as ZnO-hv. In this way, two lines of samples, with and without ALT treatment, were obtained. Part of the material of the resulting powders was annealed in a muffle furnace at a temperature of 400 °C for 4 h. For heat-treated powders, the index 400 was added to the designation (for example, ZnO-05Ag-400). Schematically, the preparation of series of samples used in this study is presented in Figure 11. More details on material preparation and experimental setups, including the preparation of mixed colloids and ALT, can be found elsewhere [55,65].

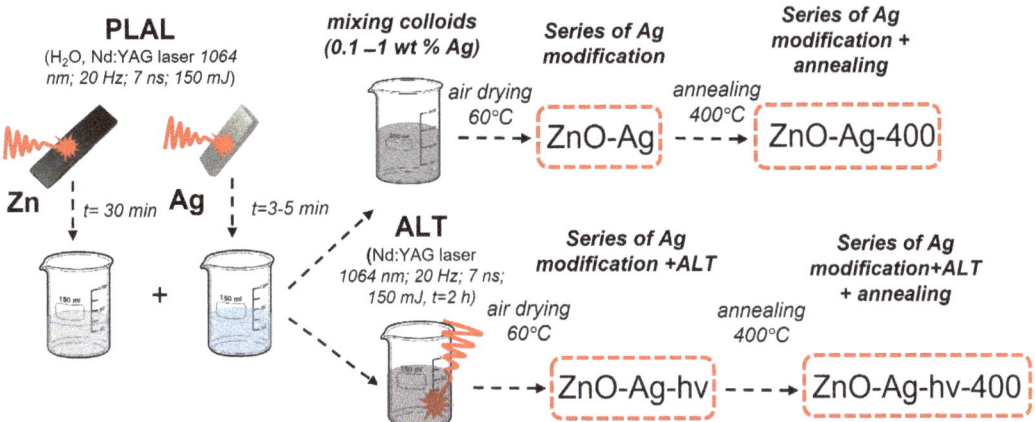

Figure 11. Schematic presentation of sample series and their preparation in this study.

3.2. Research Methods

The crystal structure of samples was studied using an XRD-7000 X-ray diffractometer (Shimadzu, Kyoto, Japan) with monochromatic CuKα radiation (1.54 Å) in the 2θ range of 20–90° and a scanning speed of 0.02 °/s using Bragg-Brentano geometry. Crystalline Si (a = 5.4309 Å, λ = 1.540562 Å) was used as an external standard to calibrate the diffractometer. The phase composition of collected patterns was analyzed using the PDF-4 database (Release 2022). To refine the parameters of the crystal lattice and determine the regions of coherent scattering (CSR), the full-profile analysis program POWDER CELL 2.4 was used.

The Ag content in the samples was estimated using an XRF-1800 sequential spectrometer (Shimadzu, Japan). To ensure accurate determination of low Ag loadings, preliminary calibration was performed.

The thermogravimetry analysis and differential scanning calorimetry (TG/DSC) were performed using an STA 409 PC Luxx analyzer (Netzsch, Selb, Germany) in a dry air atmosphere at a heating rate of 10 °C/min in the temperature range of 25–1000 °C.

Morphology and chemical composition of produced samples were also studied by transmission electron microscopy (TEM) using a JEOL JEM-2100 instrument (Tokyo Boeki Ltd., Tokyo, Japan) equipped with an energy-dispersive X-ray (EDX) analysis system at an accelerating voltage of 200 kV. Samples for TEM studies were prepared by depositing NPs dispersed in ethanol on copper grids coated with a carbon film.

Specific surface area and pore size distribution were determined by means of a TriStar II 3020 gas adsorption analyzer (Micromeritics, Norcross, GA, USA) using low-temperature nitrogen sorption. Before analysis, samples that were not subjected to heat treatment were degassed in a vacuum (10^{-2} Torr) at room temperature. Samples subjected to calcination were degassed in a vacuum (10^{-2} Torr) at 200 °C for 2 h using a laboratory degassing station or with a VacPrep Degasser (Micromeritics, USA) tool. The specific surface area was determined by the Brunauer-Emmett-Teller (BET) method.

Electrokinetic properties of dispersions were examined on an Omni S/N analyzer (Brookhaven, Upton, NY, USA). For this, powders of samples were dispersed in distilled water by means of sonicating for 6 min. The concentration of the prepared dispersions was 0.25 mg/mL. When studying the dependence of the zeta potential of dispersed particles on pH, the pH of the medium was adjusted by adding 0.1 and 0.001 M solutions of potassium hydroxide.

Optical properties of materials in the UV-Vis range were studied by diffuse reflection spectroscopy (DRS) on a Cary 100SCAN spectrophotometer with a DRA-CA-30I module (from Labsphere, North Sutton, NH, USA) in the wavelength range 230–800 nm. MgO powder was used as a reference for measurements. Then, the obtained DRSs were transformed using the Kubelka-Munk function, and hence the optical band gap of ZnO (as a direct-gap semiconductor) was estimated using the Tauck method from the curves plotted in the coordinates $(F(R)h\nu)^2 - E(eV)$. The calculated band gap obtained using the Tauck method was compared with the values obtained by the DASF method [36].

3.3. Photocatalytic Experiment

Photocatalytic activity of the prepared nanocomposites was assessed by the decomposition of the model dye Rhodamine B (with a concentration of 5×10^{-6} M), the broad-spectrum bacteriostatic antibiotic tetracycline (with a concentration of 5×10^{-5} M) and the organic pollutant phenol (with a concentration of 5×10^{-5} M). The concentration of tested catalysts was 0.5 g/L (15 mg of sample per 30 mL of aqueous medium, with no acid or alkali added). Before photocatalysis, a dark stage was carried out for 60 min to establish adsorption-desorption equilibrium. Then, the reactor was irradiated with LEDs with wavelengths of 375 nm (soft UV-A) and 410 nm (visible region). The wavelength of 410 nm corresponds to the surface plasmon resonance (SPR) of Ag NPs. The total radiation power incident on the reactor from the LEDs with 375 and 410 nm was 50 and 320 mW, respectively. The change in the concentration of Rh B and TC was monitored through their absorption spectra using an SF-56 spectrophotometer (OKB SPECTR LLC, Saint-Petersburg, Russia), while the changes in the concentration of Phen was determined from its fluorescence spectra using an RF-5031PC spectrofluorimeter (Shimadzu, Japan). To monitor the concentration of organics spectroscopically, at certain time intervals, aliquot samples were taken from the reactor, centrifuged (10 min, 12,000 rpm, 8 °C) and analyzed spectrally, after which the samples were returned to the reactor. Next, C/C_0 curves were plotted versus irradiation time, where C and C_0 are the current and initial concentrations of organic compound, respectively. Finally, the rate constants of the photocatalytic reaction were calculated from the corresponding kinetic curves, assuming the first-order reaction kinetics.

4. Conclusions

In this work, using laser techniques, four series of ZnO-Ag composite nanoparticles with silver content of 0.1–1 wt% were prepared. The composites were prepared from individual colloids first generated by pulsed laser ablation (PLA) of metallic Zn and Ag targets in water. According to X-ray diffraction, the as-prepared ZnO nanoparticles contained a zinc hydroxycarbonate phase $Zn_2(CO_3)_2(OH)_6$ as well as $Zn(OH)_2$ hydroxide and the main wurtzite phase ZnO (>90 wt%). Additional laser treatment of the mixed colloids with focused laser irradiation allows for better dispersion of silver clusters on the surface of ZnO particles, while annealing at 400 °C destroys hydroxycarbonates in the samples. Nevertheless, some traces of $Zn_2(CO_3)_2(OH)_6$ and $Zn(OH)_2$ are still seen in the SAED patterns of annealed samples. Adding silver was found not to affect the absorption band edge (and, consequently, the band gap energy) of the composite particles, and the broad SPR band of Ag clusters indicates a strong interfacial electronic interaction between Ag and ZnO.

At the same time, the addition of silver is shown to enhance both the photocatalytic properties of the composite particles and their stability as catalysts. This can be explained by the better charge separation in composite particles and by the changes in the kinetics of electrons due to the formation of the Schottky barrier and, possibly, the SPR effect in silver clusters. The obtained composite photocatalysts with low content of loaded Ag were demonstrated to effectively degrade the persistent model dye Rh B, as well as the well-known ecotoxicant phenol and the common antibiotic tetracycline (under low-power irradiation from an LED source with a wavelength of 375 nm).

Supplementary Materials: The following supporting information can be downloaded at: https://www.mdpi.com/article/10.3390/ma17020527/s1, Figure S1: X-ray diffraction patterns of a series of samples: (**a**) ZnO-Ag; (**b**) ZnO-Ag-400; (**c**) ZnO-Ag-hv; (**d**) ZnO-Ag-hv-400; Figure S2: Nitrogen adsorption-desorption isotherms for BET of samples: (**a**) ZnO-Ag; (**b**) ZnO-Ag-400; (**c**) ZnO-Ag-hv; (**d**) ZnO-Ag-hv-400; Figure S3: TG-DSC (Thermogravimetry-differential scanning calorimetry) curve of ZnO; Figure S4: Diffuse reflectance spectra of a series of samples: (**a**) ZnO-Ag; (**b**) ZnO-Ag-400; (**c**) ZnO-Ag-hv; (**d**) ZnO-Ag-hv-400; and estimation of the band gap using the Tauc method insert; Figure S5: Example spectra for calculating E_g using the DASF method for sample (**a**) ZnO and (**b**) ZnO-hv; Figure S6: Absorption spectra of (**a**) Rh B, (**b**) TC and (**c**) Phen before and after 480 min LED λ = 375 nm irradiation; Table S1: Reaction rate constants for the decomposition of RhB under irradiation with LED λ = 375 nm; Table S2: Reaction rate constants for the decomposition of TC under irradiation with LED λ = 375 nm; Table S3: Reaction rate constants for the decomposition of Phen under irradiation with LED λ = 375 nm.

Author Contributions: Conceptualization, E.D.F. and V.A.S.; methodology, S.A.K.; validation, S.A.K.; formal analysis, T.S.K.; investigation, A.V.V., D.A.G. and T.S.K.; data curation, D.A.G.; writing—original draft preparation, A.V.V. and E.D.F.; writing—review and editing, V.A.S. and S.A.K.; visualization, E.D.F. and A.V.V.; supervision, V.A.S.; project administration, E.D.F. All authors have read and agreed to the published version of the manuscript.

Funding: This study was supported by the Tomsk State University Development Programme (Priority-2030).

Institutional Review Board Statement: Not applicable.

Informed Consent Statement: Not applicable.

Data Availability Statement: The data presented in this study are available upon request from the corresponding authors.

Acknowledgments: The authors thank Dr. Ivan Smirnov (TSU) for TEM studies. The studies (low-temperature N_2 adsorption and UV-Vis spectroscopy) were carried out with equipment from Tomsk Regional Core Shared Research Facilities Center of National Research Tomsk State University.

Conflicts of Interest: The authors declare no conflicts of interest.

Sample Availability: Samples of the ZnO-Ag photocatalysts are available from the authors.

References

1. Vargas-Berrons, K.; Bernal-Jacome, L.; de Leon-Martinez, L.D.; Flores-Ramirez, R. Emerging pollutants (EPs) in Latin América: A critical review of under-studied EPs, case of study -Nonylphenol-. *Sci. Total Environ.* **2020**, *726*, 138493. [CrossRef] [PubMed]
2. Lellis, B.; Favaro-Polonio, C.Z.; Pamphile, J.A.; Polonio, J.C. Effects of textile dyes on health and the environment and bioremediation potential of living organisms. *Biotechnol. Res. Innov.* **2019**, *3*, 275–290. [CrossRef]
3. Mohod, A.V.; Momotko, M.; Shah, N.S.; Marchel, M.; Imran, M.; Kong, L.; Boczkaj, G. Degradation of Rhodamine dyes by Advanced Oxidation Processes (AOPs)–Focus on cavitation and photocatalysis–A critical review. *Water Resour. Ind.* **2023**, *30*, 100220. [CrossRef]
4. Ore, O.T.; Adeola, A.O.; Bayode, A.A.; Adedipe, D.T.; Nomngongo, P.N. Organophosphate pesticide residues in environmental and biological matrices: Occurrence, distribution and potential remedial approaches. *JECE* **2023**, *5*, 9–23. [CrossRef]
5. Yadav, G.; Ahmaruzzaman, M. Recent development of novel nanocomposites for photocatalysis mediated remediation of phenolic derivatives: A comprehensive review. *J. Ind. Eng. Chem.* **2023**, *127*, 18–35. [CrossRef]
6. Pokharel, S.; Shrestha, P.; Adhikari, B. Antimicrobial use in food animals and human health: Time to implement 'One Health' approach. *Antimicrob. Resist. Infect. Control* **2020**, *9*, 181. [CrossRef] [PubMed]
7. Ganguly, P.; Byrne, C.; Breen, A.; Pillai, S.C. Antimicrobial activity of photocatalysts: Fundamentals, mechanisms, kinetics and recent advances. *Appl. Catal. B* **2018**, *225*, 51–75. [CrossRef]
8. Mashra, S.; Sundaram, B. A review of the photocatalysis process used for wastewater treatment. *Mater. Today Proc.* **2023**. [CrossRef]
9. Al-Mamun, M.R.; Kader, S.; Islam, M.S.; Khan, M.N. Photocatalytic activity improvement and application of UV-TiO$_2$ photocatalysis in textile wastewater treatment: A review. *J. Environ. Chem. Eng.* **2019**, *7*, 103248. [CrossRef]
10. Dhiman, P.; Rana, G.; Kumar, A.; Sharma, G.; Vo, D.-V.N.; Naushad, M. ZnO-based heterostructures as photocatalysts for hydrogen generation and depollution: A review. *Environ. Chem. Lett.* **2020**, *20*, 1047–1081. [CrossRef]
11. Li, Z.; Wang, S.; Wu, J.; Zhou, W. Recent progress in defective TiO$_2$ photocatalysts for energy and environmental applications. *Renew. Sustain. Energy Rev.* **2022**, *156*, 111980. [CrossRef]

12. Rajaram, P.; Jeice, A.R.; Jayakumar, K. Review of green synthesized TiO$_2$ nanoparticles for diverse applications. *Surf. Interfaces* **2023**, *39*, 102912. [CrossRef]
13. Weldegebrieal, G.K. Synthesis method, antibacterial and photocatalytic activity of ZnO nanoparticles for azo dyes in wastewater treatment: A review. *Inorg. Chem. Commun.* **2020**, *120*, 108140. [CrossRef]
14. Jiang, Z.; Lui, B.; Yu, L.; Tong, Y.; Yan, M.; Znang, R.; Han, W.; Hao, Y.; Shanggoun, L.; Zhang, R.; et al. Research progresses in preparation methods and applications of zinc oxide nanoparticles. *J. Alloys Compd.* **2023**, *956*, 170316. [CrossRef]
15. Nachit, W.; Ait Ahsaine, H.; Ramzi, Z.; Touhtouh, S.; Goncharova, I.; Benkhouja, K. Photocatalytic activity of anatase-brookite TiO$_2$ nanoparticles synthesized by sol gel method at low temperature. *Opt. Mater.* **2022**, *129*, 112256. [CrossRef]
16. Wu, Z.-Y.; Xu, Y.-J.; Huang, L.-J.; Znang, Q.-X.; Tang, D.-L. Fullerene-cored star-shaped polyporphyrin-incorporated TiO$_2$ as photocatalysts for the enhanced degradation of rhodamine B. *J. Environ. Chem. Eng.* **2021**, *9*, 106142. [CrossRef]
17. Andronic, L.; Lelis, M.; Enesca, A.; Karazhanov, S. Photocatalytic activity of defective black-titanium oxide photocatalysts towards pesticide degradation under UV/VIS irradiation. *Surf. Interfaces* **2022**, *32*, 102123. [CrossRef]
18. Nair, P.R.; Ramirez, C.R.S.; Pinilla, M.A.G.; Krishnan, B.; Avellaneda, D.A.; Pelaes, R.F.C.; Shaji, S. Black titanium dioxide nanocolloids by laser irradiation in liquids for visible light photo-catalytic/electrochemical applications. *Appl. Surf. Sci.* **2023**, *623*, 157096. [CrossRef]
19. Madhavi, V.; Kondaiah, P.; Mahan Rao, G. Influence of silver nanoparticles on titanium oxide and nitrogen doped titanium oxide thin films for sun light photocatalysis. *Appl. Surf. Sci.* **2018**, *436*, 708–719. [CrossRef]
20. Thongam, D.D.; Chanturvdi, H. Heterostructure charge transfer dynamics on self-assembled ZnO on electronically different single-walled carbon nanotubes. *Chemosphere* **2023**, *323*, 138239. [CrossRef]
21. Kanakkillam, S.S.; Krishnan, B.; Pelaez, R.F.C.; Martinez, J.A.A.; Avellaneda, D.A.; Shaji, S. Hybrid nanostructures of Ag/Au-ZnO synthesized by pulsed laser ablation/irradiation in liquid. *Surf. Interfaces* **2021**, *27*, 101561. [CrossRef]
22. Kumaravel, V.; Mathew, S.; Bartlett, J.; Pillai, S.C. Photocatalytic hydrogen production using metal doped TiO$_2$: A review of recent advances. *Appl. Catal.* **2019**, *244*, 1021–1064. [CrossRef]
23. Panayotov, D.A.; Frenkel, A.I.; Morris, J.R. Catalysis and photocatalysis by nanoscale Au/TiO$_2$: Perspectives for renewable energy. *ACS Energy Lett.* **2017**, *2*, 1223–1231. [CrossRef]
24. Hou, W.; Cronin, S.B. A review of surface plasmon resonance-enhanced photocatalysis. *Adv. Funct. Mater.* **2013**, *23*, 1612–1619. [CrossRef]
25. He, W.; Kim, H.K.; Wamer, W.G.; Melka, D.; Callahan, J.H.; Yin, J.J. Photogenerated charge carriers and reactive oxygen species in ZnO/Au hybrid nanostructures with enhanced photocatalytic and antibacterial activity. *J. Am. Chem. Soc.* **2014**, *136*, 750–757. [CrossRef] [PubMed]
26. Li, H.; Ding, J.; Cai, S.; Zhang, W.; Zhang, X.; Wu, T.; Wang, C.; Foss, M.; Yang, R. Plasmon-enhanced photocatalytic properties of Au/ZnO nanowires. *Appl. Surf. Sci.* **2022**, *583*, 152539. [CrossRef]
27. Liu, H.; Liu, H.; Yang, J.; Zhai, H.; Liu, X.; Jia, H. Microwave-assisted one-pot synthesis of Ag decorated flower-like ZnO composites photocatalysts for dye degradation and NO removal. *Ceram. Int.* **2019**, *45*, 20133–20140. [CrossRef]
28. Adhikari, S.; Banerjee, A.; Eswar, N.K.; Sarkar, D.; Madras, G. Photocatalytic inactivation of E. coli by ZnO–Ag nanoparticles under solar radiation. *RSC Adv.* **2015**, *5*, 51067–51077. [CrossRef]
29. Forsythe, R.C.; Cox, C.P.; Wilsey, M.K.; Müller, A.M. Pulsed Laser in Liquids Made Nanomaterials for Catalysis. *Chem. Rev.* **2021**, *121*, 7568–7637. [CrossRef]
30. Mussin, A.; AlJulaih, A.A.; Mintcheva, N.; Aman, D.; Iwamori, S.; Gurbatov, S.O.; Bhardwaj, A.K.; Kulinich, S.A. PLLA nanosheets for wound healing: Embedding with iron-ion-containing nanoparticles. *Nanomanufacturing* **2023**, *3*, 401–415. [CrossRef]
31. Amendola, V.; Amans, D.; Ishikawa, Y.; Koshizaki, N.; Scire, S.; Compagnini, G.; Reichenberger, S.; Barcikowski, S. Room-Temperature Laser Synthesis in Liquid of Oxide, Metal-Oxide Core-Shells, and Doped Oxide Nanoparticles. *Chem. Eur. J.* **2020**, *26*, 9206–9242. [CrossRef] [PubMed]
32. Shabalina, A.V.; Fakhrutdinova, E.D.; Golubovskaya, A.G.; Kuzmin, S.M.; Koscheev, S.V.; Kulinich, S.A.; Svetlichnyi, V.A.; Vodyankina, O.V. Laser-assisted preparation of highly-efficient photocatalytic nanomaterial based on bismuth silicate. *Appl. Surf. Sci.* **2022**, *575*, 151722. [CrossRef]
33. Yudasari, N.; Hardiansyh, A.; Herbani, Y.; Isnaeni; Auliyanti, M.M.; Djuhana, D. Single-step laser ablation synthesis of ZnO–Ag nanocomposites for broad-spectrum dye photodegradation and bacterial photoinactivation. *J. Photochem. Photobiol. A* **2023**, *441*, 114717. [CrossRef]
34. Lapin, I.N.; Svetlichnyi, V.A. Features of the synthesis of nanocolloid oxides by laser ablation of bulk metal targets in solutions. *Proc. SPIE* **2015**, *9810*, 98100T. [CrossRef]
35. Svetlichnyi, V.; Shabalina, A.; Lapin, I.; Goncharova, D. Metal Oxide Nanoparticle Preparation by Pulsed Laser Ablation of Metallic Targets in Liquid. In *Applications of Laser Ablation-Thin Film Deposition, Nanomaterial Synthesis and Surface Modification*, 1st ed.; Yang, D., Ed.; InTech: London, UK, 2016; Chapter 11; pp. 245–263, 426p, ISBN 978-953-51-2812-0. [CrossRef]
36. Ali, H.; Ismail, A.M.; Menazea, A.A. Multifunctional Ag/ZnO/chitosan ternary bio-nanocomposites synthesized via laser ablation with enhanced optical, antibacterial, and catalytic characteristics. *Water Process. Eng.* **2022**, *49*, 102940. [CrossRef]
37. Yudasari, N.; Anugrahwidya, R.; Tahir, D.; Suliyanti, M.M.; Herbani, Y.; Imawan, C.; Khalil, M.; Djuhana, D. Enhanced photocatalytic degradation of rhodamine 6G (R6G) using ZnO–Ag nanoparticles synthesized by pulsed laser ablation in liquid (PLAL). *J. Alloys Compd.* **2021**, *886*, 161291. [CrossRef]

38. Anugrahwidya, R.; Yudasari, N.; Tahir, D. Optical and structural investigation of synthesis ZnO/Ag Nanoparticles prepared by laser ablation in liquid. *Mater. Sci. Semicond. Process.* **2020**, *105*, 1044712. [CrossRef]
39. Elsayed, K.A.; Alomari, M.; Drmosh, Q.A.; Alheshibri, M.; Al Baroot, A.; Kayed, T.S.; Manda, A.A.; Al-Alotaibi, A.L. Fabrication of ZnO-Ag bimetallic nanoparticles by laser ablation for anticancer activity. *Alex. Eng. J.* **2022**, *61*, 1449–1457. [CrossRef]
40. Jung, H.J.; Koutavarapu, R.; Lee, S.; Kim, J.H.; Choi, H.C.; Choi, M.Y. Enhanced photocatalytic degradation of lindane using metal–semiconductor Zn@ZnO and ZnO/Ag nanostructures. *J. Environ. Sci.* **2018**, *74*, 107–115. [CrossRef]
41. Whang, T.-J.; Hsieh, M.-T.; Chen, H.-H. Visible-light photocatalytic degradation of methylene blue with laser-induced Ag/ZnO nanoparticles. *Appl. Surf. Sci.* **2012**, *258*, 2796–2801. [CrossRef]
42. Ohtsuka, T.; Matsuda, M. In Situ Raman Spectroscopy for Corrosion Products of Zinc in Humidified Atmosphere in the Presence of Sodium Chloride Precipitate. *Corros. Sci.* **2003**, *59*, 407–413. [CrossRef]
43. Svetlichnyi, V.A.; Lapin, I.N. Structure and properties of nanoparticles fabricated by laser ablation of Zn metal targets in water and ethanol. *Russ. Phys. J.* **2013**, *56*, 581–587. [CrossRef]
44. Singh, S.B.; De, M. Room temperature adsorptive removal of thiophene over zinc oxide-based adsorbents. *J. Mater. Eng. Perform.* **2018**, *27*, 2661–2667. [CrossRef]
45. Souri, D.; Tahan, Z.E. A new method for the determination of optical band gap and the nature of optical transitions in semiconductors. *Appl. Phys. B* **2015**, *119*, 273–279. [CrossRef]
46. Gavrilenko, E.A.; Goncharova, D.A.; Lapin, I.N.; Gerasimova, M.A.; Svetlichnyi, V.A. Photocatalytic activity of zinc oxide nanoparticles prepared by laser ablation in a decomposition reaction of rhodamine B. *Russ. Phys. J.* **2020**, *63*, 1429–1437. [CrossRef]
47. Gavrilenko, E.A.; Goncharova, D.A.; Lapin, I.N.; Nemoykina, A.L.; Svetlichnyi, V.A.; Aljulaih, A.A.; Mintcheva, N.; Kulinich, S.A. Comparative study of physicochemical and antibacterial properties of ZnO nanoparticles prepared by laser ablation of Zn target in water and air. *Materials* **2019**, *12*, 186. [CrossRef] [PubMed]
48. Yanase, I.; Konno, S. Photoluminescence of $Zn_5(CO_3)_2(OH)_6$ nanoparticles synthesized by utilizing CO_2 and ZnO water slurry. *J. Lumin.* **2019**, *213*, 326–333. [CrossRef]
49. Gündogdu, Y.; Dursun, S.; Gezgin, S.Y.; Kiliç, H.S. Femtosecond laser-induced production of ZnO@Ag nanocomposites for an improvement in photocatalytic efficiency in the degradation of organic pollutants. *Opt. Laser Technol.* **2024**, *170*, 110291. [CrossRef]
50. Lee, K.-C.; Lin, S.-J.; Lin, C.-H.; Tsai, C.-S.; Lu, Y.-J. Size effect of Ag nanoparticles on surface plasmon resonance. *Surf. Coat. Technol.* **2008**, *202*, 5339–5342. [CrossRef]
51. Ramya, E.; Ramya, E.; Rao, M.V.; Rao, D.N. Nonlinear optical properties of Ag-enriched ZnO nanostructures. *J. Nonlinear Opt. Phys. Mater.* **2019**, *28*, 1950027. [CrossRef]
52. Patil, S.S.; Patil, R.H.; Kale, S.B.; Tamboli, M.S.; Ambekar, J.D.; Gade, W.N.; Kolekar, S.S.; Kale, B.B. Nanostructured microspheres of silver@zinc oxide: An excellent impeder of bacterial growth and biofilm. *J. Nanopart. Res.* **2014**, *16*, 2717. [CrossRef]
53. Hu, X.; Mohamood, T.; Ma, W.; Chen, C.; Zhao, J. Oxidative decomposition of Rhodamine B dye in the presence of VO_2^+ and/or Pt(IV) under visible light irradiation: N-deethylation, chromophore cleavage, and mineralization. *J. Phys. Chem. B* **2006**, *110*, 26012–26018. [CrossRef] [PubMed]
54. Tian, L.; Guan, X.; Zong, S.; Dai, A.; Qu, J. Cocatalysts for Photocatalytic Over all Water Splitting: A Mini Review. *Catalysts* **2023**, *13*, 355. [CrossRef]
55. Fakhrutdinova, E.; Reutova, O.; Maliy, L.; Kharlamova, T.; Vodyankina, O.; Svetlichnyi, V. Laser-based Synthesis of TiO_2-Pt Photocatalysts for Hydrogen Generation. *Materials* **2022**, *15*, 7413. [CrossRef] [PubMed]
56. Muñoz-Fernandez, L.; Gomez-Villalba, L.S.; Milošević, O.; Rabanal, M.E. Influence of nanoscale defects on the improvement of photocatalytic activity of Ag/ZnO. *Mater. Charact.* **2022**, *185*, 111718. [CrossRef]
57. Ziashahabi, A.; Prato, M.; Dang, Z.; Poursalehi, R.; Naseri, N. The effect of silver oxidation on the photocatalytic activity of Ag/ZnO hybrid plasmonic/metal-oxide nanostructures under visible light and in the dark. *Sci. Rep.* **2019**, *9*, 11839. [CrossRef] [PubMed]
58. Quan, Y.; Liu, M.; Wu, H.; Tian, X.; Dou, L.; Wang, Z.; Ren, C. Rational design and construction of S-scheme CeO_2/AgCl heterojunction with enhanced photocatalytic performance for tetracycline degradation. *Appl. Surf. Sci.* **2024**, *642*, 158601. [CrossRef]
59. Jin, C.; Li, W.; Chen, Y.; Li, R.; Huo, J.; He, Q.; Wang, Y. Efficient Photocatalytic Degradation and Adsorption of Tetracycline over Type-II Heterojunctions Consisting of ZnO Nanorods and K-Doped Exfoliated g-C_3N_4 Nanosheets. *Ind. Eng. Chem. Res.* **2020**, *59*, 2860–2873. [CrossRef]
60. Wang, X.Q.; Wang, F.; Chen, B.; Cheng, K.; Wang, J.L.; Zhang, J.J.; Song, H. Promotion of phenol photodecomposition and the corresponding decomposition mechanism over g-C_3N_4/TiO_2 nanocomposites. *Appl. Surf. Sci.* **2018**, *453*, 320–329. [CrossRef]
61. Blažeka, D.; Radičić, R.; Maletić, D.; Živković, S.; Momčilović, M.; Krstulović, N. Enhancement of Methylene Blue Photodegradation Rate Using Laser Synthesized Ag-Doped ZnO Nanoparticles. *Nanomat.* **2020**, *12*, 2677. [CrossRef]
62. Liu, Q.; Liu, E.; Li, J.; Qiu, Y.; Chen, R. Rapid ultrasonic-microwave assisted synthesis of spindle-like Ag/ZnO nanostructures and their enhanced visible-light photocatalytic and antibacterial activities. *Catal. Today* **2020**, *339*, 391–402. [CrossRef]
63. Li, T.; Liu, Y.; Li, M.; Jiang, J.; Gao, J.; Dong, S. Fabrication of oxygen defect-rich pencil-like ZnO nanorods with CDots and Ag co-enhanced photocatalytic activity for tetracycline hydrochloride degradation. *Sep. Purif. Technol.* **2021**, *266*, 118605. [CrossRef]

64. Kononenko, V.V.; Ashikkalieva, K.K.; Arutyunyan, N.R.; Romshin, A.M.; Kononenko, T.V.; Konov, V.I. Femtosecond laser-produced plasma driven nanoparticle formation in gold aqueous solution. *J. Photochem. Photobiol. A* **2022**, *426*, 113709. [CrossRef]
65. Fedorovich, Z.P.; Gerasimova, M.A.; Fakhrutdinova, E.D.; Svetlichnyi, V.A. Effect of laser and temperature treatment on the optical properties of titanium dioxide nanoparticles prepared via pulse laser ablation. *Rus. Phys. J.* **2022**, *64*, 2115–2122. [CrossRef]

Disclaimer/Publisher's Note: The statements, opinions and data contained in all publications are solely those of the individual author(s) and contributor(s) and not of MDPI and/or the editor(s). MDPI and/or the editor(s) disclaim responsibility for any injury to people or property resulting from any ideas, methods, instructions or products referred to in the content.

Article

The Influence of Different Aggregates on the Physico-Mechanical Performance of Alkali-Activated Geopolymer Composites Produced Using Romanian Fly Ash

Adrian-Victor Lăzărescu [1], Andreea Hegyi [1,2,*], Alexandra Csapai [1,2,*] and Florin Popa [2]

[1] NIRD URBAN-INCERC Cluj-Napoca Branch, 117 Calea Florești, 400524 Cluj-Napoca, Romania; adrian.lazarescu@incerc-cluj.ro

[2] Faculty of Materials and Environmental Engineering, Technical University of Cluj-Napoca, 103-105 Muncii Boulevard, 400641 Cluj-Napoca, Romania; florin.popa@stm.utcluj.ro

* Correspondence: andreea.hegyi@incerc-cluj.ro (A.H.); alexandra.csapai@stm.utcluj.ro (A.C.)

Abstract: In light of the urgent need to develop environmentally friendly materials that, at some point, will allow the reduction of concrete and, consequently, cement consumption—while at the same time allowing the reuse of waste and industrial by-products—alkali-activated fly ash (AAFA) geopolymer composite emerges as a material of great interest. The aim of this study was to investigate the physico-mechanical performance of composites based on AAFA binders and the effect of different types of aggregates on these properties. The experimental results indicate variations in flexural and compressive strength, which are influenced both by the nature and particle size distribution of aggregates and the binder-to-aggregate ratio. The analysis of the samples highlighted changes in porosity, both in distribution and pore size, depending on the nature of the aggregates. This supports the evolution of physico-mechanical performance indicators.

Keywords: geopolymer composites; fly ash; micronized quartz; glass aggregates

Citation: Lăzărescu, A.-V.; Hegyi, A.; Csapai, A.; Popa, F. The Influence of Different Aggregates on the Physico-Mechanical Performance of Alkali-Activated Geopolymer Composites Produced Using Romanian Fly Ash. *Materials* **2024**, *17*, 485. https://doi.org/10.3390/ma17020485

Academic Editor: Carlos Leiva

Received: 17 November 2023
Revised: 11 January 2024
Accepted: 17 January 2024
Published: 19 January 2024

Copyright: © 2024 by the authors. Licensee MDPI, Basel, Switzerland. This article is an open access article distributed under the terms and conditions of the Creative Commons Attribution (CC BY) license (https://creativecommons.org/licenses/by/4.0/).

1. Introduction

Cement production contributes at least 5–8% of global carbon dioxide emissions [1]. Such a massive output has a significant impact on the environment. A sustainable alternative to cement-intensive concrete is geopolymer binders [2–4]. These are currently under development, and research is focused on meeting the imperative to reduce global CO_2 emissions. With excellent mechanical properties and durability in challenging environments, these materials offer an opportunity for both environmental and engineering considerations, providing an alternative to conventional technology [5,6]. Geopolymer concrete is considered a third-generation binder after lime and cement. Some studies using the European life cycle database, Ecoinvent, suggest that using it may lead to a potential reduction in greenhouse gas emissions ranging from 25 to 45% [7], or even up to 70% [8].

Geopolymer binders are essentially formed through chemical reactions. The specific category is created by the alkaline activation of materials abundant in SiO_2 and Al_2O_3 [9]. The ash from thermal power plants contains significant proportions of aluminium and amorphous silica, making it a suitable source for geopolymer production [10]. The steps of the chemical process used to obtain geopolymers by alkaline activation of the thermal power plant ash, as outlined by Buchwald et al. [11], can be described by the chemical reaction. The overall chemical reaction of thermal power plant ash is expressed by Equation (1). As a result of this repolymerization mechanism, a distinct spatial, three-dimensional arrangement can be observed at a microstructural level in geopolymers; this is in contrast to the non-spatial arrangement of Si and Ca oxides typically found in cementitious composites [12].

$$SiO_2 \cdot \alpha Al_2O_3 \cdot \beta CaO \cdot \gamma Na_2O \cdot \delta Fe_2O_3 \cdot \varepsilon TiO_2 + (\beta + \gamma + 3\delta)H_2O + (2 + 2\alpha + \varepsilon)OH^- \rightarrow$$
$$SiO_3^{2-} + 2\alpha AlO_2^- + \beta Ca^{2+} + 2\gamma Na^+ + \delta Fe^{3+} + \varepsilon HTiO_3^- + (1 + \alpha)H_2O + 2(\beta + \gamma + 3\delta) \quad (1)$$

Therefore, as reported in the literature, the microstructural characteristics of geopolymer composites are closely related to the specific raw materials and the production technology [4,9,13–18]. These significantly influence the physico-mechanical performance and durability of the cured and matured composite [19–72].

Several studies have documented a wide variation in the oxide composition of fly ash; the main oxides are Al_2O_3, SiO_2, Fe_2O_3 and CaO. The prevailing consensus is that fly ash consists predominantly of spherical, amorphous particles, but it also contains crystalline structures that dissolve more gradually and often only partially during the geopolymerization process. In addition, it is acknowledged in the literature that these oxides are present in fly ash in different mineralogical phases such as anhydrite, quartz, portlandite, hematite, calcite or, to a lesser extent, mullite. The reactivity of fly ash tends to be higher when the content of crystalline phases is lower [10].

When produced from various raw materials, geopolymers typically consist of a blend of crystalline aluminosilicate particles and semicrystalline and amorphous aluminosilicate gels. A detailed characterization of their structural constitution is challenging because of their complex composition and the difficulty in separating the crystalline aluminosilicate particles from the semi-crystalline and amorphous gel phases; however, this is vital to understand their mechanical strength. Therefore, a comprehensive understanding of both the structural composition and the gel phases in geopolymers is essential [73]. Scanning electron microscopy (SEM) coupled with energy-dispersive spectroscopy (EDS) stands out as a primary method employed for the analysis of microstructures. These methods have been previously employed for the characterization of clay, zeolites, fly ash, cement and concrete [73–81].

Consequently, the literature points to a significant variation in the mechanical strength of both the geopolymer binder and geopolymer composites (which consists of the binder matrix containing the aggregate skeleton). The variability is influenced by factors such as the characteristics and percentage content of Si and Al in the fly ash, the Si/Al ratio, the type of alkaline activator used, the molarity of the NaOH or KOH solution used, the mass ratio between the Na_2SiO_3 and NaOH (or KOH) solutions in the preparation of the alkaline activator, the duration and temperature of the heat treatment or the age at testing. Compressive strengths in the range of 10 MPa to 95 MPa have been reported for both the geopolymer binder and the geopolymer composite, and aggregate granulation up to 4 mm has been observed [19–72]. Research has shown that heat treatment temperature is critical in formulating and producing geopolymer materials. The ideal temperature for heat treatment falls between 60 °C and 75 °C [61].

There are limited studies about the use of different sands as aggregates in the geopolymer mortar. Aggregates greatly influence the characteristics of mortar or concrete, both in fresh and hardened states. Their grading, shape and texture greatly affect properties of concrete in a fresh state (i.e., workability, finishability, bleeding or segregation). Moreover, when hardened characteristics are considered, density, mechanical strength, porosity or water absorption are also highly affected by aggregate features [82].

Several researchers have studied the influence of different types of aggregates in terms of mechanical properties of geopolymer materials [82–86]. Mechanical properties of geopolymer concrete with different fine aggregate content and grading (sand and granite slurry) were mixed together in different proportions (100:0, 80:20, 60:40 and 40:60) using fly ash and granulated slag as raw material, with a 50:50 aggregate:binder ratio [83]. The mechanical properties of the geopolymer materials (compressive strength, flexural strength) were studied after 7, 28 and 90 days of curing at ambient room temperature. The results show that the mechanical properties increased up to a fine aggregate proportion of 60:40; a decreasing trend has been observed at a proportion of 40:60.

Other studies have investigated the influence of aggregate mass percentages in the geopolymer binder to determine the impact of the ratio of binder:aggregate on the synthesized material properties [84]. The study shows that the incorporation of aggregates in the reaction mixtures changes the aspect of the materials due to interactions between binders and aggregates. Several other parameters that could influence adhesion between the aggregates and the binder are the porosity, the roughness of the aggregates and the chemical composition at the interface [85,86]. The optimal aggregate choice may vary depending on the specific application, desired properties and local material availability. Additionally, proper mix design and testing should be conducted to ensure the desired performance of geopolymer-based materials with selected aggregates.

The aim of this study is to analyse the effects of mixing different local source aggregates with an alkali-activated fly ash-based geopolymer binder in composite materials. These aggregates are sourced from either recycled waste (i.e., glass waste, spent garnet) or quartz aggregates; each is characterized by a different granulation that influences the basic physico-mechanical properties of the material.

2. Materials and Methods

2.1. Preparation of the Geopolymer Binder

The geopolymer binder was obtained by alkaline activation of fly ash (FA) obtained from Rovinari power plant in Romania. Fly ash was chosen because previous studies [87] have shown that its chemical composition and particle size distribution make it suitable for the production of geopolymer binders with increased mechanical performances. Characterization of the fly ash was carried out prior to its use in the preparation of the geopolymer binder. This included the determination of the chemical composition by XRF analysis and the assessment of the $R_{0.045}$ fineness (Table 1 and Figure 1). The chemical composition and the particle size distribution of the fly ash were investigated using X-ray fluorescence (XRF) analysis, using a HELOS RODOS/L, R5 instrument (Sympatec GmbH, Clausthal-Zellerfeld, Germany) (Table 1). The mineralogical composition of the fly ash was investigated using an X-ray diffraction (XRD) analysis (Figure 2), Bruker D8 ADVANCE X-ray diffractometer (Bruker, Karlsruhe, Germany). Scans were collected in the range of 5–60° (2θ) with a step size of 0.02° and a scan speed of 10 s per step.

Table 1. Characterization of fly ash.

	SiO_2	Al_2O_3	Fe_2O_3	CaO	MgO	SO_3	Na_2O	K_2O	P_2O_5
Oxide composition (%)	46.94	23.83	10.08	10.72	2.63	0.45	0.62	1.65	0.25
	TiO_2	Cr_2O_3	Mn_2O_3	ZnO	SrO	CO_2	P.C.	$SiO_2+Al_2O_3$	
	0.92	0.02	0.06	0.02	0.03	-	2.11	70.77	
$R_{0.045}$ (%)					31.40				

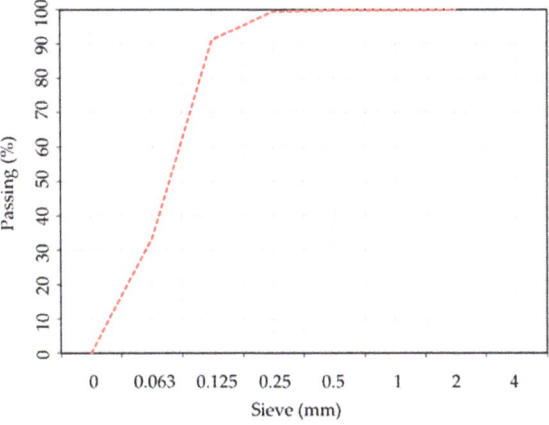

Figure 1. Fly ash particle size distribution.

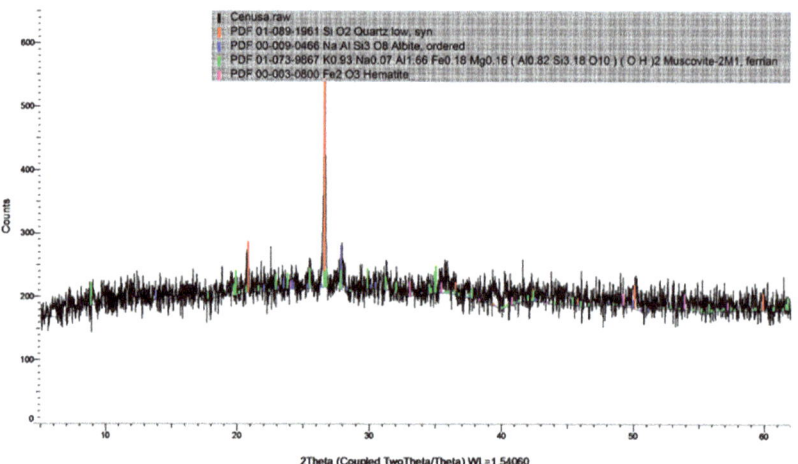

Figure 2. Screenshot of the XRD spectra for the fly ash powder.

The diffuse reflection peak for quartz was predominantly at 26.59° (2θ), with additional smaller peaks at 20.82 and 50.14° 2θ. Albite was detected at 13.88, 23.53 and 27.90° (2θ). Muscovite M1 was identified at various angles: 19.91, 22.93, 25.50, 26.78, 27.90, 29.84 and 35.06° (2θ). Hematite was identified by X-ray diffraction at angles of 24.02, 33.09, 35.47, 40.82, 49.28, 53.96 and 57.29° (2θ). The peak with the highest intensity at 2θmax, 26.59° (2θ) was assigned to quartz.

Based on previous studies regarding the production of alkali-activated fly ash-based geopolymer binders [87], the alkaline activator solution (AA) was prepared by mixing a sodium hydroxide (NaOH) solution with a molar concentration of 8 M in a 1:1 constant mass ratio with a commercially purchased sodium silicate (Na_2SiO_3) solution at room temperature. After preparation and before use, the alkaline activator solution was stored in a sealed container under laboratory conditions (23 °C) for 24 h to mature.

The mixing of the components and the preparation of the geopolymer binder were carried out in a laboratory environment (23 °C and 65% RH) using an ELE laboratory mixer, with a stainless-steel beater (ELE International, Milton Keynes, UK), according to EN 196-1 [88]. The mass ratio between the amount of fly ash and the amount of alkaline activator used in the preparation was kept constant at 0.9.

The geopolymer binder sample, identified by code P1, was used as a control sample in comparison to all subsequent composite samples and was prepared using only fly ash and alkaline activator. The control sample, i.e., the binder sample, was prepared without aggregates, because the aim of the study was to comparatively analyse both the influence of different types of aggregates on the performance of the geopolymer material and the influence of introducing aggregates into the geopolymer binder, thus obtaining the transition to geopolymer mortar composites.

2.2. Preparation of the Geopolymer Composite Samples

For the formulation of geopolymer composites, different types of aggregates with a maximum particle size of 8 mm have been incorporated into the geopolymer binder during the preparation process, with specific identification codes (P2–P12) for each mixture. These include polygranular CEN- NORMSAND EN 196-1 sand (P2), granular class 0/4 mm natural aggregates (P3), granular class 4/8 mm natural aggregate (P4), granular class 0/4 mm recycled glass aggregate (P5), granular class 4/8 mm recycled glass aggregate (P6), micronized quartz (P7), 0/0.3 mm granulated quartz (P8), 0/0.5 mm granulated quartz (P9), 0/0.6 mm granulated quartz (P10), 0.3/0.7 mm granulated quartz (P11) and spent garnet (P12). For each type of aggregate used in the production of the samples, bulk apparent density was analysed in accordance with EN 1097-3 [89] (Table 2); and particle size distribution (Figure 3) was analysed using a sieving method. This range of aggregates was chosen to facilitate an analysis of their influence on the physico-mechanical properties of the geopolymer composite, considering both the nature of the aggregate and the granularity characteristics.

Table 2. Bulk apparent density of the aggregates used in production of the alkali-activated fly ash-based geopolymer samples.

Aggregate Type	Polygranular Sand	Natural Aggregates 0/4 mm	Natural Aggregates 4/8 mm	Glass Aggregate Reciclată 0/4 mm	Glass Aggregate Reciclată 4/8 mm	Micronized Quartz	Granulated Quartz 0/0.3 mm	Granulated Quartz 0/0.5 mm	Granulated Quartz 0/0.6 mm	Granulated Quartz 0.3/0.7 mm	Spent Garnet
Bulk apparent density (kg/m³)	1660	1580	1530	1350	1430	1030	1430	1440	1450	1420	2690

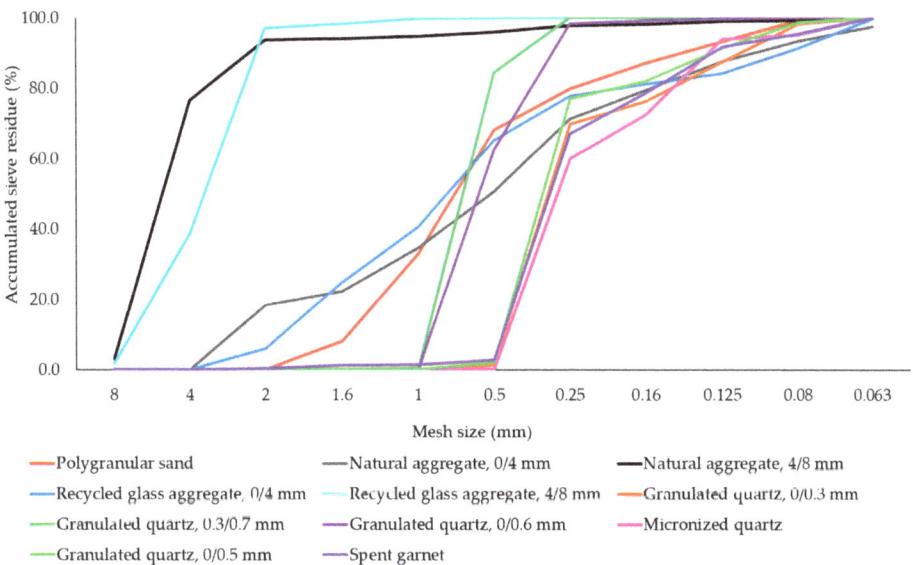

Figure 3. Graphical representation of the particle size distribution of the aggregates used in the production of alkali-activated fly ash-based geopolymer samples.

To analyse both the influence of the type of aggregates used in the production of the geopolymer samples and the influence of mass ratio between geopolymer binder and aggregates on the physico-mechanical properties samples, three binder:aggregate ratios were used (1:1, 0.75:1 and 1.25:1). The components were mixed and the geopolymer composites were prepared under the same temperature, relative humidity and equipment conditions as those described in Section 2.1. A minimum of 3 sets of samples were produced to analyse the influence of the type, specific granulation and geopolymer binder/aggregate ratio on the physico-mechanical performances of the alkali-activated fly ash-based geopolymer mixtures.

After being cast into 40 × 40 × 160 mm moulds, with the corresponding vibration, the samples were subjected to a (70 °C for 24 h) heat treatment using a thermostatic MEMMERT ULE 500 chamber (MEMMERT GmbH+Co.KG, Schwabach, Germany). After demoulding, the geopolymer samples were stored in laboratory conditions (T = 23 °C and

RH = 65%). Testing to determine their mechanical properties was carried out after 7 days. The experimental research methodology is graphically outlined in Figure 4.

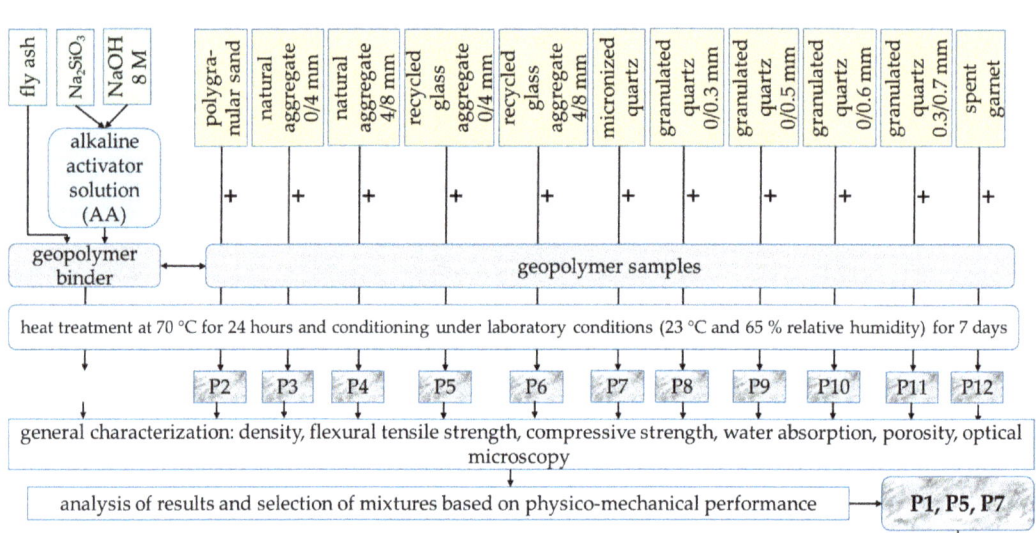

Figure 4. Graphical representation of the work flow for the preparation and analysis of the geopolymer recipes.

2.3. Physico-Mechanical Analysis of the Alkali-Activated Geopolymer Samples

Prior to the physico-mechanical performance evaluation tests, the density of the geopolymer samples was determined as the ratio between the mass of the samples, determined by weighing using a precision balance (KERN FKB 36K0.1, KERN & SOHN, Albstadt, Ebingen, Germany) and their volume, by direct measurement of the real dimensions using an electronic calliper with an accuracy of 0.01 mm.

To obtain the flexural and compressive strength results for all of the alkali-activated fly ash based geopolymer samples, tests were conducted according to the (EN 196-1:2006 [88]) standard method for evaluation of mechanical performances of OPC paste and standard type mortar.

Total water absorption of the alkali-activated fly ash-based geopolymer samples was determined by submerging the test samples in water, at a constant temperature of (20 ± 5) °C, until they reached constant mass. The minimum immersion period of the samples was 3 days. After reaching the saturated state constant mass, they were weighed. Subsequently, samples were placed in an oven and dried at a temperature of (105 ± 5) °C, to reach the constant mass. The water absorption of each sample was expressed as mass percentage loss.

Prior to the water absorption test, porosity was evaluated using a calculation for all alkali-activated fly ash-based geopolymer samples. This method was used by assimilation with standardized methods for characterizing the porosity of concrete, which allows the determination of this parameter based on the apparent and real density of the material. The porosity of the alkali-activated fly ash-based geopolymer samples was determined using the pycnometer method. The concrete samples were crushed and a representative amount was collected. The obtained material was placed in the mill and then passed through a 0.02 mm sieve. After sieving, it was dried at constant mass in the oven. The porosity of the samples was calculated by means of apparent density and real density, as a percentage.

The porosity of geopolymer materials was monitored and quantified to study the influence of this parameter on the mechanical properties, for the use of different types of aggregates.

For each physico-mechanical indicator experimentally determined for the alkali-activated fly ash-based geopolymer samples (density, flexural strength, compressive strength, porosity), a variation of the specified indicator was calculated and expressed as a percentage difference to the value obtained in the control sample (P1).

2.4. Optical, SEM and EDS Analysis of Samples

Aggregate distribution and porosity analysis were carried out on each sample by microscopic examination using a Leica SAPO optical stereomicroscope (LEICA, Wetzlar, Germany). From the overall sample set, three mixtures were selected for scanning electron microscopy (SEM) and energy-dispersive spectroscopy (EDS): P1—control sample, P5—granular class 0/4 mm recycled glass aggregates and P7—micronized quartz. The selection criteria included the mixtures with the most favourable physico-mechanical performances for recycled glass aggregates and the mixture with micronized quartz, which showed superior physico-mechanical performance compared to all quartz-type aggregate mixtures. Considering that a binder-aggregate ratio greater than 1:1 (1.25:1) typically results in an increased water absorption, it was important to balance the amount of available binder in the composite matrix to effectively incorporate the aggregates to achieve good mechanical strength performances; therefore, samples with the average binder to aggregate ratio of 1:1 were selected for the optical, SEM and EDS analyses.

The SEM and EDS images were acquired with a JEOL/JSM 5600-LV scanning electron microscope (JEOL Ltd., Tokyo, Japan) using the secondary electron imaging (SEI) mode at an acceleration voltage of 15 kV. As part of the preparation process, to increase the electrical conductivity for electron microscopy analysis, the samples were coated with gold by plasma sputtering.

The aim of using these methods was both to demonstrate the good conditions of the geopolymerization reaction process, with the formation of specific compounds, and to highlight their homogeneous distribution in the geopolymer matrix, with direct effects on the mechanical behaviour.

3. Results and Discussions

3.1. Physico-Mechanical Properties of the Samples

The experimental results regarding the physico-mechanical properties of the alkali-activated fly ash-based geopolymer samples are presented in Figures 5–9, as the arithmetic mean of the individual values for each situation.

As observed in Figure 5, as expected, the density of the geopolymer samples increased when aggregates were incorporated in the geopolymer binder matrix, regardless of their type, compared to the control sample. However, results show that both the type of aggregate and the geopolymer binder to aggregate ratio influence the behaviour in the variation of this parameter. It can, therefore, be stated that irrespective of the geopolymer binder to aggregate ratio the maximum percentage increase in bulk density compared to the control sample was obtained when incorporating granular size 4/8 mm natural aggregates (Mixture P4). This increase was 19.55% for the geopolymer samples with a binder: aggregate ratio 1:1, 39.72% for 0.75:1 and 27.33% for 1.25:1. The lowest increase in bulk density, regardless of the geopolymer binder to aggregate ratio, was obtained for samples produced using spent garnet as the aggregate: 4.91% (binder: aggregate ratio 1:1), 15.65% (binder: aggregate ratio 0.75:1) and 6.41% (binder: aggregate ratio 1.25:1).

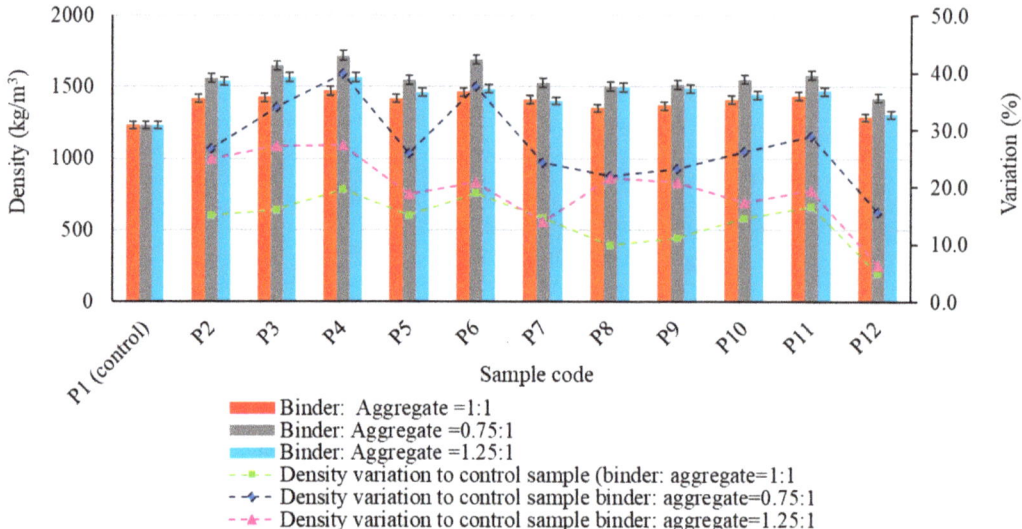

Figure 5. The influence of aggregates on the density of the alkali-activated fly ash-based geopolymer samples.

The impact on flexural strength of using different aggregates is shown in Figure 6. Results show that adding aggregates into the geopolymer binder matrix can have both a positive and a negative effect on this parameter, depending on the type of aggregate and the binder:aggregate ratio.

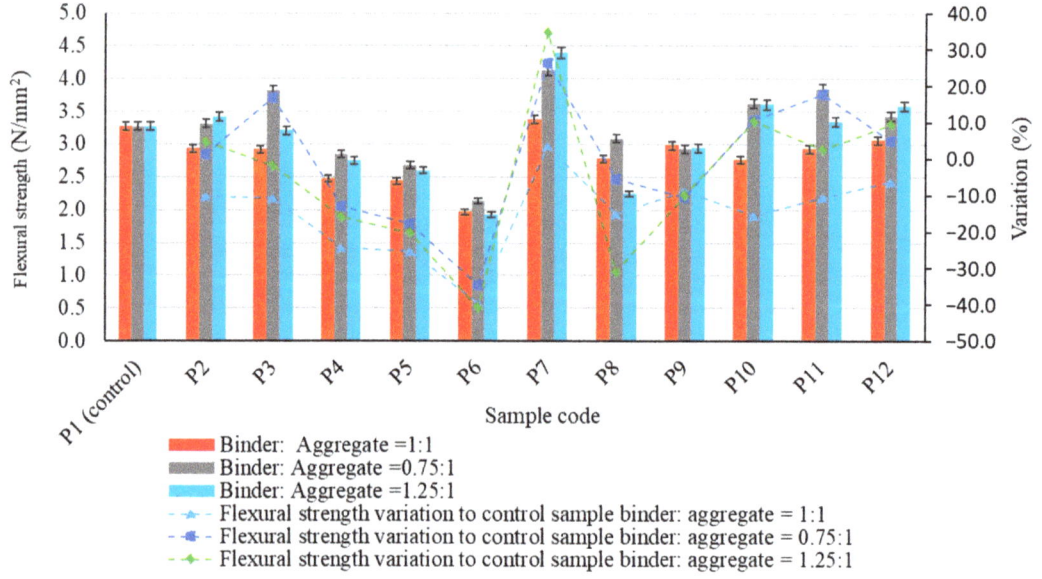

Figure 6. The influence of aggregates on the flexural strength of the alkali-activated fly ash-based geopolymer samples.

Thus, except for micronized quartz (P7), which showed a slight increase (3.47% with respect to the control sample), a reduction in flexural tensile strength was observed for all the types of aggregate used when their respective quantities were equal to that of the geopolymer matrix in the composite matrix.

The maximum reduction in flexural tensile strength of geopolymer binder composites compared to the control sample was 37.72%. This was observed when using recycled glass aggregates of granular class 4/8 mm (P6), at a geopolymer binder to aggregate ratio of 1:1. As the amount of geopolymer binder in the composite matrix decreased (geopolymer binder to aggregate ratio of 0.75:1), the influence of the aggregate type and its granular class on the flexural tensile strength became more apparent. The parameter increases with respect to the control sample were recorded when polygranular sand (P2), natural aggregates granular class 0/4 mm (P3), micronized quartz (P7), granulated quartz 0/0.6 mm (P10), 0.3/0.7 mm (P11) or spent garnet (P12) were used. The maximum increase was 26.41% compared to the control sample; this was observed in the case of micronized quartz. Conversely, the use of coarse aggregates, natural aggregates of granular class 4–8 mm (P4) or recycled glass aggregates of granular class 4/8 mm (P6), and the effect of using recycled glass aggregates—including small-sized recycled glass aggregates of granular class 0/4 mm (P5)—leads to a decrease in flexural strength. The peak value is 34.80% (P6) compared to the control sample. As the amount of geopolymer binder in the composite matrix was increased (geopolymer binder to aggregate ratio of 1.25:1), different effects were also observed, both in terms of increase and decrease in flexural strength, depending on the type and granulation of the aggregate. In general, the positive effect, which results in an increase in the parameter studied, is maintained using granular sand or quartz. This is consistent with the previous scenario (geopolymer binder to aggregate ratio of 0.75:1). The most significant improvement was observed when micronized quartz (P7) was used. This resulted in an increase of 34.51% compared to the control sample. Similarly, the diminishing effect on flexural strength due to the incorporation of recycled glass aggregates into the geopolymer binder matrix is again evident, with the most significant reduction of 41.01% being observed for large-sized recycled glass aggregates of granular class 4/8 mm (P6).

When analysing the influence of the type of aggregate, its granulation and the geopolymer binder:aggregate ratio on the compressive strength of the geopolymer composites (Figure 7), several similarities can be observed when compared with the control sample (geopolymer binder without aggregates). Regardless of the geopolymer binder to aggregate ratio, the most significant reduction in compressive strength compared to the control is observed when using spent garnet (P12): 34.46%, 30.99% and 11.64%. In addition, the use of micronized quartz (P7) increased the compressive strength of the samples regardless of the geopolymer binder to aggregate ratio, with values of 21.47%, 14.21% and 52.38%. The use of quartz aggregates does not guarantee a consistent effect on compressive strength. Depending on the granulation of the aggregates, compressive strength may be improved or reduced. Even when using recycled glass aggregates, an increase in the compressive strength was observed. It should be noted that the granular class of the aggregates is a key element that can have either positive or negative effects. In some cases, the compressive strength of the composite can be increased by more than 18%, especially when these aggregates with a large grain size (4/8 mm) are introduced into the geopolymer binder matrix. However, this is at a reduced quantity (geopolymer binder to aggregate ratio = 0.75:1).

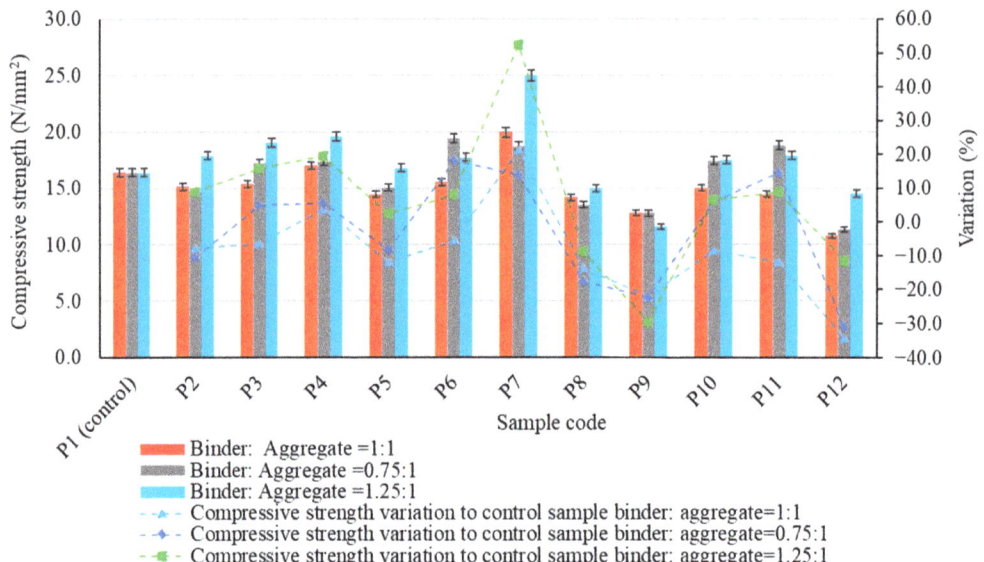

Figure 7. The influence of aggregates on the compressive strength of the alkali-activated fly ash-based geopolymer samples.

A cumulative analysis of the mechanical strength in both flexural and compressive strength tests for all geopolymer binder composites considering all types of aggregate used suggests that in most cases a geopolymer binder to aggregate ratio of 1:1 is not the most favourable alternative. Thus, in most situations, reducing the amount of binder by 25% (geopolymer binder to aggregate ratio = 0.75:1) resulted in an improvement in flexural strength, while maintaining the aggregate type.

However, there were instances where increasing the amount of binder by 25% (geopolymer binder to aggregate ratio = 1.25:1) resulted in an improvement in mechanical performances, particularly natural aggregates with a bigger particle size (4/8 mm). In the case of compressive strength, it was observed that an increase of 25% in the amount of binder (geopolymer binder to aggregate ratio = 1.25:1) can be a factor, with the potential to improve this performance, both for natural aggregates, recycled glass aggregates or quartz aggregates. It can, therefore, be seen that in terms of mechanical strength performance, similar to cementitious composites, both the binder matrix and the aggregate matrix structure of the composite, as well as their interaction and the bond between them, contribute to the overall result. Furthermore, as will be shown indirectly by the analysis of porosity and water absorption of the analysed samples, all these aspects are closely related to the porosity and distribution of pores within the composite mass.

The experimental results indicate the influence of introducing aggregates into the geopolymer binder matrix, but there is also an influence from the characteristics of these aggregates on the apparent porosity of the composites (Figure 8). As observed, in general, the porosity has a decreasing trend with the introduction of aggregates into the geopolymer binder matrix. This decreasing trend of porosity is generalized for the situations binder:aggregate = 1:1, respectively 0.75:1, regardless of the nature of the aggregates. In the case of a binder:aggregate ratio = 1.25:1, exceptions, i.e., increases in open porosity compared to the control sample, are observed for the situations P4 (natural aggregates granular class 4/8 mm), P5 (glass granular class 0/4 mm) and P7 (micronized quartz).

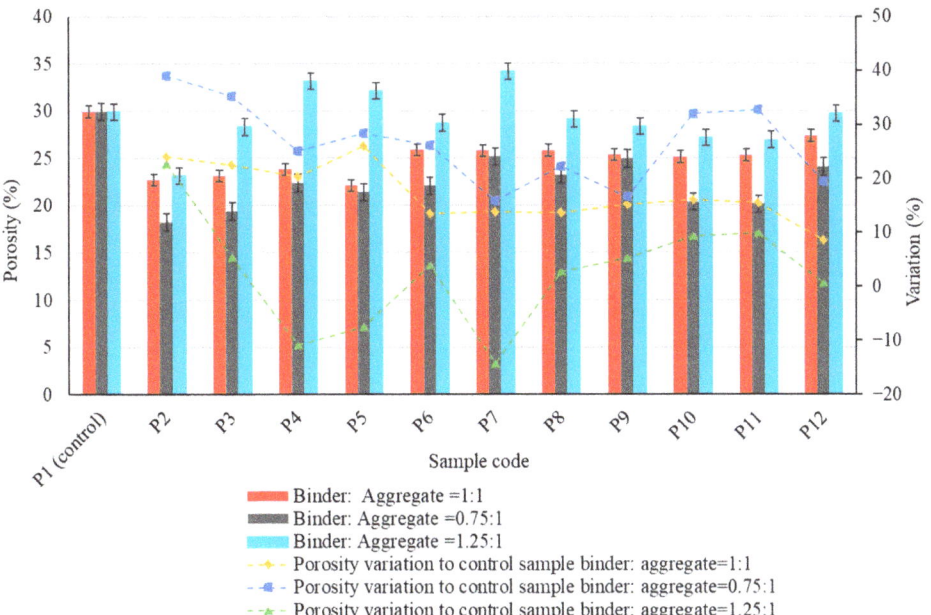

Figure 8. The influence of aggregates on the porosity of the alkali-activated fly ash-based geopolymer samples.

The increase in open porosity is probably due to various causes: in the case of natural aggregates, this is due to the larger grain size; in the case of glass aggregates, it is due to the shape of the grains and the lack of adhesion in the contact zone; and in the case of micronized quartz, it is probably due to the additional SiO_2 input involved in the geopolymerization mechanism. For each type of aggregate, the same trend of porosity evolution is observed as a function of the binder:aggregate ratio, i.e., a higher amount of binder results in a higher open porosity, which is in correlation with the experimental results recorded for water absorption (a phenomenon that occurs through the capillarity of the material and is directly related to the open porosity).

Research clearly evidences that the water to solids ratio of the mixtures plays an important role in the pore size distribution of the cured geopolymers. In the case of geopolymerization, water is consumed only marginally during the alkali activation process of fly ash. For this reason, the volume fraction of the liquid activator governs the final open porosity of the geopolymers. Curing temperature and curing time also play an important role in the definition of geopolymer porosity. In general, a systematic increase in the total volume of pores is observed when the curing temperature increases. Inconsistent results have also been found in relation to the variation of porosity and pore size distribution during the curing time [90,91].

In terms of water absorption of the geopolymer binder composites (Figure 9), it is most evident that for all types of aggregates used, no specific behaviour could be observed for the results obtained when compared to the control sample. It can also be observed that a 25% reduction in the amount of geopolymer binder in the composite (geopolymer binder to aggregate ratio = 0.75:1) results in a reduction in water absorption, whereas an increase in the amount of geopolymer binder in the composite (geopolymer binder to aggregate ratio = 1.25:1) results in an increase in this parameter compared to the value obtained for the same binder to aggregate ratio.

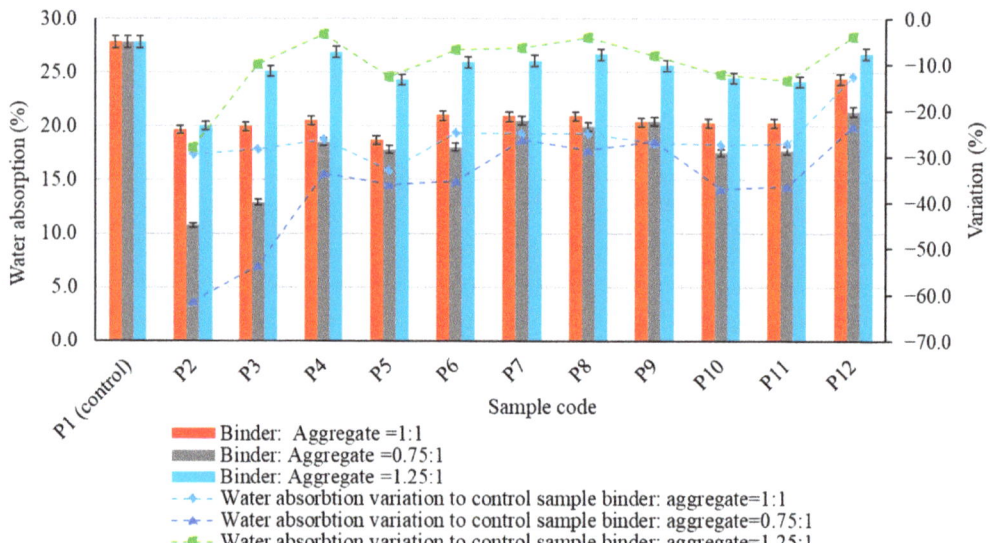

Figure 9. The influence of aggregates on water absorption of the alkali-activated fly ash-based geopolymer samples.

The most significant changes were observed for the samples produced using polygranular sand (P2), in which a reduction in the quantity of geopolymer binder resulted in a reduction in water absorption of over 45%. Similarly, when using natural aggregates with a large grain size (P4), an increase in the amount of geopolymer binder resulted in a reduction in water absorption of more than 31% compared to the situation with a binder to aggregate ratio of 1:1. Analysis of the experimental data obtained for a constant binder to aggregate ratio shows the influence of the nature and granulometry of the aggregates on water absorption. Thus, for a binder to aggregate ratio of 1:1, the most significant reduction in water absorption compared to the control sample was recorded for the composite with recycled glass aggregates of granular size 0–4 mm (P5)—33.03%. However, this trend was not as pronounced when using granular class 4–8 mm recycled glass aggregates (P6), indicating that their distribution in the geopolymer binder matrix modifies the open porosity. This is a factor closely related to water absorption. It is probable that recycled glass aggregates with a small grain size allow a better distribution in the mass, together with a lower open porosity; or even the possibility of the fine part of the aggregate closing some of the open pores formed in the geopolymer binder during the geopolymerization process. This hypothesis is supported by the behaviour of composites produced using polygranular sand (P2), natural aggregates of granular class 0–4 mm (P3) or quartz aggregates; these are all characterized by a higher proportion of fine particles and a significant reduction in water absorption. Furthermore, when examining the results in which the amount of geopolymer binder in the composite was reduced (geopolymer binder to aggregate ratio = 0.75:1), there was an enhanced effect in reducing water absorption. This exceeded a 60% reduction compared to the control sample when using polygranular sand (P2). Conversely, when the amount of geopolymer binder in the composite was increased (geopolymer binder to aggregate ratio = 1.25:1), the effect on reducing water absorption was less pronounced. These results underline both the importance and the feasibility of reducing water absorption in composites through key factors: the type of aggregate, granularity and the amount of binder available in the matrix.

3.2. Optical, SEM and EDS Analysis of Samples

In order to assess the behaviour of the alkali-activated fly ash-based geopolymer samples, the variation of the observed physico-mechanical indicators was further supported by means of optical analysis and microstructural characterization of the samples.

Microscopic analysis of the structure of the geopolymers (Figure 10) has highlighted mixtures in which the distribution of the aggregates is homogeneous and uniform. An exception is seen when using natural or recycled glass aggregates with large particle sizes (P4) and (P6), which show a tendency towards segregation. In addition, the presence of pores formed during the setting process was observed in the mass, with a variable distribution and size depending on the type and granular class of the aggregates. The control sample has a porosity characterized by an even distribution of numerous small pores. In contrast, the composite samples show a non-uniform distribution of porosity, with the presence of larger pores (maximum diameter, 1140 µm, for sample P7) but in smaller quantities. The use of recycled glass aggregates tends to reduce porosity, both in terms of pore size and frequency. In this case, the behaviour of the geopolymer samples is attributed to a combination of effects: improvement due to reduced porosity, but also lack of improvement due to the tendency for segregation, with the aggregates migrating towards the lower part of the specimens; this is probably due to a reduced degree of cohesion with the geopolymer binder. Quartz aggregates generally tend to reduce porosity but depending on their granulometry the composite matrix may have areas of clustered pores along with more compact areas of few and small pores. Spent garnet has a comparable effect on porosity, such as the patterns observed with coarse sand and natural aggregates, where the composite matrix presents a combination of larger and smaller pores which are randomly distributed.

Figure 10. Optical microscopy images of the control sample and geopolymer binder composites for binder:aggregate ratio = 1:1 (4× measurement).

The morphology and microstructure of the alkali-activated fly ash geopolymer composites were investigated by analysing sections of samples P1, P5 and P7 using SEM and EDS techniques. Figure 11 P1a, P5a and P7a show the SEM images at ×50 magnification, while Figure 11 P1b, P5b and P7b show the elemental distribution maps in the samples investigated. Figure 11 P1a shows the SEM image of the sample corresponding to the control composition. Visual assessment of the region examined suggests the formation of a dense structure with uniform micro-pore distribution within the sample. The average micro-pore size measured within the region analysed was 61.2 μm. A comparable structure was observed in sample P7. However, in this instance, the average micro-pore size measured approximately 80.8 μm. However, in Figure 11 P5a, which illustrates the sample containing recycled glass aggregates (granular class 0/4 mm), a distinct scenario emerges. The glass aggregates appear embedded in the geopolymer binder mass, which appears less cohesive and more porous.

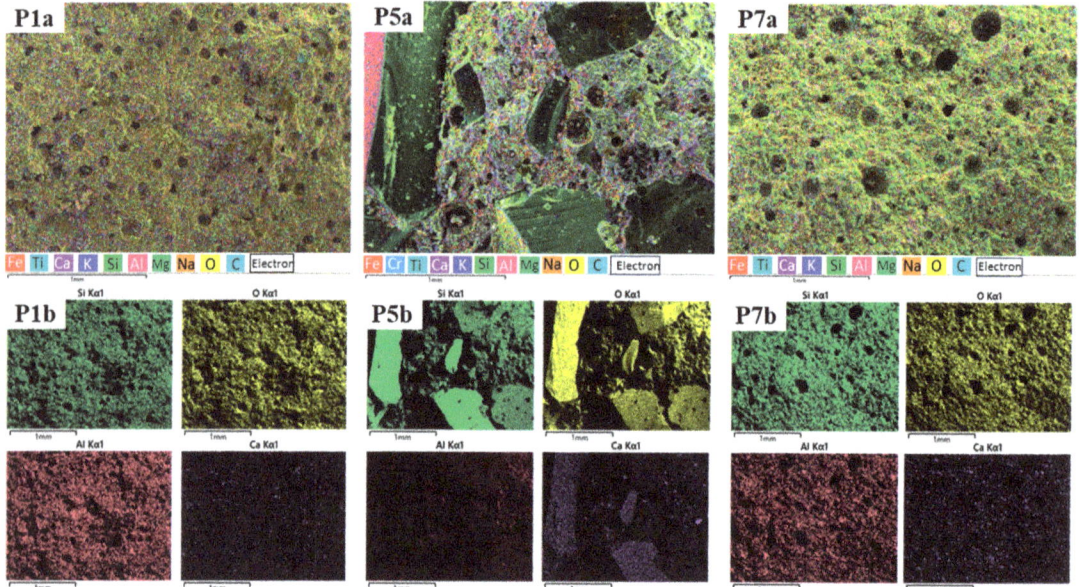

Figure 11. Elemental distribution map of Si, O, Ca and Al in the control sample (P1(a,b)) and the samples with recycled glass aggregates granular class 0/4 mm (P5(a,b)) and micronized quartz (P7(a,b)) for binder:aggregate ratio = 1:1.

The comprehensive EDS spectra (Figure 12 P1b, P5b and P7b and Table 3) as well as the EDS elemental distribution maps (Figure 11 P1b, P5b and P7b) reveal a uniform distribution of Si, Na and Al over the entire area examined in each specimen. This is consistent across all specimens. These elements play a direct role in the geopolymerization reactions that form robust Si-Al and Na-Al-Si bonds and, therefore, provide a basis for the strength of the geopolymer composite. The elevated presence of O, Si, Al and Na in distinct regions of the samples, correlated with the observed microstructures, suggests the presence of sodium alumino-silicate hydrate (N-A-S-H) gel. Conversely, in the case of P5, concentrations of Si and O are heightened in regions where glass aggregates are present (attributed to the specific composition of glass: CaO, Al_2O_3 and SiO_2), while Al distribution is observable.

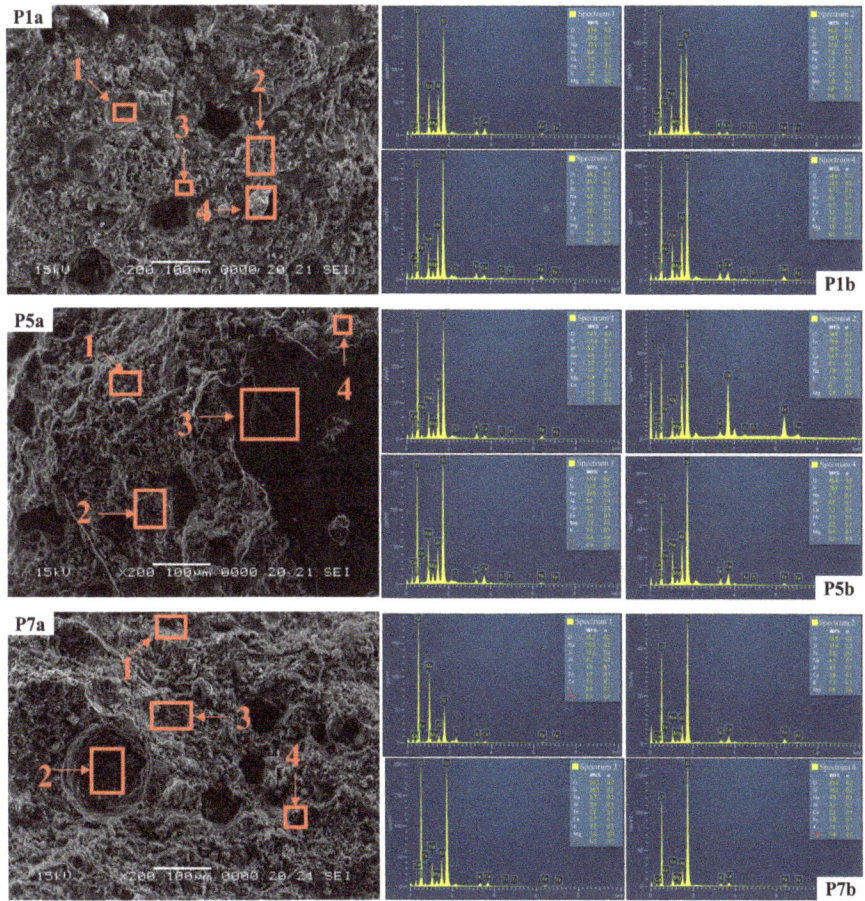

Figure 12. SEM micrographs and EDS spectra, recorded in the selected marked area (1–4), of the P1a (control composition), P5a (recycled glass) and P7a (micronized quartz) samples; for binder:aggregate ratio = 1:1.

Table 3. Spectra values of EDS for various distributions of elements in the examined regions (1–4).

Mixture	Spectra	O %	S %i	Na %	Al %	Ca %	Fe %	K %	Mg %
P1a	S1	49.8	23.8	10.3	8.4	3	2.3	1.8	0.6
	S2	48	20.3	7.6	17	1.4	2.3	1.1	1
	S3	44.3	27.7	6.9	9.2	1.7	5.6	2	1.4
	S4	44.6	24.3	8.2	9.7	3.2	6.1	2.2	1
P5a	S1	53.3	22.3	6.4	9.2	1.2	3.8	1.6	1.4
	S2	34.9	16.6	3.9	8.1	14.7	19.3	1.2	0.5
	S3	51.4	23.5	10	6	3.1	2	1.4	1.9
	S4	46.4	26.2	8.7	8	5.1	2.5	2.2	0.9
P7a	S1	55.1	12.2	26	2.2	1	1.2	0.5	0.3
	S2	45.8	31.9	4.5	4.1	3.8	8	1.3	0.6
	S3	52.3	28.5	6.7	5.9	2	2.2	1.2	1
	S4	49.2	36.2	4.9	4.2	2.2	1.8	1.2	0.4

The SEM images were also used to determine the influence of glass and micronized quartz aggregates on the morphology and surface topography of the geopolymer composites obtained from alkali-activated fly ash, as shown in Figure 13 control sample, at magnifications of ×200 (Figure 13 P1a), ×1000 (Figure 13 P1b) and ×5000 (Figure 13 P1c), the sample containing glass aggregates (0/4 mm) at magnifications of ×200 (Figure 13 P5a), ×1000 (Figure 13 P5b) and ×5000 (Figure 13 P5c), and the sample containing micronized quartz at magnifications of ×200 (Figure 13 P7a), ×1000 (Figure 13 P7b) and ×5000 (Figure 13 P7c).

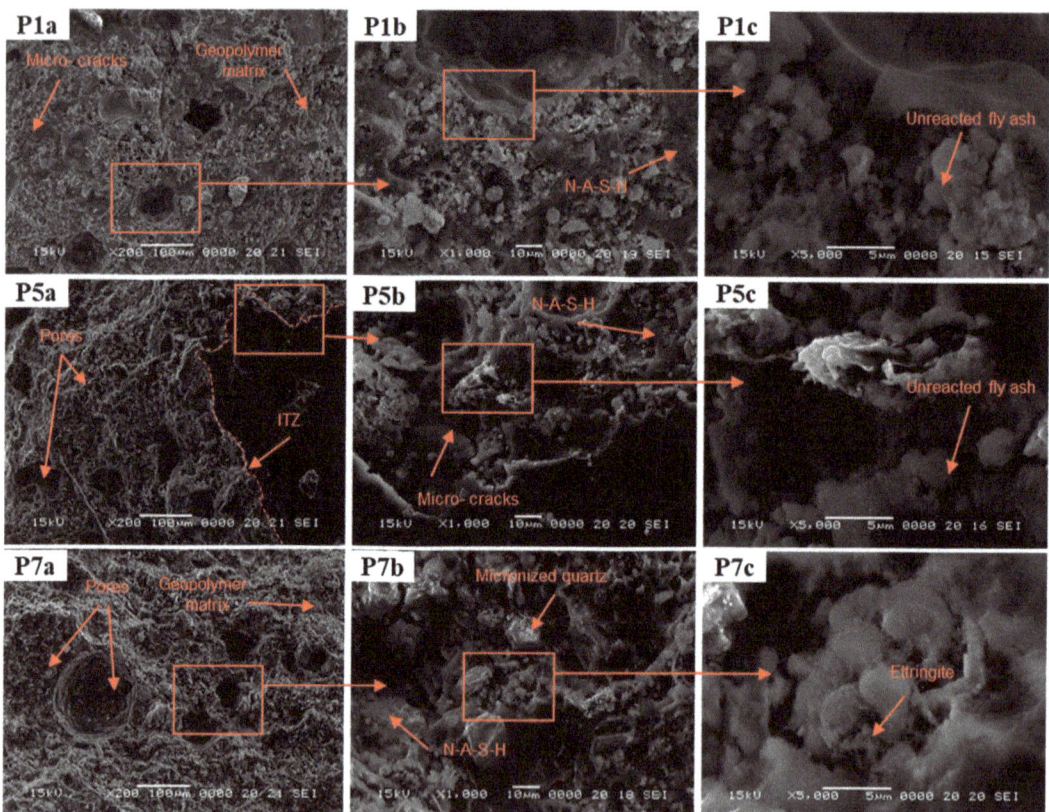

Figure 13. SEM images of the control sample (P1) and the samples containing glass aggregate (P5) and micronized quartz (P7), respectively, at magnifications of ×200 (P1a, P5a and P7a), ×1000 (P1b, P5b and P7b) and at ×5000 (P1c, P5c and P7c); for binder:aggregate ratio = 1:1. The areas recoded at higher magnification are marked in the low magnification image. ITZ—Interfacial Transition Zone.

The main features observed in the SEM images of the three samples analysed include a compact distribution of fully reacted, partially reacted or unreacted fly ash particles, a coherent and homogeneous mass encapsulating and binding the particles (the geopolymer binder), and networks of intergranular and intragranular pores [81]. Thus, the SEM images reveal that sample P7 exhibits a more granular appearance, which is characterized by larger-diameter pores that are fewer in number compared to specimens P1 and P5. Although a well-developed geopolymer matrix is observed for sample P5, it appears to be less homogeneous and uniform than that observed for P1 and P7, which can be attributed to the inclusion of the glass aggregate. Furthermore, the glass aggregate appears to be well embedded and encapsulated within the geopolymer binder, with little to no porosity or voids surrounding the aggregate particles. In addition, all samples show small cracks at low magnification.

Upon closer examination of all three samples (Figure 13 P1c, P5c and P7c), the images indicate the presence of particles with smooth surfaces that correspond to the morphology of barely reactive class F fly ash, which appears to be surrounded by a layer of a fine needle-shaped material presumably composed of unreacted alkali that precipitated during the process and formed these stripe-like structures. The general morphology of the geopolymer composites shows densely packed particles with most of the small cenospheres likely dissolved and the amorphous phase, identified as alkaline aluminium silicate hydrate gel (N-A-S-H), surrounding the embedded aggregates. Figure 13 P7b details the aspect of the sample containing micronized quartz at a magnification of $\times 1000$. Two types of angular and irregular shaped particles can be observed. The bulky, isolated crystals correspond to the micronized quartz that is utilized as an aggregate in this specimen, while the hexagonal prism crystals and partially acicular particles can be attributed to the incomplete dissolution of the fly ash.

4. Conclusions

The choice of aggregates in geopolymer formulations depends on various factors, including local availability, the desired properties of the final product, and the specific application of the geopolymer. Experimentation and optimization of mix designs are often necessary to achieve the desired balance of strength, workability and durability in geopolymer concrete. The use of recycled aggregates or industrial by-products as alternative aggregate aligns with sustainability goals by reducing the environmental impact of construction materials.

This study investigated the effect of different aggregates on alkali-activated fly ash geopolymer composites, produced using Romanian local materials, and shows that density increases with the addition of aggregates, especially with 4–8 mm natural aggregates; this is influenced by both aggregate type and geopolymer binder to aggregate ratio. The introduction of aggregates has variable effects on flexural tensile and compressive strength, with results dependent on the type and particle size distribution of the aggregate and the binder to aggregate ratio. Flexural tensile strength varies, decreasing with recycled glass aggregates and increasing with micronized quartz. A 25% reduction in binder generally improves flexural strength, while a 25% increase in binder can improve compressive strength. This demonstrates the interplay between binder matrix, aggregate structure and their interaction. In addition, water absorption in geopolymer composites decreases significantly as the amount of binder is reduced, highlighting the key role of these factors in shaping composite properties.

These results are supported by the structural and microstructural characterization of the samples by optical, SEM and EDS images. These images show a largely homogeneous distribution of Si, Na and Al in the geopolymer composites, and a relatively uniform distribution of macro- and micro-pores; therefore, this contributes to the physico-mechanical properties of alkali-activated fly ash geopolymer composites.

The particle size distribution of aggregates can affect the workability, strength and density of geopolymer-based materials. A well-graded aggregate mix is generally preferred to optimize packing density and improve mechanical properties. The type of aggregate can also influence the porosity and permeability of the geopolymer composite. Porous aggregates may increase water absorption, which affects the durability of the material. Therefore, selecting aggregates with low porosity is desirable for improved performance.

Aggregates contribute significantly to the mechanical properties of geopolymer composites. The type and quality of the aggregate can impact the compressive and flexural strength, as well as the hardness of the final material.

Considering environmental aspects, recycled aggregates or industrial by-products may be used as alternatives to traditional natural aggregates, aligning with sustainable construction practices.

Therefore, optimizing the combination of aggregates and other components is essential to achieve the desired properties in geopolymer materials. Trial-and-error programmes and testing are often required to fine-tune mix designs for specific applications and performance requirements. Further studies will be conducted with specific mix-design optimization with variations in molar concentration of the NaOH solution, $Na_2SiO_3/NaOH$ solutions ratio and binder to aggregate ratio.

Author Contributions: Conceptualization, A.-V.L. and A.H.; methodology, A.-V.L., A.H., A.C. and F.P.; validation, A.-V.L. and A.H.; formal analysis, A.-V.L., A.H., A.C. and F.P.; investigation, A.-V.L., A.H., A.C. and F.P.; resources, A.-V.L.; data curation, A.H. and A.C.; writing—original draft preparation, A.H. and A.C.; writing—review and editing, A.-V.L., A.H. and A.C.; visualization, A.H. and A.C.; supervision, A.-V.L. and A.H.; project administration, A.-V.L.; funding acquisition, A.-V.L. All authors have read and agreed to the published version of the manuscript.

Funding: This research was funded by the Romanian Government Ministry of Research, Innovation and Digitalization, project No. PN 23 35 05 01 "Innovative sustainable solutions to implement emerging technologies with cross- cutting impact on local industries and the environment, and to facilitate technology transfer through the development of advanced, eco- smart composite materials in the context of sustainable development of the built environment".

Institutional Review Board Statement: Not applicable.

Informed Consent Statement: Not applicable.

Data Availability Statement: Data are contained within the article.

Acknowledgments: The authors would like to extend their gratitude to the Faculty of Materials and Environmental Engineering, within the Technical University of Cluj- Napoca for the contribution of the students: Peter I.I.; Mihalca S.L.; Buta R.A.; Dunca A.C.; Puscas R.S. who carried out practical activities in the laboratories of NIRD URBAN-INCERC Cluj-Napoca Branch.

Conflicts of Interest: The authors declare no conflicts of interest.

References

1. Cembureau. Available online: https://www.cembureau.eu/library/reports/2050-carbon-neutrality-roadmap/ (accessed on 21 February 2023).
2. Aitcin, P.C. Cements of yesterday and today: Concrete of tomorrow. *Cem. Concr. Res.* **2000**, *30*, 1349–1359. [CrossRef]
3. Sandu, A.V. Obtaining and Characterization of New Materials. *Materials* **2021**, *14*, 6606. [CrossRef] [PubMed]
4. Jamaludin, L.; Razak, R.A.; Abdullah, M.M.; Vizureanu, P.; Bras, A.; Imjai, T.; Sandu, A.V.; Abd Rahim, S.Z.; Yong, H.C. The Suitability of Photocatalyst Precursor Materials in Geopolymer Coating Applications: A Review. *Coatings* **2022**, *12*, 1348. [CrossRef]
5. Warid Wazien, A.Z.; Mustafa, M.; Abdullah, A.B.; Razak, R.A.; Rozainy, M.M.A.Z.R.; Faheem, M.; Tahir, M.; Faris, M.A.; Hamzah, H.N. Review on Potential of Geopolymer for Concrete Repair and Rehabilitation. *MATEC Web Conf.* **2016**, *78*, 01065. [CrossRef]
6. Lloyd, N.; Rangan, B. Geopolymer Concrete with Fly Ash. In *Second International Conference on Sustainable Construction Materials and Technologies*; Zachar, J., Claisse, P., Naik, T., Ganjian, G., Eds.; UWM Center for By-Products Utilization: Ancona, Italy, 2010; Volume 3, pp. 1493–1504.
7. Stengel, T.; Reger, J.; Heinz, D. Life cycle assessment of geopolymer concrete—What is the environmental benefit? In Proceedings of the Concrete Solutions 09: Proceedings of the 24th Biennial Conference of the Concrete Institute of Australia, Sydney, Australia, 17–19 September 2009.
8. Weil, M.; Dombrowski, K.; Buchwald, A. Life-cycle analysis of geopolymers. In *Geopolymers: Structures, Processing, Properties and Industrial Applications*; Provis, J.L., Van Deventer, J.S.J., Eds.; Woodhead Publishing Limited: Cambridge, UK, 2009; pp. 194–210.
9. Al Bakri Abdullah, A.M.; Kamarudin, H.; Binhussain, M.; Nizar, K.; Mastura, W.I.W. Mechanism and Chemical Reaction of Fly Ash Geopolymer Cement—A Review. *Asian J. Sci. Res.* **2011**, *1*, 247–253.
10. Amran, M.; Fediuk, R.; Murali, G.; Avudaiappan, S.; Ozbakkaloglu, T.; Vatin, N.; Karelina, M.; Klyuev, S.; Gholampour, A. Fly Ash-Based Eco-Efficient Concretes: A Comprehensive Review of the Short-Term Properties. *Materials* **2021**, *14*, 4264. [CrossRef] [PubMed]
11. Buchwald, A. What are geopolymers? Current State of Research and Technology, The Opportunities They Offer, and Their Significance for the Precast Industry. *Concr. Precast. Plant Technol. (Betonw. Fert.-Tech.)* **2006**, *72*, 42–49.
12. Tian, L.; He, D.; Zhao, J.; Wang, H. Durability of geopolymers and geopolymer concretes: A review. *Rev. Adv. Mater. Sci.* **2021**, *60*, 1–14.

13. Fernandez-Jimenez, A.; Palomo, A. Composition and microstructure of alkali activated fly ash binder: Effect of the activator. *Cem. Concr. Res.* **2005**, *35*, 1984–1992. [CrossRef]
14. Panagiotopoulou, C.; Kontori, E.; Perraki, T.; Kakali, G. Dissolution of aluminosilicate minerals and by-products in alkaline media. *J. Mater. Sci.* **2007**, *42*, 2967–2973. [CrossRef]
15. Adewuyi, Y.G. Recent Advances in Fly-Ash-Based Geopolymers: Potential on the Utilization for Sustainable Environmental Remediation. *ACS Omega* **2021**, *6*, 15532–15542. [CrossRef] [PubMed]
16. Abbas, R.; Khereby, M.A.; Ghorab, H.Y.; Elkhoshkhany, N. Preparation of Geo-polymer Concrete Using Egyptian Kaolin Clay and the Study of Its Environmental Effects and Economic Cost. *Clean Technol. Environ. Policy* **2020**, *22*, 669–687. [CrossRef]
17. Albidah, A.; Alghannam, M.; Abbas, H.; Almusallam, T.; Al-Salloum, Y. Charac-teristics of Metakaolin-Based Geopolymer Concrete for Different Mix Design Parameters. *J. Mater. Res. Technol.* **2021**, *10*, 84–98. [CrossRef]
18. Ionescu, B.A.; Lăzărescu, A.-V.; Hegyi, A. The Possibility of Using Slag for the Production of Geopolymer Materials and Its Influence on Mechanical Performances—A Review. *Proceedings* **2020**, *63*, 30.
19. Xu, H.; van Deventer, J. The effect of alkali metals on the formation of geopolymeric gels from alkali-feldspats. *Colloids Surf. A Physicochem. Eng. Asp.* **2013**, *216*, 27–44. [CrossRef]
20. Weng, L.; Sagoe-Crentsil, K. Dissolution processes, hydrolysis and condensation reactions during geopolymer synthesis: Part I—Low Si/Al ratio systems. *J. Mater. Sci.* **2007**, *42*, 2997–3006. [CrossRef]
21. Andini, S.; Cioffi, R.; Colangelo, F.; Grieco, T.; Montagnaro, F.; Santoro, L. Coal fly ash as raw material for the manufacture of geopolymer-based product. *J. Waste Manag.* **2008**, *28*, 416–423. [CrossRef]
22. Duxon, P.; Fernande-Jimenez, A.; Provis, J.L.; Lukey, G.C.; Palomo, A.; van Deventer, J.S.J. Geopolymer technology: The current state of the art. *J. Mater. Sci.* **2007**, *42*, 2917–2933. [CrossRef]
23. Moreno, N.; Querol, X.; Andrés, J.M.; Stanton, K.; Towler, M.; Jurcovicova, M.; Jones, R. Physico-chemical characteristics of European pulverized coal combustion fly ashes. *Fuel* **2005**, *84*, 1351–1563. [CrossRef]
24. Chen-Tan, N.W.; Van Riessen, A.; Ly, C.V.; Southam, D. Determining the reactivity of a fly ash for production of geopolymer. *J. Am. Ceram. Soc.* **2009**, *92*, 881–887. [CrossRef]
25. Nath, P.; Sarker, P.K.; Rangan, V.B. Early Age Properties of Low-calcium Fly Ash Geopolymer Concrete Suitable for Ambient Curing. *Procedia Eng.* **2015**, *125*, 601–607. [CrossRef]
26. Swanepoel, J.C.; Strydom, C.A. Utilisation of fly ash in a geopolymeric material. *Appl. Geochem.* **2002**, *17*, 114–148. [CrossRef]
27. Goretta, K.C.; Gutierrez-Mora, F.; Singh, D.; Routbort, J.L.; Lukey, G.C.; van Deventer, J.S.J. Erosion of geopolymers made from industrial waste. *J. Mater. Sci.* **2007**, *42*, 3066–3072. [CrossRef]
28. Puertas, F.; Martinez-Ramirez, S.; Alonso, S.; Vazquez, T. Alkali activated fly ash/slag cements: Strength behavior and hydration products. *Cem. Concr. Res.* **2000**, *30*, 1625–1632. [CrossRef]
29. Farhana, Z.; Kamarudin, H.; Rahmat, A.; Al Bakri, A.M. The Relationship between Water Absorption and Porosity for Geopolymer Paste. *Mater. Sci. Forum* **2014**, *803*, 166–172. [CrossRef]
30. Aly, M.; Hashmi, M.S.; Olabi, A.G.; Messeiry, M. Effect of colloidal nano-silica on the mechanical and physical behavior of waste-glass cement mortar. *Mater. Des.* **2012**, *33*, 127135. [CrossRef]
31. Khater, M.H. Effect of nano-silica on microstructure formation of low-cost geopolymer binder. *Nanocomposites* **2016**, *2*, 84–97. [CrossRef]
32. Khater, M.H. Physicomechanical properties of nano-silica effect on geopolymer composites *J. Build. Mater. Struct.* **2016**, *3*, 1–14. [CrossRef]
33. Assaedi, H.; Shaikh, F.U.; Low, I.M. Effect of nanoclay on durability and mechanical properties of flax fabric reinforced geopolymer composites. *J. Asian Ceram. Soc.* **2017**, *5*, 62–70. [CrossRef]
34. Adak, D.; Sarkar, M.; Mandal, S. Effect of nano-silica on strength and durability of fly ash based geopolymer mortar. *Construct. Build. Mater.* **2014**, *70*, 453–459. [CrossRef]
35. Shaikh, F.U.; Supit, S.W.; Sarker, P.K. A study on the effect of nano silica on compressive strength of high volume fly ash mortars and concretes. *Mater. Des.* **2014**, *60*, 433–442. [CrossRef]
36. Li, Z.; Zhang, W.; Wang, R.; Chen, F.; Jia, X.; Cong, P. Effects of Reactive MgO on the Reaction Process of Geopolymer. *Materials* **2019**, *12*, 526. [CrossRef] [PubMed]
37. Hu, M.; Zhu, X.; Long, F. Alkali-activated fly ash-based geopolymers with zeolite or bentonite as additives. *Cem. Concr. Compos.* **2009**, *31*, 762–768. [CrossRef]
38. Bakharev, T. Geopolymeric materials prepared using Class F fly ash elevated temperature curing. *Cem. Concr. Res.* **2005**, *35*, 1224–1232. [CrossRef]
39. Atis, C.D.; Görür, E.B.; Karahan, O.; Bilim, C.; Ilkentapar, S.; Luga, E. Very high strength (120 MPa) Class F fly ash geopolymer mortar activated at different NaOH amount, heat curing temperature and heat curing duration. *Constr Build. Mater.* **2015**, *96*, 673–678. [CrossRef]
40. Al Bakri, M.M.; Mohammed, H.; Kamarudin, H.; Niza, K.; Zarina, Y. Review of Fly Ash-Based Geopolymer Concrete Without Portland Cement. *J. Eng. Technol.* **2011**, *3*, 1–4.
41. Hardjito, D.; Rangan, B.V. *Development and Properties of Low-Calcium Fly Ash-Based Geopolymer Concrete*; Technical Report GC1; Civil Engineering Faculty, Technical University: Perth, Australia, 2005.

42. Al Bakri Mustafa, A.M.; Kamarudin, H.; Binhussain, M.; Niza, I.K. The effect of curing temperature on physical and chemical properties of geopolymers. *Phys. Procedia* **2011**, *22*, 286–291.
43. Al Bakri Mustafa, A.M.; Kamarudin, H.; Bnhussain, M.; Nizar, I.K.; Rafiza, A.R.; Zarina, Y. The processing, characterization, and properties of fly ash based geopolymer concrete. *Rev. Adv. Mater. Sci.* **2012**, *30*, 90–97.
44. Chindaprasirt, P.; Chareerat, T.; Sirivivatnano, V. Workability and strength of coarse high calcium fly ash geopolymer. *Cem. Conc. Comp.* **2007**, *29*, 224–229. [CrossRef]
45. Morsy, M.S.; Alsaye, S.H.; Al-Salloum, Y.; Almusallam, T. Effect of sodium silicate to sodium hydroxide ratios on strength and microstructure of fly ash geopolymer binder. *Arab. J. Sci. Eng.* **2014**, *39*, 4333–4339. [CrossRef]
46. Álvarez-Ayuso, E.; Querol, X.; Plana, F.; Alastuey, A.; Moreno, N.; Izquierdo, M.; Font, O.; Moreno, T.; Diez, S.; Vasquez, K.; et al. Environmental, physical and structural characterisation of geopolymer matrixes synthesised from coal (co-)combustion fly ashes. *J. Hazard. Mater.* **2008**, *154*, 175–183. [CrossRef]
47. Hardjito, D.; Rangan, B.V. *Development and Properties of Low-Calcium Fly Ash-Based Geopolymer Concrete*; Technical Report GC2; Civil Engineering Faculty, Technical University: Perth, Australia, 2005.
48. Provis, J.L.; Yong, C.Z.; Duxson, P.; van Deventer, J. Correlating mechanical and thermal properties of sodium silicate-fly ash geopolymers. *Colloids Surf. A Physicochem. Eng.* **2009**, *336*, 57–63. [CrossRef]
49. Sumajouw, D.; Hardjito, D.; Wallah, S.; Rangan, B. Fly ash-based geopolymer concrete: Study of slender reinforced columns. *J. Mater. Sci.* **2007**, *42*, 3124–3130. [CrossRef]
50. Vora, P.; Dave, U. Parametric Studies on Compressive Strength of Geopolymer Concrete. *Procedia Eng.* **2013**, *51*, 210–219. [CrossRef]
51. Sindhunata; van Deventer, J.S.J.; Lukey, G.C.; Xu, H. Effect of Curing Temperature and Silicate Concentration on Fly-Ash-Based Geopolymerization. *Ind. Eng. Chem. Res.* **2006**, *45*, 3559–3569. [CrossRef]
52. Raijiwala, D.B.; Patil, H.S. Geopolymer concrete: A green concrete. In Proceedings of the 2nd International Conference on Chemical, Biological and Environmental Engineering (ICBEE 2010), Cairo, Egypt, 2–4 November 2010.
53. Alonso, S.; Palomo, A. Alkaline activation of metakaolin and calcium hydroxide mixtures: Influence of temperature, activator concentration and solids ratio. *Mater. Lett.* **2001**, *47*, 55–62. [CrossRef]
54. Memon, F.; Nuruddin, M.F.; Khan, S.H.; Shafiq, N.R. Effect of sodium hydroxide concentration on fresh properties and compressive strength of self-compacting geopolymer concrete. *J. Eng. Sci. Technol.* **2013**, *8*, 44–56.
55. Barbosa, V.; Mackenzie, K.; Thaumaturgo, C. Synthesis and characterisation of sodium polysialate inorganic polymer based on alumina and silica. In Proceedings of the Geopolymer'99 International Conference, Saint-Quentin, Geopolymer Institute, Saint-Quentin, France, 30 June–2 July 1999.
56. Luhar, S.; Dave, U. Investigations on mechanical properties of fly ash and slag based geopolymer concrete. *Ind. Concr. J.* **2016**, 34–41.
57. Ma, Y.; Hu, J.; Ye, G. The effect of activating solution on the mechanical strength, reaction rate, mineralogy, and microstructure of alkali-activated fly ash. *J. Mater. Sci.* **2012**, *47*, 4568–4578. [CrossRef]
58. Xie, J.; Yin, J.; Chen, J.; Xu, J. Study on the geopolymer based on fly ash and slag. *Energy Environ.* **2009**, *3*, 578–581.
59. van Jaarsveld, J.G.S.; van Deventer, J.S.J.; Lukey, G.C. The effect of composition and temperature on the properties of fly ash-and kaolinite-based geopolymers. *J. Chem. Eng.* **2002**, *89*, 63–73. [CrossRef]
60. Rovnaník, P. Effect of curing temperature on the development of hard structure of metakaolin-based geopolymer. *Constr. Build. Mater.* **2010**, *24*, 1176–1183. [CrossRef]
61. Chindaprasirt, P.; Chareerat, T.; Hatanaka, S.; Cao, T. High-strength geopolymer using fine high-calcium fly ash. *Mater. Civ. Eng.* **2010**, *23*, 264–270. [CrossRef]
62. Fernández-Jiménez, A.; Garcia-Lodeiro, I.; Palomo, A. Durability of alkali-activated fly ash cementitious materials. *J. Mater. Sci.* **2007**, *42*, 3055–3065. [CrossRef]
63. Phoo-ngernkham, T.; Sinsiri, T. Workability and compressive strength of geopolymer mortar from fly ash containing diatomite. *Eng. J.* **2011**, *38*, 11–26.
64. Kong, D.; Sanjayan, J. Effect of elevated temperatures on geopolymer paste, mortar and concrete. *Cem. Concr. Res.* **2010**, *40*, 334–339. [CrossRef]
65. Guo, X.; Shi, H. Self-solidification/stabilization of heavy metal wastes of class C fly ash-based geopolymers. *J. Mater. Civ. Eng.* **2012**, *25*, 491–496. [CrossRef]
66. Rashad, A.; Zeedan, S. The effect of activator concentration on the residual strength of alkali-activated fly ash pastes subjected to thermal load. *Constr. Build. Mater.* **2011**, *25*, 3098–3107. [CrossRef]
67. Sukmak, P.; Horpibulsuk, S.; Shen, S. Strength development in clay-fly ash geopolymer. *Contr. Build. Mater.* **2013**, *40*, 566–574. [CrossRef]
68. Taebuanhuad, S.; Rattanasak, U.; Jenjirapanya, S. Strength behavior of fly ash geopolymer with microwave pre-radiation curing. *J. Ind. Technol.* **2012**, *8*, 1–8.
69. Lăzărescu, A.V.; Szilagyi, H.; Baeră, C.; Ioani, A. Parameters Affecting the Mechanical Properties of Fly Ash-Based Geopolymer Binders–Experimental Results. *IOP Conf. Ser. Mater. Sci. Eng.* **2018**, *374*, 012035. [CrossRef]
70. Hardjito, D.; Rangan, B.V. *Development and Properties of Low-Calcium Fly Ash-Based Geopolymer Concrete*; Technical Report GC3; Civil Engineering Faculty, Technical University: Perth, Australia, 2006.

71. Omar, O.M.; Heniegal, A.M.; Abd Elhameed, G.D.; Mohamadien, H.A. Effect of Local Steel Slag as a Coarse Aggregate on Properties of Fly Ash Based-Geopolymer Concrete. *Int. J. Civ. Environ.* **2015**, *3*, 1452–1460.
72. Perera, D.S.; Uchida, O.; Vance, E.R.; Finnie, K.S. Influence of curing schedule on the integrity of geopolymers. *J. Mater. Sci.* **2007**, *42*, 3099–3106. [CrossRef]
73. Xu, H.; Van Deventer, J.S.J. Microstructural characterisation of geopolymers synthesised from kaolinite/stilbite mixtures using XRD, MAS-NMR, SEM/EDX, TEM/EDX, and HREM. *Cem. Concr. Res.* **2002**, *32*, 1705–1716. [CrossRef]
74. Kutchko, B.G.; Kim, A.G. Fly ash characterization by SEM–EDS. *Fuel* **2006**, *85*, 2537–2544. [CrossRef]
75. Kumari, N.; Mohan, C. Basics of clay minerals and their characteristic properties. *Clay Clay Miner.* **2021**, *24*, 1–29.
76. Velde, B. Electron microprobe analysis of clay minerals. *Clay Miner.* **1984**, *19*, 243–247. [CrossRef]
77. Gomes, S.; Francois, M. Characterisation of mullite in silicoaluminous fly ash by XRD, TEM, and 29Si MAS NMR. *Cem. Concr. Res.* **2020**, *30*, 175–181. [CrossRef]
78. Derbe, T.; Temesgen, S.; Bitew, M. A Short Review on Synthesis, Characterization, and Applications of Zeolites. *Hindawi Adv. Mater. Sci. Eng.* **2021**, *2021*, 6637898. [CrossRef]
79. Chen, X.; Wang, G.; Dong, Q.; Zhao, X.; Wang, Y. Microscopic characterizations of pervious concrete using recycled Steel Slag Aggregate. *J. Clean. Prod.* **2020**, *254*, 120149. [CrossRef]
80. Lyu, K.; She, W.; Miao, C.; Chang, H.; Gu, Y. Quantitative characterization of pore morphology in hardened cement paste via SEM-BSE image analysis. *Constr. Build. Mater.* **2019**, *202*, 589–602. [CrossRef]
81. Galiano, L.Y.; Pereira, F.C.; Izquierdo, M. Contributions to the study of porosity in fly ash-based geopolymers. Relationship between degree of reaction, porosity and compressive strength. *Mater. Construcción.* **2016**, *66*, 324.
82. Hudson, B.P. Modification to the fine aggregate angularity test investigation into the way we measure fine aggregate angularity. In Proceedings of the 7th Annual Symposium, International Center for Aggregate Research (ICAR) Symposium, Austin, TX, USA, 19–21 April 1999.
83. Sreenivasulu, C.; Jawahar, J.G.; Reddy, M.V.S.; Kumar, D.P. Effect of fine aggregate blending on short-term mechanical properties of geopolymer concrete. *Asian J. Civ. Eng. (BHRC)* **2016**, *17*, 537–550.
84. Nuaklong, P.; Sata, V.; Chindaprasirt, P. Influence of recycled aggregate on fly ash geopolymer concrete properties. *J. Clean. Prod.* **2016**, *112*, 2300–2307. [CrossRef]
85. Jahromi, S.G. Estimation of resistance to moisture destruction in asphalt mixtures. *Constr. Build. Mater.* **2009**, *23*, 2324–2331. [CrossRef]
86. Bagampadde, U.; Isacsson, U.; Kiggundu, B.M. Impact of bitumen and aggregate composition on stripping in bituminous mixtures. *Mater. Struct.* **2006**, *39*, 303–315. [CrossRef]
87. Lăzărescu, A.-V.; Ionescu, B.A.; Hegyi, A.; Florean, C. Analysis Regarding the Mechanical Properties of Alkali-Activated Fly Ash-Based Geopolymer Concrete Containing Spent Garnet as Replacement for Sand Aggregates. *EJMSE* **2023**, *8*, 11–21. [CrossRef]
88. SR EN 196-1:2016; Methods of Testing Cement—Part 1: Determination of Strength. ASRO: Bucharest, Romania, 2016. (In Romanian)
89. SR EN 1097-3:2002; Tests for Mechanical and Physical Properties of Aggregates Determination of Loose Bulk Density and Voids. ASRO: Bucharest, Romania, 2003. (In Romanian)
90. Lloyd, R.R.; Provis, J.L.; Smeaton, K.J.; Van Deventer, J.S.J. Spatial distribution of pores in fly ash-based inorganic polymer gels visualised by Wood's metal intrusion. *Micropor. Mesopor. Mater.* **2009**, *126*, 32–39. [CrossRef]
91. Zhang, Z.; Xiao, Y.; Huajun, Z. Potential application of geopolymers as protection coatings for marine concrete. II. Microstructure and anticorrosion mechanism. *Appl. Clay. Sci.* **2010**, *49*, 7–12. [CrossRef]

Disclaimer/Publisher's Note: The statements, opinions and data contained in all publications are solely those of the individual author(s) and contributor(s) and not of MDPI and/or the editor(s). MDPI and/or the editor(s) disclaim responsibility for any injury to people or property resulting from any ideas, methods, instructions or products referred to in the content.

Article

DyMnO₃: Synthesis, Characterization and Evaluation of Its Photocatalytic Activity in the Visible Spectrum

Miguel Ángel López-Álvarez [1,*], Pedro Ortega-Gudiño [2,*], Jorge Manuel Silva-Jara [3], Jazmín Guadalupe Silva-Galindo [3], Arturo Barrera-Rodríguez [4], José Eduardo Casillas-García [5], Israel Ceja-Andrade [6], Jesús Alonso Guerrero-de León [1] and Carlos Alberto López-de Alba [1]

[1] Departamento de Ingeniería Mecánica, Centro Universitario de Ciencias Exactas e Ingenierías, Universidad de Guadalajara, Blvd. Marcelino García Barragán 1421, Guadalajara 44430, Jalisco, Mexico; alonso.guerrero@academicos.udg.mx (J.A.G.-d.L.); carlos.ldealba@academicos.udg.mx (C.A.L.-d.A.)

[2] Departamento de Ingeniería Química, Centro Universitario de Ciencias Exactas e Ingenierías, Universidad de Guadalajara, Blvd. Marcelino García Barragán 1421, Guadalajara 44430, Jalisco, Mexico

[3] Departamento de Farmacobiología, Centro Universitario de Ciencias Exactas e Ingenierías, Universidad de Guadalajara, Blvd. Marcelino García Barragán 1421, Guadalajara 44430, Jalisco, Mexico; jorge.silva@academicos.udg.mx (J.M.S.-J.); jazmin.silva@alumnos.udg.mx (J.G.S.-G.)

[4] Centro de Investigación en Nanocatálisis Ambiental y Energías Limpias CUCIENEGA, Universidad de Guadalajara, Av. Universidad 1115, Ocotlán 47820, Jalisco, Mexico; arturo.barrera@academicos.udg.mx

[5] Departamento de Ciencias Tecnológicas, Centro Universitario de la Ciénega (CUCIENEGA), Universidad de Guadalajara, Av. Universidad 1115, Ocotlán 47820, Jalisco, Mexico; jose.casillas2569@academicos.udg.mx

[6] Departamento de Física, Centro Universitario de Ciencias Exactas e Ingenierías, Universidad de Guadalajara, Blvd. Marcelino García Barragán 1421, Guadalajara 44430, Jalisco, Mexico; israel.ceja@academicos.udg.mx

* Correspondence: miguel.lalvarez@academicos.udg.mx (M.Á.L.-Á.); pedro.ogudino@academicos.udg.mx (P.O.-G.)

Abstract: DyMnO₃ is a p-type semiconductor oxide with two crystal systems, orthorhombic and hexagonal. This material highlights its ferroelectric and ferromagnetic properties, which have been the subject of numerous studies. Nevertheless, its photocatalytic activity has been less explored. In this work, the photocatalytic activity of DyMnO₃ is evaluated through the photodegradation of MG dye. For the synthesis of this oxide, a novel and effective method was used: polymer-decomposition. The synthesized powders contain an orthorhombic phase, with a range of absorbances from 300 to 500 nm and a band gap energy of 2.4 eV. It is also highlighted that, when using this synthesis method, some of the main diffraction lines related to the orthorhombic phase appear at 100 °C. Regarding its photocatalytic activity, it was evaluated under visible light (λ = 405 nm), reaching a photodegradation of approximately 88% in a period of 30 min. Photocurrent tests reveal a charge carrier separation (e^-, h^+) at a 405 nm wavelength. The main reactive oxygen species (ROS) involved in the photodegradation process were radicals, OH•, and photo-holes (h^+). These results stand out because it is the first time that the photodegradation capability of this oxide in the visible spectrum has been evaluated.

Keywords: DyMnO₃; photocatalysis; malachite green dye; visible-light photocatalyst

1. Introduction

Today's landscape is contaminated with heavy metals and inorganic and organic compounds in seas, rivers, and lakes. Organic contaminants are the dyes used substantially in the textile, paper, and pharmaceutical industries, and in our environment, they can be found in products from plastics to toys [1]. Among the dyes that stand out for their toxic effect are azo dyes. They are characterized by the presence of one or more azo groups (-N=N-) linked to aromatic rings such as benzene and naphthalene. They are normally resistant to biodegradation. These dyes are widely used in the textile, leather, food, cosmetics, and paper products industries due to their fastness and variety of colors compared to natural ones. The annual global production of azo dyes is estimated at around one million tons, and more than 2000 azo dyes are currently used [2]. Because many of them are soluble in water and are related to various carcinogenic processes, different physicochemical techniques have been proposed for their degradation, such as adsorption, coagulation, electrodeionization, precipitation, photolysis, or membrane filtration [3]. Sometimes, these techniques are inefficient, expensive, and consume a large amount of energy. Therefore, photocatalysis has been proposed as an alternative technique for the degradation of azo dyes [4].

Of the group of azo dyes, one of the most representative is malachite green (MG) dye. MG is a cationic dye used as a fungicide and antiseptic in aquaculture and the industries of textile and paper [5]. However, MG dye represents a risk to human health due to its negative effects on the immune and reproductive systems, as well as its potential genotoxicity and carcinogenic properties [6]. Thus, diverse oxides have been used as photocatalysts for the degradation of MG dye, including: Fe_2O_3, TiO_2, ZnO, CeO_2, WO_3, V_2O_5, Cu_2O, CuO, NiO, Mn_3O_4, Ag_2O, and $CuMn_2O_4$ [4]. Furthermore, the photodegradation of MG dye with an yttrium perovskite photocatalyzed in the visible spectrum was recently observed [7]. Nevertheless, there are other groups of oxides that have been used in the photodegradation of azo dyes at wavelengths in the visible spectrum. These are oxides with a perovskite-type structure (ABO_3) (where A is a rare earth or alkaline metal, B is a transition metal, and O is oxygen), such as: $XTaO_3$ (X = Na, K, Ag, Li), $XNbO_3$ (X = Na, K, Ag, Cu), $XTiO_3$ (X = Sr, Ba, Ca, Ni), $XFeO_3$ (X = Y, La, Bi), and $AgVaO_3$ [8]. The reasons why oxides with a perovskite-type structure are considered potential candidates for the photodegradation of organic dyes are the following: they are formed from a wide variety of compositional and constituent elements; their valency, stoichiometry, and vacancy can be varied widely; and there is much information on their physical and solid-state chemical properties [9].

On the other hand, the oxides with a perovskite-type structure, whose photocatalytic activity has been the subject of recent research, are the rare earth manganites. Of this group of perovskites, $DyMnO_3$ stands out, mainly, for its ferroelectric, ferromagnetic [10–12], and magnetocaloric properties [13,14], which have contributed to its technological applications [15,16]. However, its optical applications (for example, its use as a photocatalyst) have been little explored.

Considering the above, the synthesis of $DyMnO_3$ and its use as a photocatalyst in the degradation of MG dye are proposed in this work. Furthermore, its photocatalytic activity will be evaluated at a wavelength in the visible spectrum (λ = 405 nm). Regarding its characterization, this will be carried out by XRD, TEM, UV-Vis, and XPS spectroscopics. It should be noted that it is the first time that the photocatalytic activity of $DyMnO_3$ has been evaluated at a wavelength of the visible spectrum, as well as the first time that the polymer-decomposition method has been used for the synthesis of this perovskite.

2. Materials and Methods

2.1. Synthesis

2.1.1. Precursor Optimization

In the synthesis, the concentrations of dysprosium nitrate hydrate (Sigma Aldrich, Saint Louis, MO, USA, 99.9%) and manganese nitrate tetrahydrate (Sigma Aldrich, 97%) were 1.388 and 1 g, respectively (The concentration would be 4×10^{-3} M for each reagent). Both were mixed simultaneously and diluted with 20 mL of bi-distilled water. Subsequently, 0.200 g of polyvinyl alcohol (PVA) (Sigma Aldrich, 99.8%, Mw = 89,000–98,000), diluted in 20 mL of bi-distilled water, was added to the solution containing the metal cations. The final mixture was stirred moderately for 30 min and placed in a drying oven for 1 h. This caused almost 70% of the water volume to evaporate, and a viscous solution was obtained. Finally, this solution was placed in a conventional microwave oven for 2 min. During that period, an exothermic reaction occurred, releasing an abundant amount of gases. The product of this reaction was a porous-looking black powder.

2.1.2. Temperature Optimization

The precursor powder obtained from the PVA decomposition reaction was calcined at 800, 1000, and 1200 °C. These temperatures were selected because, in previous research, it was observed that, from 800 °C onwards, the orthorhombic phase of this material begins to be obtained. All calcinations were for 3 h.

2.2. Characterization of $DyMnO_3$ Powders

The crystallinity evaluation of $DyMnO_3$ powders was carried out using an Empyrean diffractometer (PANalytical, Westborough, MA, USA) with CuK1 radiation. The morphology analysis was performed using transmission electron microscopy (TEM) with a Jeol JEM1010 microscope (Tokyo, Japan). The absorbance spectrum of the dysprosium manganate powders was obtained using a UV-Vis spectrophotometer (Cary 100 Agilent Technologies, Santa Clara, CA, USA).

2.3. Photocatalytic Activity

To evaluate the photodegradation of the malachite green dye (MG) using $DyMnO_3$ as a photocatalyst, two solutions with 40 mL of this dye at a concentration of 1.5×10^{-5} M were prepared. To one solution, 20 mg of $DyMnO_3$ was added, while 40 mg was added to the other. Both solutions were exposed to visible light (λ = 405 nm) for 180 min. An LED (Light Emitting Diode) with an optical irradiance (E_e) of 100 mW/cm^2 was used as the light source. Then, aliquots were extracted at different intervals and analyzed using a NanoDrop 2000 spectrophotometer (Thermo Scientific, Waltham, MA, USA). The degradation percentage was estimated using Equation (1).

$$Degradation = \frac{A_0 - A}{A_0} = \frac{C_0 - C}{C_0} \times 100 \qquad (1)$$

where A_0 and A correspond to the solutions' absorbance values before and after exposure to visible light, respectively. Similarly, C_0 and C represent the concentrations of the solutions before and after irradiation, respectively.

2.4. Transient Photocurrent Tests

To carry out this test, a pellet with a diameter of 1.2 cm and a thickness of 2 mm of $DyMnO_3$ was fabricated. The pellet was sintered at 600 °C for 12 h. Subsequently, it was placed on two NiAg foils with a thickness of approximately 1 mm. These foils served as electrodes, applying a direct voltage (DC) of 5 V to the pellet. Then, the pellet was exposed to visible light (λ = 405 nm) using an LED with the same optical irradiance as in the photocatalysis experiments. Finally, the photocurrent measurements were obtained using a DMM 6500 multimeter (Keithley, Portland, OR, USA) under the conditions described above.

The emission spectrum of the LED was obtained using a CCS200 Spectrometer (Thorlabs, NJ, USA).

3. Results and Discussion

3.1. X-ray Diffraction

Figure 1 shows the diffraction pattern of the calcined precursor powder at 800 °C. As can be observed, each diffraction line was identified and associated with the orthorhombic phase of $DyMnO_3$ (JCPDF #25-0330). It is worth noting that no diffraction lines associated with the hexagonal phase or with the oxides Dy_2O_3, MnO_2, or Mn_2O_3 were observed. Moreover, when this synthesis method is compared with some of those used for obtaining the orthorhombic phase, it was observed that the polymer-decomposition method achieves this phase at a lower temperature and without the use of acidic or organometallic reagents [17,18].

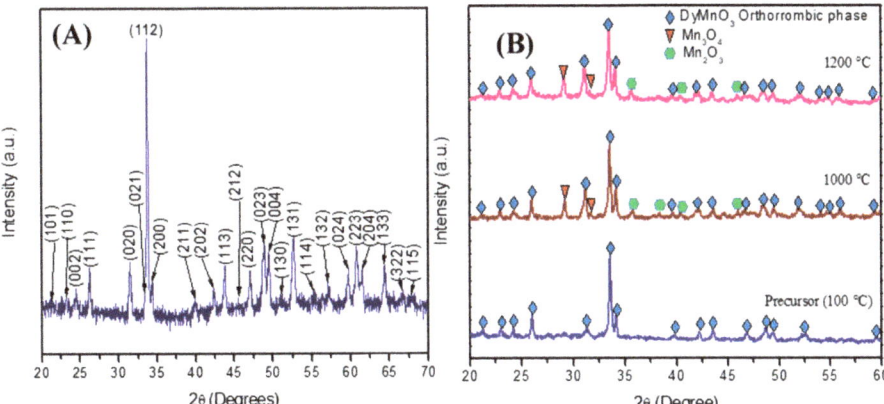

Figure 1. (**A**) Diffractograms of precursor powder calcined at 800 °C and (**B**) its calcinations a 100, 1000, and 1200 °C.

In order to evaluate the crystalline evolution and propose a formation mechanism for DyMnO$_3$, diffraction patterns of the precursor powder and its calcinations at 1000 and 1200 °C were analyzed by XRD. The diffractogram of the precursor powder calcined at 100 °C shows some of the main diffraction lines related to the orthorhombic phase. According to this result, it is observed that the synthesis method used in this work promotes obtaining the orthorhombic phase, even when using a relatively low calcination temperature. On the other hand, the XRD patterns of the calcinations of the precursor powder at 1000 °C and 1200 °C show diffraction lines with greater intensities. Furthermore, more diffraction peaks associated with the orthorhombic phase appear after the calcinations. However, these calcination temperatures also promote the formation of impurities associated with manganese, particularly the presence of the oxides Mn$_2$O$_3$ (JCPDF #41-1442) and Mn$_3$O$_4$ (JCPDF #24-0734).

Based on the XRD patterns of the calcinations carried out on the precursor (Figure 1B), the following reaction to obtain DyMnO$_3$ is proposed:

$$Dy(NO_3)_3 \times xH_2O + Mn(NO_3)_2 \times 4H_2O + (CH_2 - CHOH)_n + H_2O \xrightarrow{800°C} DyMnO_3 + \alpha(NO_x) + \beta(H_2O) + \varepsilon(CO_?) \quad (2)$$

According to Equation (2), the mechanism to obtain the orthorhombic phase can be described as follows: The metal cations are bound to the hydroxyl groups of the polymer. This causes the cations to stabilize in the polymer structure via interactions with the hydroxyl groups [19]. Then, a precipitation process does not occur. When the solution obtained is placed inside a conventional microwave oven, a polymer-decomposition reaction at approximately 100 °C occurs. During the reaction, the manganese cation is oxidized and the chemical structure of PVA is decomposed. Thus, an abundant amount of gases is released, as shown by Equation (2). At the end of the reaction, the metal cations adopt the architecture of a polymer network.

3.2. Transmission Electron Microscopy

Figure 2A,B show the TEM micrographs of the DyMnO$_3$ powders. In both images, flake-shaped nanostructures with porosity can be observed. Notably, the presence of porosity may be related to the release of NO$_x$ gases, which are produced during the thermal decomposition of the manganese and dysprosium nitrates used in the synthesis process. It should be mentioned that this type of morphology has already been observed in other oxides synthesized using the polymer-decomposition method [19].

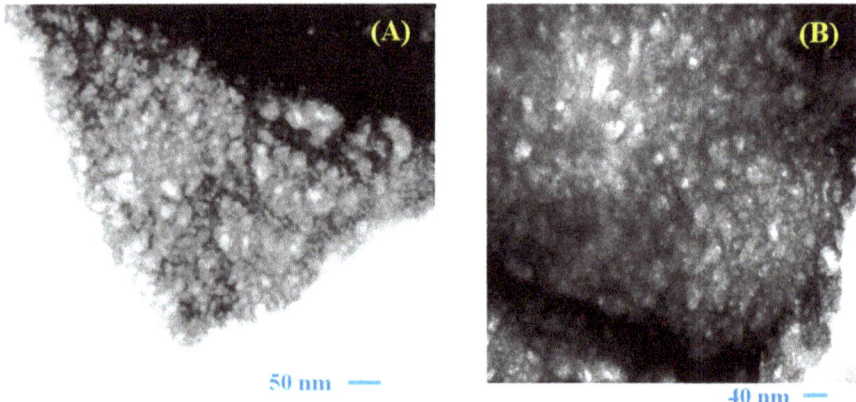

Figure 2. TEM micrographs of DyMnO$_3$ powders at (**A**) 80,000 and (**B**) 100,000 magnifications.

Based on other research, it is known that micro- or nanostructures containing porosity can contribute to the photodegradation of dyes because they promote their absorption and, also, interactions between reactive oxygen species (ROS) and the dye molecule occur more efficiently [20,21].

3.3. Porosity Analysis

Figure 3 shows the nitrogen adsorption—desorption isotherm of DyMnO$_3$ obtained at 800 °C. This isotherm can be considered II-type, which is related to macroporous material [22]. The specific surface area (SBET) was 7.3 m^2 g^{-1}, which is similar to that corresponding to lanthanide oxides. Its porosity is corroborated by its total specific pore volume (0.0113 cm^3 g^{-1}) and its micropore volume (0.000128 cm^3 g^{-1}) calculated by the BJH and DR methods [22], respectively. These results are summarized in Table 1.

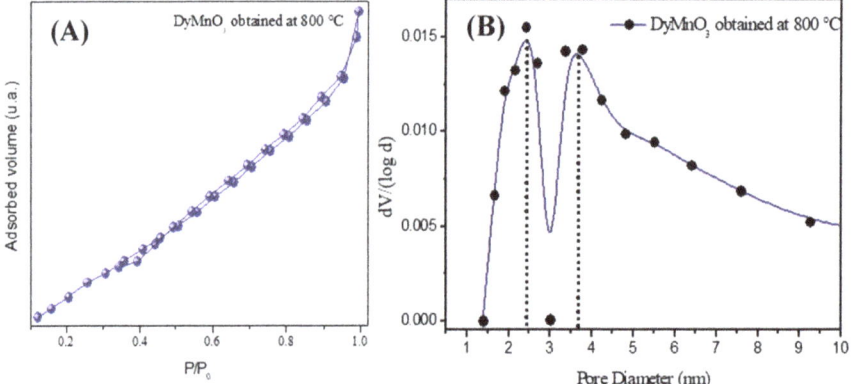

Figure 3. (**A**) Nitrogen adsorption—desorption isotherm and (**B**) pore size distribution of DyMnO$_3$. The blue circles correspond to the data acquired from the experiment. The black circles dotted line represents two modal peaks corresponding to two more frequent pore sizes.

Table 1. Textural properties of DyMnO$_3$ prepared by the polymer-decomposition method.

Material	SBET (m^2 g^{-1})	BJH Total Pore Volume (cm^3 g^{-1})	DR Micropore Volume (cm^3 g^{-1})	BJH Pore Diameter (nm)
DyMnO$_3$	7.3	0.0113	0.000128	2.4

The pore diameter of DyMnO$_3$ calculated with the BJH method was 2.2 nm, and it is at the limit of the mesopore range according to the IUPAC classification [23]. Its pore size distribution is bimodal since it exhibits two modal peaks corresponding to two more frequent pore sizes at 2.4 and 3.7 nm.

3.4. UV-Vis Spectroscopy

Figure 4 shows the graph of $(\alpha h \nu)^n$ vs. $h\nu$ (Tauc plot), where α represents the absorption coefficient, h is the Planck constant, and ν is the frequency of the light. The exponent, n, value is associated with the type of electronic transition the material exhibits, with n = 2 for direct and n = 1/2 for indirect transitions. For DyMnO$_3$, the value is n = $\frac{1}{2}$ [24]. In the graph, the intersection of the dashed line with the x-axis indicates the band gap energy (E_g). In this case, the approximate value of E_g is 2.4 eV. Based on other research, it is known that the valence band (VB) of rare-earth manganates (such as DyMnO$_3$) consists of O 2p and lanthanide 4f orbitals, while the conduction band (CB) consists of Mn 3d orbitals [24].

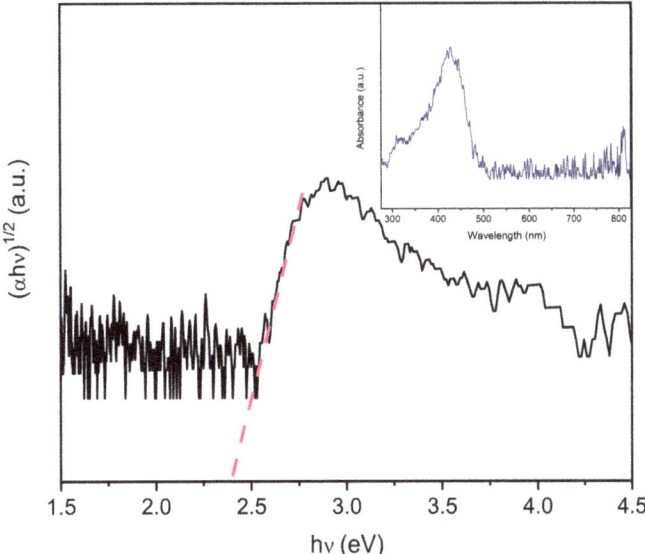

Figure 4. Tauc plot of DyMnO$_3$. Within the graph is the absorption spectrum of this oxide. The pink line represent the intersection with the x-axis indicates the band gap energy. The blue lines correspond to the absorbance spectrum of DyMnO$_3$, and the black lines are the graphic results of the Tauc-plot.

Within the Tauc plot, the absorbance spectrum of the DyMnO$_3$ powders is also shown. As can be observed, this oxide exhibits an absorbance range that extends from 300 nm to 500 nm, making it a potential candidate for evaluating its photocatalytic activity at wavelengths in the visible spectrum.

By applying Equation (3) to the band gap energy obtained in the Tauc plot, it is possible to estimate the minimum wavelength required to promote electrons from the VB to migrate to the CB. In this case, the value obtained is 516 nm. Based on this result, DyMnO$_3$ can be considered a potential candidate to be used as a photocatalyst at wavelengths in the visible spectrum.

$$\lambda \text{ (nm)} = \frac{1240}{E_g \text{ (eV)}} \quad (3)$$

3.5. Transient Photocurrent Measurements

Figure 5A shows the photocurrent vs. time graph obtained from transient photocurrent measurements. During the light exposure period, the electric current in the material increases, while in the absence of light, it decreases. The increase in photocurrent was approximately 2.44×10^{-6} A. This reveals that the incident photons (with energy $h\nu$ = 3.1 eV) promote the creation of electron–hole pairs (e^-, h^+) on the surface of this material, preventing their recombination. Subsequently, the applied voltage contributes to the mobility of the charge carriers, generating an electric current.

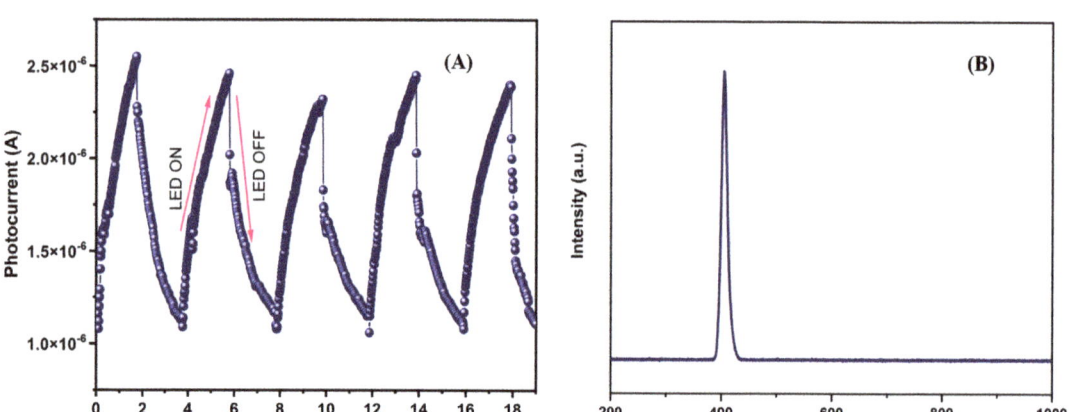

Figure 5. (**A**) Graphical results of transient photocurrent measurements using a wavelength in the visible spectrum (λ = 405 nm) at an optical irradiance of 100 mW/cm². (**B**) Emission spectrum of LED.

Additionally, Figure 5B shows the emission spectrum of the LED used in the transient photocurrent tests and photocatalysis experiments. As observed, the light emitted by the LED has a wavelength of 405 nm. Furthermore, it can be considered monochromatic because no contributions from other wavelengths that could influence the photocatalytic activity of this oxide are detected.

3.6. Degradation of Malachite Green (MG) Dye

Figure 6A,B show the absorbance spectra of the aliquots extracted from MG solutions containing 20 mg and 40 mg of $DyMnO_3$, respectively. A significant decrease in the absorbance peak located at approximately 617 nm can be observed. Specifically, it is observed that 40 mg of this oxide can promote the photodegradation of the dye by 30.8 and 88.6% in 5 and 30 min, respectively (Figure 7). Meanwhile, with 20 mg, the percentages achieved in the same time intervals are 3.1 and 71.1%, respectively. This increase in photocatalytic activity can be attributed to increasing the amount of $DyMnO_3$.

During the photodegradation process, the absorbance spectra of the MG dye show a slight blue shift of the major absorbance peak, from 617 to 598 nm. This shift has been observed in other research and was related to a process of N-demethylation that occurred in the degradation of the dye [25]. Additionally, a decrease in the intensity of the absorbance peaks located at 425 and 315 nm is also observed, indicating that the chromophore groups of the MG dye have been destroyed. These results reveal that reactive oxygen species (ROS) attack the C-N bonds of the MG dye, causing its degradation.

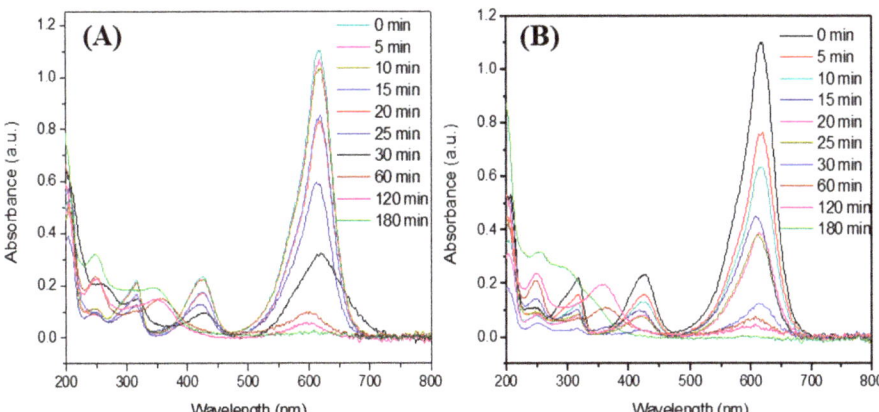

Figure 6. UV-Vis absorbance spectra obtained during the degradation of the MG dye using (**A**) 20 and (**B**) 40 mg of DyMnO$_3$ as a photocatalyst.

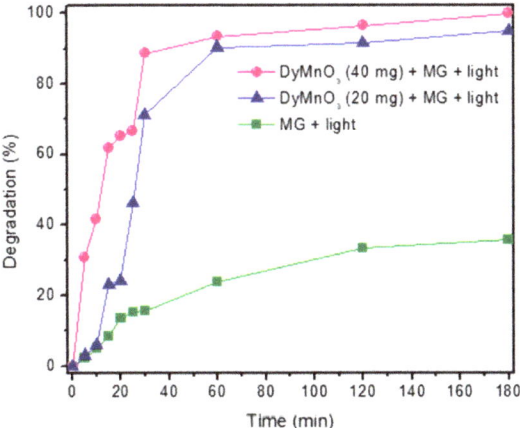

Figure 7. Percentages of photocatalytic degradation of MG dye using 20 and 40 mg of DyMnO$_3$. The peak of highest absorbance located at 617 (associated with N-methyl groups) was selected as a reference to estimate the percentage of degradation.

To evaluate the contribution of the light (λ = 405 nm, E_e = 100 mW/cm^2) to the photodegradation of this dye, 40 mL of the MG dye was exposed to this wavelength of light without adding DyMnO$_3$. The results revealed that after 180 min, a photodegradation of 35.6% occurred. However, this value can be considered negligible compared to those obtained using 20 and 40 mg of photocatalyst.

Due to the conditions in which these photodegradation experiments were carried out (E_e = 100 mW/cm^2, λ = 405 nm), it is difficult to compare the photocatalytic efficiency of DyMnO$_3$ with other oxides used as photocatalysts. However, Table 2 shows some photocatalyst oxides used in the degradation of MG dye at an optical irradiance of 100 mW/cm^2.

Table 2. Some photocatalyst oxides used in the degradation of MG dye at $E_e = 100$ mW/cm^2.

Photocatalyst	Wavelength (nm)	Amount of Photocatalyst (mg)	Degradation Obtained after 60 min (%)
YMnO$_3$	405	20	20 [7]
LaCoO$_3$	365	10	50 [26]
DyCoO$_3$	365	20	90 [27]
Ga$_2$O$_3$	365	30	10 [28]

The first three have a perovskite-type structure. However, in all cases, DyMnO$_3$ shows greater efficiency in degrading this dye.

Additionally, the contribution of DyMnO$_3$ absorption to the degradation process of the MG dye was evaluated. In this test, 40 mg of the photocatalyst powder was added to a solution containing 40 mL of the MG dye. Figure 8 shows the graphic results. As can be observed, the greatest degradation occurs in the 60 min period, at approximately 32%. After 240 min, the dye degrades by approximately 40%. These results confirm that the degradation of the dye can be attributed mainly to the formation of ROS rather than to the absorption process that occurred on the surface of the photocatalyst.

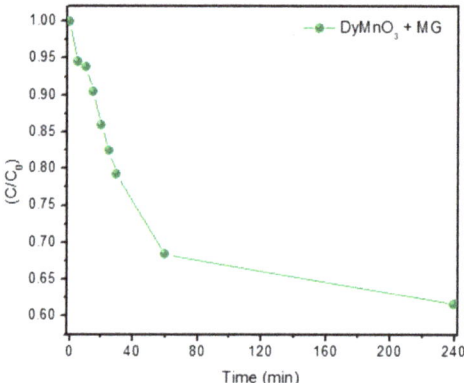

Figure 8. Graphic results of the DyMnO$_3$ absorption tests on the MG dye.

3.6.1. Photodegradation Kinetics

To identify and associate a kinetic model with the photocatalytic degradation process of MG dye using DyMnO$_3$ as a photocatalyst, the pseudo-first-order kinetic model was selected as the kinetic model (Equation (4)).

$$ln\left(\frac{C_0}{C}\right) = kt \qquad (4)$$

where C_0 and C represent the concentration values of the solutions at the beginning and after a time t of light exposure, respectively; the photochemical degradation rate is represented by k.

Figure 9 shows the graphical results obtained using the pseudo-first-order kinetic model, with the absorbance values recorded in the spectra of Figure 5. The straight line in each graph corresponds to the linear fit applied to the results after using the pseudo-first-order kinetic model. As can be seen, the correlation coefficients (R^2) were 0.93 (20 mg) and 0.88 (40 mg). These results indicate that the absorbance values recorded during the photocatalytic degradation process of the MG dye closely relate to the proposed kinetic model. It should be mentioned that other kinetic models were also used, such as: the zero-order ($R^2 < 0.50$), second-order ($R^2 < 0.70$), pseudo-second-order ($R^2 < 0.70$), parabolic diffusion ($R^2 < 0.70$) and modified Freundlich models ($R^2 < 0.60$). However, their correlation coefficients were less than 0.70. Additionally, the photocatalytic degradation rate (k) was also obtained. The apparent values for k were 0.020 min^{-1} (20 mg) and 0.031 min^{-1} (40 mg).

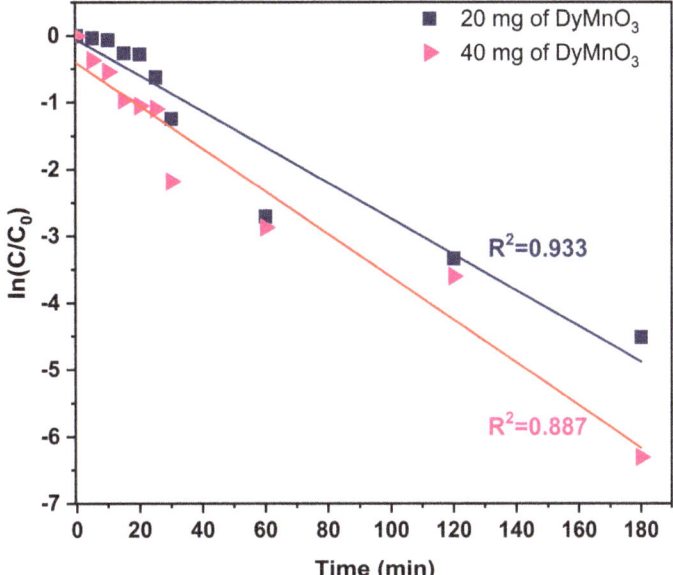

Figure 9. Linear fit corresponds to the pseudo-first-order kinetic model using 20 and 40 mg of DyMnO$_3$.

Although there are many studies based on the photodegradation of MG dye with photocatalyzing oxides, only some have reported the estimated value of the k constant. Among the values of the constant k reported for Cu/ZnO [29], ZnFeO$_4$ [30], and ZnO [31] were 10×10^{-3}, 7.1×10^{-3}, and 20.4×10^{-3}, respectively. However, it is difficult to compare the values of k obtained in this research with those recorded previously. This is because the values obtained in the other works were in different concentrations, and, in addition, their linear fits present an R^2 close to 0.99.

It is evident that the proposed kinetic model (pseudo-first-order kinetics) does not completely fit the photodegradation process of the MG dye with DyMnO$_3$. Furthermore, when the zero-order, second-order, pseudo-second-order, parabolic diffusion, and modified Freundlich models were applied in this photodegradation process, they showed a correlation coefficient of less than 0.70. One of the possible causes associated with this chemical behavior may be the following: In most kinetic models, the absorption rate is greater than the photodegradation rate. However, in this case, the photodegradation rate exceeds the absorption rate.

3.6.2. Recycling Tests and pH Influence

With the purpose of evaluating the reuse of DyMnO$_3$ as a photocatalyst, recycling tests were carried out. In four cycles, a decrease in the degradation rate was observed (Figure 10A). It is possible that fragments of the dye molecule are absorbed on the surface of the photocatalyst, causing a decrease in its degradation rate.

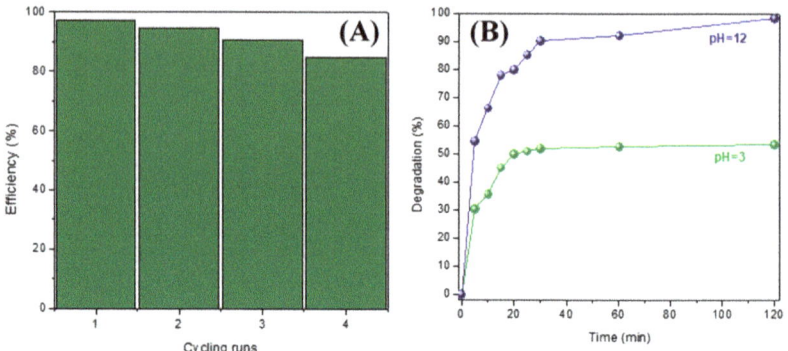

Figure 10. (**A**) Reusability of DyMnO$_3$ photocatalyst for MG dye degradation. (**B**) Influence of pH on the degradation of the MG dye using DyMnO$_3$ as a photocatalyst.

Additionally, the influence of pH on the surface of the photocatalyst was evaluated. For this, two solutions with pH = 3 and pH = 12 were prepared. Then, 40 mg of DyMnO$_3$ was added to each solution and exposed to visible light (λ = 405 nm; E_e = 100 mW/cm^2). Figure 10B shows the graphical results. In 120 min, degradations of approximately 53% (pH = 3) and 98% (pH = 12) were achieved. Although it is difficult to interpret the influence of pH on the photocatalytic reaction due to its multiple roles, a possible interpretation of this result could be the following: by increasing the pH, the surface charge of the photocatalyst would be negative. This may be possible because DyMnO$_3$ is a p-type semiconductor; therefore, it electrostatically attracts negative charges. This causes the dye to be absorbed more effectively on the surface of the photocatalyst and its degradation to occur more effectively.

3.6.3. XPS Analysis

To identify and associate the chemical species present on the DyMnO$_3$ powder surface, the O 1s and Mn 2p core levels (Figure 11) were analyzed through XPS.

Before being used as a photocatalyst, the O 1s XPS spectrum (Figure 11A) was deconvoluted into two Gaussians centered at approximately 529.5 and 531.4 eV. The former was associated with Dy-O and Mn-O bonds, while the latter was attributed to the presence of hydroxyl groups (-OH) [32]. Additionally, the area of each Gaussian curve was quantified: 60.24% for the curve associated with the Dy-O and Mn-O bonds (O Lattice) and 39.76% for the one associated with -OH. After being used as a photocatalyst, the new O 1s spectrum (Figure 11C) was also deconvoluted. The results show two Gaussians associated with the same chemical species described previously. However, the area covered by each was 54.04% (O Lattice) and 45.96% (-OH). Comparing these results with those obtained before being used as a photocatalyst, an approximately 6.2% increase in -OH was quantified.

The XPS spectrum of Mn 2p was also analyzed before DyMnO$_3$ was used as a photocatalyst (Figure 11B). Two peaks located at approximately at 642 eV (Mn 2p$^{3/2}$) and 653.7 eV (Mn 2p$^{1/2}$) appear in its spectrum, which are associated with the +3 oxidation state of manganese (Mn^{+3}) [30]. However, after being used as a photocatalyst (Figure 11D), the Mn 2p spectrum was deconvolved. The Gaussians located at 641.9 and 653.5 eV were associated with the +3 state of manganese (Mn^{+3}), while those located at 644 and 655.5 eV were associated with the +4 oxidation state (Mn^{+4}) [32]. The presence of the +4 oxidation state in the Mn 2p spectrum may be related to the creation of photo-holes, which contribute to the change in its oxidation state, as described in Equation (5).

$$Mn^{+3} + h^{+}_{VB} \rightarrow Mn^{+4} \tag{5}$$

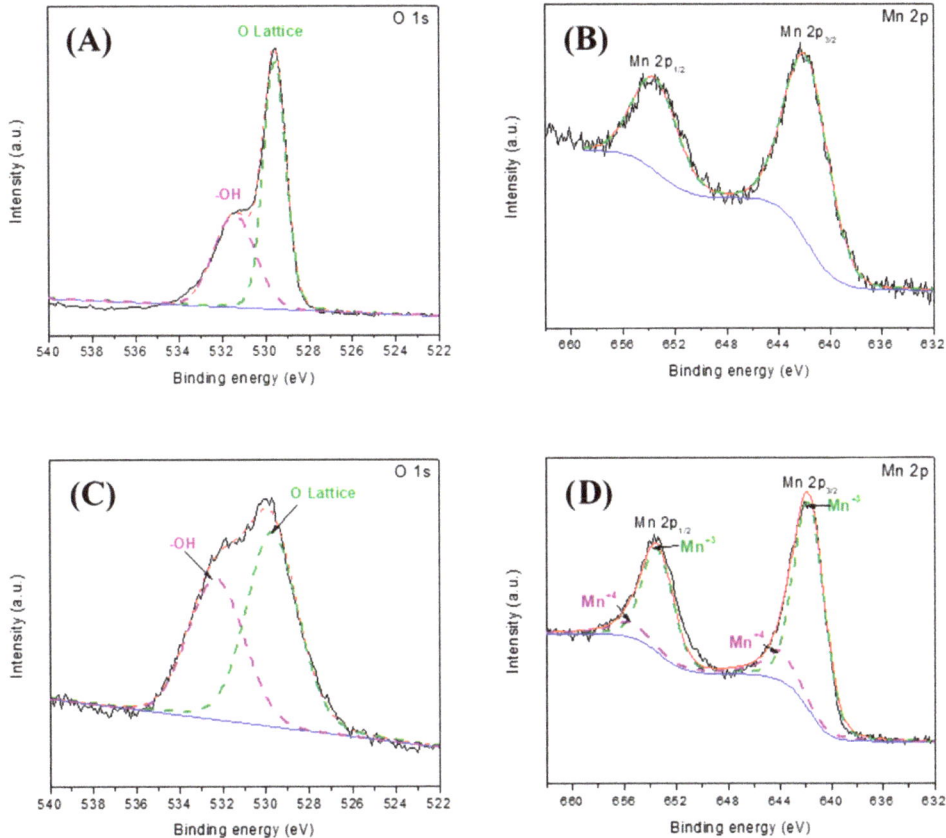

Figure 11. XPS spectrum of the O 1s and Mn 2p core levels before (**A,B**) and after (**C,D**) DyMnO$_3$ was used as a photocatalyst, respectively. The dashed lines represent the deconvolutions.

3.6.4. Contribution of Reactive Oxygen Species to the Photodegradation of the MG Dye

To evaluate the contribution of reactive oxygen species (ROS) involved in the photocatalytic degradation process of MG dye, disodium ethylenediamine tetra acetic acid (EDTA), isopropyl alcohol (ISPA), and p-benzoquinone (p-BZQ) were used as scavengers of h^+, $OH^\bullet$, and O_2^-, respectively. Three solutions with 40 mL of MG and 40 mg of DyMnO$_3$ each were prepared for this. A scavenger (EDTA, ISPA, or p-BZQ) at a concentration of 0.5×10^{-3} M was added to each solution. The results are shown in Figure 12.

Comparing the photocatalytic efficiency of DyMnO$_3$ obtained with each scavenger, it is observed that with EDTA, a photodegradation of 50% is achieved after 180 min. On the other hand, using p-BZQ and ISPA, 65% and 98% photodegradations are achieved within the same time interval, respectively. According to these results, holes (h^+) and hydroxyl radicals ($OH^\bullet$) are the main ROS involved in the photocatalytic degradation process of MG dye. The superoxide radical (O_2^-) had the least contribution to the photodegradation of the dye.

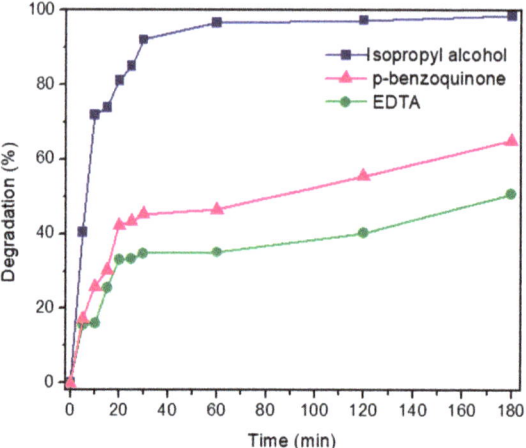

Figure 12. Percent degradation of MG dye using DyMnO$_3$ as a photocatalyst and isopropyl alcohol, p-benzoquinone, and disodium ethylenediamine tetraacetic acid (EDTA) as scavengers.

3.6.5. Photodegradation Mechanism

According to the results obtained from the scavengers and photocurrent tests, a possible photocatalytic degradation mechanism based on the formation of ROS was proposed (Figure 13) [33]. This mechanism can be explained as follows: Photons with energy hυ = 3.01 eV are emitted on the surface of DyMnO$_3$ (E_g = 2.40 eV). Since the energy of the incident photons, hυ, is greater than the E_g of the oxide, electrons from the valence band (VB) can migrate to the conduction band (CB), forming electron–hole pairs (Equation (6)). The electrons in the conduction band promote the creation of superoxide radicals (O$_2^-$) (Equation (7)), which in turn contributes to the formation of hydroperoxyl (OOH$^\bullet$) and hydroxyl (OH$^\bullet$) radicals (Equations (9)–(11)). On the other hand, the interaction of photo-holes (h$^+$) with water molecules and hydroxyl groups (−OH) also promotes the formation of OH$^\bullet$ (Equations (8) and (12)). As it is known, OH$^\bullet$ are strong oxidizing agents that can degrade organic molecules, in this case, allowing for the degradation of the MG dye, as shown in Equation (13).

$$DyMnO_{3(surface)} + h\upsilon_{(photon)} \rightarrow e^-_{(CB)} + h^+_{(VB)} \tag{6}$$

$$O_2 + e^-_{(CB)} \rightarrow O_2^- \tag{7}$$

$$H_2O + h^+_{(VB)} \rightarrow OH^\bullet + H^+ \tag{8}$$

$$O_2^- + H^+ \rightarrow OOH^\bullet \tag{9}$$

$$2(OOH^\bullet) \rightarrow O_2 + H_2O_2 \tag{10}$$

$$H_2O_2 + O_2^- \rightarrow -OH + OH^\bullet + O_2 \tag{11}$$

$$-OH + h^+_{VB} \rightarrow OH^\bullet \tag{12}$$

$$OH^\bullet + MG_{dye} \rightarrow Degradation\ products \tag{13}$$

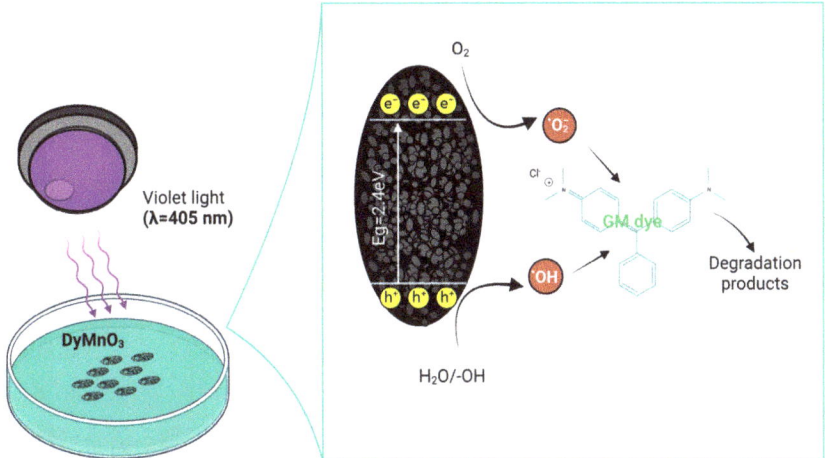

Figure 13. Graphical photodegradation mechanism of MG dye using DyMnO$_3$ as photocatalyst under visible light (λ = 405 nm).

3.7. Photocatalyst Oxides Used in the Degradation of MG Dye

Many oxides have been used as photocatalysts in the degradation of MG dye. Table 3 shows some of them as well as the conditions under which their photocatalytic activity was evaluated.

Table 3. Photocatalyst oxides comparing the amount used, light type, irradiation time, and efficiency.

Photocatalyst	Photocatalyst Amount	Light Type	Irradiation Time (min)	Efficiency
ZnO and Dy/ZnO [33]	80 and 30 mg, respectively	Visible light	60	90% and 100%, respectively
Bi$_2$WO$_6$ [34]	1 g/L (pH = 2)	Visible light	30	87%
Fe$_2$O$_3$ (Hematite) [35]	0.1 g	UV	180	45%
NaNbO$_3$ and Au/NaNbO$_3$ [36]	0.4 g, respectively	Visible light	60	25% and 85%, respectively
WO$_3$, TiO$_2$/WO$_3$ [37]	0.1 g, respectively	Visible light	60	50% and 100%, respectively
GO, CuFe$_2$O$_4$, GO/CuFe$_2$O$_4$ [38]	0.01 g, respectively	Visible light	210	90%, 30% and 63%, respectively
WO$_3$ and Pd/WO$_3$ [39]	150 mg/L, respectively	Solar light	300	50% and 80%, respectively
Bi$_2$O$_3$, CaFe$_2$O$_4$, B$_2$O$_3$/CaFe$_2$O$_4$ [40]	0.05 g, respectively	Visible light	240	70%, 58% and 89%, respectively
Fe$_2$O$_3$/SnO$_2$ [41]	40 mg	Solar light	240	86%
ZnO [42]	0.2 g/L	Solar light	100	85%
Co$_3$O$_4$ [43]	50 mg	Visible light	100	90%
Cu$_2$O [44]	10 mg	Visible light	45	92%
DyMnO$_3$	40 mg	Visible light	60	93%

It is difficult to make a comparison regarding the photocatalytic efficiency of each one with that obtained with DyMnO$_3$. This is because in many cases, the optical irradiance of the light-emitting source is not described, and in others, its photocatalytic activity was evaluated under basic or acidic conditions. Despite this, there are oxides that are more efficient than DyMnO$_3$ in the degradation of this dye, such as Dy/ZnO or TiO$_2$/WO$_3$. However, they need to be mixed with other materials to achieve efficiencies greater than or similar to those obtained with dysprosium manganate.

4. Conclusions

DyMnO$_3$ was synthesized using a novel and efficient polymer-decomposition method. Unlike other methods used in the synthesis of this oxide (such as hydrothermal, combustion, solid-state reactions), the method proposed in this work promotes the orthorhombic phase at 100 °C. Finally, at 800 °C, it was possible to obtain its orthorhombic phase without the presence of residual oxides (Mn$_2$O$_3$, Mn$_3$O$_4$, or Dy$_2$O$_3$). However, at 1000 and 1200 °C, diffraction lines associated with manganese oxides were observed.

Regarding its photocatalytic activity, a photodegradation of 80% of the MG dye was achieved in approximately 60 min under visible light (λ = 405 nm). The main reactive oxygen species (ROS) involved in the photodegradation process were h$^+$ and OH$^\bullet$. Due to the conditions under which the photodegradation experiments were carried out (λ = 405 nm; E_e = 100 mW/cm^2), it is difficult to compare the efficiency obtained. However, under conditions similar to those described, the efficiency of this material is greater than that achieved with other oxides with a perovskite-type structure, particularly YMnO$_3$, DyMnO$_3$, and LaMnO$_3$. A key factor associated with the photocatalytic efficiency of DyMnO$_3$ was its reduced band gap energy (2.4 eV). This value allows us to approximate the minimum wavelength at which the separation of charge carriers (h$^+$, e$^-$) can occur in a photocatalyst. In this material, it was λ~516 nm.

In summary, there are diverse photocatalyst oxides, and among the most used are TiO$_2$, ZnO, and CeO$_2$. However, those oxides are very efficient under UV light (λ = 365 nm). Due to this, various synthesis methods have been developed that allow the band gap energy to be reduced. Moreover, they have been used in heterostructures in order to increase their photocatalytic activity at wavelengths in the visible spectrum. In this work, a new oxide photocatalyst with a perovskite-type structure (DyMnO$_3$) was developed, with a degradation capability greater than 90% under visible light. However, its photocatalytic efficiency is lower than that of many heterostructures based on semiconductor oxides. For this reason, its use in heterostructures is considered future work with the purpose of further improving its photocatalytic efficiency.

Author Contributions: Conceptualization, M.Á.L.-Á. and P.O.-G.; methodology, J.M.S.-J., J.G.S.-G. and I.C.-A.; validation, M.Á.L.-Á., A.B.-R. and J.E.C.-G.; data curation, J.A.G.-d.L. and C.A.L.-d.A.; writing—review, J.M.S.-J. and M.Á.L.-Á.; Project administration M.Á.L.-Á. and P.O.-G.; investigation, M.Á.L.-Á. All authors have read and agreed to the published version of the manuscript.

Funding: This research received no external funding.

Institutional Review Board Statement: Not applicable.

Informed Consent Statement: Not applicable.

Data Availability Statement: The data presented in this study are available upon request from the corresponding author.

Acknowledgments: The authors thank José Antonio Rivera Mayorga and the team at the XPS Laboratory of CUCEI Universidad de Guadalajara for the XPS measurements.

Conflicts of Interest: The authors declare no conflict of interest.

References

1. Žužić, A.; Ressler, A.; Macan, J. Perovskite oxides as active materials in novel alternatives to well-known technologies: A review. *Ceram. Int.* **2022**, *48*, 27240–27261. [CrossRef]
2. Stolz, A. Basic and applied aspects in the microbial degradation of azo dyes. *Appl. Microbiol. Biotechnol.* **2001**, *56*, 69–80. [CrossRef]
3. Rauf, M.A.; Ashraf, S.S. Fundamental principles and application of heterogeneous photocatalytic degradation of dyes in solution. *Chem. Eng. J.* **2009**, *151*, 10–18. [CrossRef]
4. Saeed, M.; Muneer, M.; Haq, A.u.; Akram, N. Photocatalysis: An effective tool for photodegradation of dyes—A review. *Environ. Sci. Pollut. Res.* **2022**, *29*, 293–311. [CrossRef]
5. Pérez-Estrada, L.A.; Agüera, A.; Hernando, M.D.; Malato, S.; Fernández-Alba, A.R. Photodegradation of malachite green under natural sunlight irradiation: Kinetic and toxicity of the transformation products. *Chemosphere* **2008**, *70*, 2068–2075. [CrossRef] [PubMed]
6. Srivastava, S.; Sinha, R.; Roy, D. Toxicological effects of malachite green. *Aquat. Toxicol.* **2004**, *66*, 319–329. [CrossRef] [PubMed]
7. López-Alvarez, M.Á.; Silva-Jara, J.M.; Silva-Galindo, J.G.; Reyes-Becerril, M.; Velázquez-Carriles, C.A.; Macías-Rodríguez, M.E.; Macías-Lamas, A.M.; García-Ramírez, M.A.; López de Alba, C.A.; Reynoso-García, C.A. Determining the Photoelectrical Behavior

and Photocatalytic Activity of an h-YMnO$_3$ New Type of Obelisk-like Perovskite in the Degradation of Malachite Green Dye. *Molecules* **2023**, *28*, 3932. [CrossRef] [PubMed]
8. Grabowska, E. Selected perovskite oxides: Characterization, preparation and photocatalytic properties—A review. *Appl. Catal. B Environ.* **2016**, *186*, 97–126. [CrossRef]
9. Silva, G.R.O.; Santos, J.C.; Martinelli, D.M.H.; Pedrosa, A.M.G.; de Souza, M.J.B.; Melo, D.M.A. Synthesis and Characterization of LaNixCo1-xO$_3$ Perovskites via Complex Precursor Methods. *Mater. Sci. Appl.* **2010**, *1*, 39–45. [CrossRef]
10. Bhoi, K.; Dam, T.; Mohapatra, S.R.; Patidar, M.M.; Singh, D.; Singh, A.K.; Vishwakarma, P.N.; Babu, P.D.; Siruguri, V.; Pradhan Dillip, K. Studies of magnetic phase transitions in orthorhombic DyMnO$_3$ ceramics prepared by acrylamide polymer gel template method. *J. Magn. Magn. Mater.* **2019**, *480*, 138–149. [CrossRef]
11. Patra, M.; Midya, A.; Mandal, P. Investigation of large dielectric permittivity and relaxation behavior of DyMnO$_3$ single crystal. *Solid State Commun.* **2022**, *353*, 114845. [CrossRef]
12. Tajiri, T.; Terashita, N.; Hamamoto, K.; Deguchi, H.; Mito, M.; Morimoto, Y.; Konishi, K.; Kohno, A. Size dependences of crystal structure and magnetic properties of DyMnO$_3$ nanoparticles. *J. Magn. Magn. Mater.* **2013**, *345*, 288–293. [CrossRef]
13. Balli, M.; Mansouri, S.; Jandl, S.; Fournier, P.; Dimitrov, D.Z. Large rotating magnetocaloric effect in the orthorhombic DyMnO$_3$ single crystal. *Solid State Commun.* **2016**, *239*, 9–13. [CrossRef]
14. Midya, A.; Das, S.N.; Mandal, P.; Pandya, S.; Ganesan, V. Anisotropic magnetic properties and giant magnetocaloric effect in antiferromagnetic R MnO$_3$ crystals (R = Dy, Tb, Ho, and Yb). *Phys. Rev. B* **2011**, *84*, 235127. [CrossRef]
15. Valian, M.; Khoobi, A.; Salavati-Niasari, M. Green synthesis and characterization of DyMnO$_3$-ZnO ceramic nanocomposites for the electrochemical ultratrace detection of atenolol. *Mater. Sci. Eng. C* **2020**, *111*, 110854. [CrossRef] [PubMed]
16. Aman, S.; Tahir, M.B.; Ahmad, Z.; Znaidia, S.; Ahmad, N.; Khosa, R.Y.; Waheed, M.S.; Manzoor, S.; Abdullah, M.; Taha, T.A. Rational design of novel dysprosium manganite sandwich layered morphology for supercapacitor applications. *Chin. J. Phys.* **2022**, *79*, 531–539. [CrossRef]
17. Carp, O.; Patron, L.; Ianculescu, A.; Pasuk, J.; Olar, R. New synthesis routes for obtaining dysprosium manganese perovskites. *J. Alloys Compd.* **2003**, *351*, 314–318. [CrossRef]
18. Alonso, J.A.; Martínez-Lope, M.J.; Casais, M.T.; Fernández-Díaz, M.T. Evolution of the Jahn−Teller Distortion of MnO$_6$ Octahedra in RMnO$_3$ Perovskites (R = Pr, Nd, Dy, Tb, Ho, Er, Y): A Neutron Diffraction Study. *Inorg. Chem.* **2000**, *39*, 917–923. [CrossRef]
19. Gülgün, M.A.; Nguyen, M.H.; Kriven, W.M. Polymerized Organic-Inorganic Synthesis of Mixed Oxides. *J. Am. Ceram. Soc.* **1999**, *82*, 556–560. [CrossRef]
20. Das, B.; Das, B.; Das, N.S.; Sarkar, S.; Chattopadhyay, K.K. Tailored mesoporous nanocrystalline Ga$_2$O$_3$ for dye-selective photocatalytic degradation. *Microporous Mesoporous Mater.* **2019**, *288*, 109600. [CrossRef]
21. Takahara, Y.; Kondo, J.N.; Takata, T.; Lu, D.; Domen, K. Mesoporous Tantalum Oxide. 1. Characterization and Photocatalytic Activity for the Overall Water Decomposition. *Chem. Mater.* **2001**, *13*, 1194–1199. [CrossRef]
22. Sing, K.S.W. Reporting physisorption data for gas/solid systems with special reference to the determination of surface area and porosity (Recommendations 1984). *Pure Appl. Chem.* **1985**, *57*, 603–619. [CrossRef]
23. Lowell, S.; Shields, J.E.; Thomas, M.A.; Thommes, M. Adsorption Mechanism. In *Characterization of Porous Solids and Powders: Surface Area, Pore Size and Density*; Particle Technology Series; Springer: Dordrecht, The Netherlands, 2004; Volume 16. [CrossRef]
24. Chen, L.; Zheng, G.; Yao, G.; Zhang, P.; Dai, S.; Jiang, Y.; Li, H.; Yu, B.; Ni, H.; Wei, S. Lead-Free Perovskite Narrow-Bandgap Oxide Semiconductors of Rare-Earth Manganates. *ACS Omega* **2020**, *5*, 8766–8776. [CrossRef] [PubMed]
25. Ju, Y.; Yang, S.; Ding, Y.; Sun, C.; Zhang, A.; Wang, L. Microwave-Assisted Rapid Photocatalytic Degradation of Malachite Green in TiO$_2$ Suspensions: Mechanism and Pathways. *J. Phys. Chem. A* **2008**, *112*, 11172–11177. [CrossRef] [PubMed]
26. Michel, C.R.; López-Alvarez, M.A.; Martínez-Preciado, A.H.; Oleinikov, V. Ultraviolet Detection and Photocatalytic Activity of Nanostructured LaCoO$_3$ Prepared by Solution-Polymerization. *ECS J. Solid State Sci. Technol.* **2019**, *8*, Q9–Q14. [CrossRef]
27. Michel, C.R.; Lopez-Alvarez, M.A.; Martínez-Preciado, A.H.; Carbajal-Arízaga, G.G. Novel UV Sensing and Photocatalytic Properties of DyCoO$_3$. *J. Sens.* **2019**, *2019*, 1–12. [CrossRef]
28. Rodríguez, C.I.M.; Álvarez, M.Á.L.; Rivera, J.d.J.F.; Arízaga, G.G.C.; Michel, C.R. α-Ga$_2$O$_3$ as a Photocatalyst in the Degradation of Malachite Green. *ECS J. Solid State Technol.* **2019**, *8*, Q3180–Q3186. [CrossRef]
29. Modwi, A.; Abbo, M.A.; Hassan, E.A.; Al-Duaij, O.K.; Houas, A. Adsorption kinetics and photocatalytic degradation of malachite green (MG) via Cu/ZnO nanocomposites. *J. Environ. Chem. Eng.* **2017**, *5*, 5954–5960. [CrossRef]
30. Tsvetkov, M.; Zaharieva, J.; Milanova, M. Ferrites, modified with silver nanoparticles, for photocatalytic degradation of malachite green in aqueous solutions. *Catal. Today* **2020**, *357*, 453–459. [CrossRef]
31. Meena, P.L.; Poswal, K.; Surela, A.K. Facile synthesis of ZnO nanoparticles for the effective photodegradation of malachite green dye in aqueous solution. *Water Environ. J.* **2022**, *36*, 513–524. [CrossRef]
32. Moulder, J.F.; Chastain, J.; King, R.C. (Eds.) *Handbook of X-ray Photoelectron Spectroscopy: A Reference Book of Standard Spectra for Identification and Interpretation of XPS Data*; Physical Electronics Division, Perkin-Elmer Corp: Eden Prairie, MN, USA, 1995.
33. Suganya Josephine, G.A.; Ramachandran, S.; Sivasamy, A. Nanocrystalline ZnO doped lanthanide oxide: An efficient photocatalyst for the degradation of malachite green dye under visible light irradiation. *J. Saudi Chem. Soc.* **2015**, *19*, 549–556. [CrossRef]
34. Chen, Y.; Zhang, Y.; Liu, C.; Lu, A.; Zhang, W. Photodegradation of Malachite Green by Nanostructured Bi$_2$WO$_6$ Visible Light-Induced Photocatalyst. *Int. J. Photoenergy* **2012**, *2012*, 510158. [CrossRef]

35. Alharbi, A.; Abdelrahman, E.A. Efficient photocatalytic degradation of malachite green dye using facilely synthesized hematite nanoparticles from Egyptian insecticide cans. *Spectrochim. Acta Part A Mol. Biomol. Spectrosc.* **2020**, *226*, 117612. [CrossRef] [PubMed]
36. Baeissa, E.S. Photocatalytic degradation of malachite green dye using Au/NaNbO$_3$ nanoparticles. *J. Alloys Compd.* **2016**, *672*, 564–570. [CrossRef]
37. Wei, Y.; Huang, Y.; Fang, Y.; Zhao, Y.; Luo, D.; Guo, Q.; Fan, L.; Wu, J. Hollow mesoporous TiO$_2$/WO$_3$ sphere heterojunction with high visible-light-driven photocatalytic activity. *Mater. Res. Bull.* **2019**, *119*, 110571. [CrossRef]
38. Yadav, P.; Surolia, P.K.; Vaya, D. Synthesis and application of copper ferrite-graphene oxide nanocomposite photocatalyst for the degradation of malachite green. *Mater. Today Proc.* **2021**, *43*, 2949–2953. [CrossRef]
39. Liu, Y.; Ohko, Y.; Zhang, R.; Yang, Y.; Zhang, Z. Degradation of malachite green on Pd/WO$_3$ photocatalysts under simulated solar light. *J. Hazard. Mater.* **2010**, *184*, 386–391. [CrossRef]
40. Luo, D.; Kang, Y. Synthesis and characterization of novel CaFe$_2$O$_4$/Bi$_2$O$_3$ composite photocatalysts. *Mater. Lett.* **2018**, *225*, 17–20. [CrossRef]
41. Pradhan, G.K.; Reddy, K.H.; Parida, K.M. Facile fabrication of mesoporous α-Fe$_2$O$_3$/SnO$_2$ nanoheterostructure for photocatalytic degradation of malachite green. *Catal. Today* **2014**, *224*, 171–179. [CrossRef]
42. Saikia, L.; Bhuyan, D.; Saikia, M.; Malakar, B.; Dutta, D.K.; Sengupta, P. Photocatalytic performance of ZnO nanomaterials for self sensitized degradation of malachite green dye under solar light. *Appl. Catal. A Gen.* **2015**, *490*, 42–49. [CrossRef]
43. Verma, M.; Mitan, M.; Kim, H.; Vaya, D. Efficient photocatalytic degradation of Malachite green dye using facilely synthesized cobalt oxide nanomaterials using citric acid and oleic acid. *J. Phys. Chem. Solids* **2021**, *155*, 110125. [CrossRef]
44. Muthukumaran, M.; Gnanamoorthy, G.; Prasath, P.V.; Abinaya, M.; Dhinagaran, G.; Sagadevan, S.; Mohammad, F.; Oh, W.C.; Venkatachalam, K. Enhanced photocatalytic activity of Cuprous Oxide nanoparticles for malachite green degradation under the visible light radiation. *Mater. Res. Express* **2020**, *7*, 015038. [CrossRef]

Disclaimer/Publisher's Note: The statements, opinions and data contained in all publications are solely those of the individual author(s) and contributor(s) and not of MDPI and/or the editor(s). MDPI and/or the editor(s) disclaim responsibility for any injury to people or property resulting from any ideas, methods, instructions or products referred to in the content.

Article

Effect of Post-Washing on Textural Characteristics of Carbon Materials Derived from Pineapple Peel Biomass

Chi-Hung Tsai [1], Wen-Tien Tsai [2,*] and Li-An Kuo [3]

[1] Department of Resources Engineering, National Cheng Kung University, Tainan 701, Taiwan; ap29fp@gmail.com
[2] Graduate Institute of Bioresources, National Pingtung University of Science and Technology, Pingtung 912, Taiwan
[3] Department of Environmental Science and Engineering, National Pingtung University of Science and Technology, Pingtung 912, Taiwan; sanck112204@gmail.com
* Correspondence: wttsai@mail.npust.edu.tw; Tel.: +886-8-7703202

Abstract: Porous carbon materials have been widely used to remove pollutants from the liquid-phase streams. However, their limited pore properties could be a major problem. In this work, the effects of post-washing methods (i.e., water washing and acid washing) on the textural characteristics of the resulting biochar and activated carbon products from pineapple peel biomass were investigated in the carbonization and CO_2 activation processes. The experiments were set at an elevated temperature (i.e., 800 °C) holding for 30 min. It was found that the enhancement in pore property reached about a 50% increase rate, increasing from 569.56 m^2/g for the crude activated carbon to the maximal BET surface area of 843.09 m^2/g for the resulting activated carbon by water washing. The resulting activated carbon materials featured the microporous structures but also were characteristic of the mesoporous solids. By contrast, the enhancement in the increase rate by about 150% was found in the resulting biochar products. However, there seemed to be no significant variations in pore property with post-washing methods. Using the energy dispersive X-ray spectroscopy (EDS) and the Fourier Transform infrared spectroscopy (FTIR) analyses, it showed some oxygen-containing functional groups or complexes, potentially posing the hydrophilic characters on the surface of the resulting carbon materials.

Keywords: pineapple peel; activated carbon; biochar; post-washing; pore analysis; surface characteristics

1. Introduction

Pineapple (*Ananas comosus*) is a popular and edible fruit for drinking and food industries, but it also generates large amounts of pineapple peel in the industrial and residential sectors. The biomass residue could cause adverse impacts on living quality if dumped into the environment without the proper treatment. Although pineapple peel can be treated by the sanitary landfill and incineration, these treatment methods pose potential drawbacks due to the high contents of moisture and lignocellulose. For example, treatment by thermal combustion at municipal solid waste (MSW) incineration plants not only reduces the energy efficiency for power generation but also could generate huge amounts of air pollutants. In this regard, the traditional options for utilizing pineapple peel are to reuse it as a feedstock for animal feed and a variety of natural products (e.g., organic acids, bioethanol, bromelain enzyme, and phenolic antioxidants) or organic fertilizer by composting [1]. In order to promote the circular economy, the value-added valorization of pineapple peel (PP) has been exploited and reviewed by researchers in recent years [1–6]. In the paper by Tran et al. [6], the authors reviewed the production and application of pineapple-derived carbon adsorbents (i.e., biochar and activated carbon), which were produced by thermal processing or chemical modification to improve the surface chemistry and porosity. Other

carbon-based nanomaterials (e.g., carbon nanotube and carbon nanosphere) can be also produced from agricultural biomass like rice-based residues [7].

In general, the dried biomass is mainly composed of lignocellulosic constituents, thus containing oxygen-containing groups on the surface. These polar complexes can provide the features of complexation and ion exchange, suggesting that pineapple peel can be directly reused as a biosorbent for removing cationic pollutants from the water environment [6,8–12]. However, these biosorbents only have limited uptake capacities due to their poor pore properties like specific surface area and pore volume. In order to increase the textural characteristics of biomass-derived adsorbent, the biomass precursor must be thermally processed by pyrolysis and pyrolysis-activation methods [13,14], converting it into biochar [15–21] and activated biochar (activated carbon) [22–29], respectively. The maximal uptake capacities are highly associated with its physical properties like surface porosity and particle size. It has been found that the pore properties of the resulting biochar products derived from pineapple peel indicated relatively low values, which could be attributed to the carbonization conditions at mild temperatures. On the other hand, the production of activated carbon from pineapple peel has almost always been based on chemical activation in the literature [22–29]. It should be noted that the commercial manufacturing process for activated carbon was to employ the physical activation process due to less wastewater treatment problems in comparison with the chemical activation process [1].

In order to enhance the physical and chemical characteristics of biomass-based carbon materials, a plenitude of post-washing treatment methods have been reviewed in the literature [30–32]. In a previous study [33], the findings showed that post-washing with deionized water and/or dilute acid can enhance the pore properties of cocoa pod husk (CPH)-derived biochar due to the removal of the inorganic residues and the improvement in the accessibility of the carbon structure by lessoning the pore blockage. For example, the Brunauer–Emmet–Teller (BET) surface area of the resulting biochar produced at 400 °C holding for 30 min was significantly increased from 101 m^2/g to 342 m^2/g after post-washing with a dilute acid (0.25 M HCl). The effects of post-washing on the enhancing pore properties of biomass-based carbon materials (i.e., biochar and activated carbon) were also found by other studies [34–36].

As mentioned above, it seems that no study has reported on PP-based activated carbon produced by physical activation using carbon dioxide (CO_2). In order to upgrade the applicability of PP as a precursor for producing carbon adsorbents, the effects of post-washing on the textural characteristics of the resulting carbon materials were investigated in the present study. Based on the results in a previous study [21], the porous carbon products (including biochar and activated carbon) were produced at an elevated pyrolysis/activation temperature (i.e., 800 °C) holding for 30 min in this work because of the significant pore properties increased under the process conditions. Thereafter, the resulting carbon materials were further treated by water washing using deionized water and dilute acid (0.1 M HCl), respectively. The textural characteristics of the resulting carbon products were obtained by the nitrogen (N_2) adsorption–desorption isotherms for determining pore properties, the scanning electron microscope (SEM) for observing porous structure, and the energy dispersive X-ray spectroscopy (EDS) and Fourier infrared spectrometer (FTIR) for surveying surface elemental compositions.

2. Materials and Methods

2.1. Materials

The starting feedstock (i.e., pineapple peel) for producing biochar and activated carbon was collected from a local market (Chienchen district, Kaohsiung, Taiwan). It was first dewatered in the sun and further dried in the air-circulating oven at 105 °C. A crusher was used to transform the dried biomass into powdered particles, which were sieved to the size range of 0.841 mm (opening size of mesh No. 20) to 0.420 mm (opening size of mesh No. 40).

The dried and sieved biomass sample was used to find its thermochemical characteristics and perform the pyrolysis-activation experiments at 800 °C holding for 30 min.

2.2. Thermochemical Property Determination of Pineapple Peel

Due to the different sources, the proximate analysis, calorific value, and thermogravimetric analysis (TGA) of the pineapple biomass were also determined in this work. The adopted methods and analytical instruments refer to a previous study [21]. These thermochemical properties were relevant to the potential for producing carbon materials properly. Herein, the TGA was performed in the temperature range of 25–900 °C at a specific rate of 10 °C/min, which was close to the heating condition of the pyrolysis-activation experiments.

2.3. Experimental Methods

As referred to in a previous study [21], the pyrolysis-activation experiments were carried out by using a vertical fixed-bed reactor in this work. For the pyrolysis experiment, about 3 g of the dried pineapple (PP) biomass was used to produce biochar at 800 °C holding for 30 min, where the heating profile was set at about 10 °C/min. For the physical activation experiment, the first stage was to increase the system temperature from 25 °C (room temperature) to 500 °C under the inert atmosphere by purging a nitrogen (N_2) flow. When the system reached 500 °C, the flowing gas was changed to pass carbon dioxide (CO_2) by stopping N_2 gas. Meanwhile, the system was continuously heated to 800 °C at the same rate (about 10 °C/min) holding for 30 min. The resulting biochar (BC) and activated carbon (AC) products derived from PP were further treated by washing with deionized water (WW) and dilute acid of 0.1 M HCl (AW), as based on previous studies [21,37,38]. Therefore, these carbon products were coded as PP-BC, PP-BC-WW, PP-BC-AW, PP-AC, PP-AC-WW, and PP-AC-AW.

2.4. Characterization Analysis of Resulting Carbon Materials

As mentioned above, the adopted procedures and analytical instruments for the textural characteristics of the resulting carbon products refer to previous studies [21,39]. It should be noted that the data on specific surface area were based on the Brunauer–Emmett–Teller (BET) model [20–22], which were correlated by 4–6 points using the relative pressure (P/P_0) range of 0.05–0.10. However, the elemental compositions on the surface of the resulting carbon materials were determined by the energy dispersive X-ray spectroscopy (EDS) (model: X-stream-2; Oxford Instruments, Abingdon, UK).

3. Results and Discussion

3.1. Thermochemical Characteristics of Pineapple Peel (PP)

According to the American Society for Testing and Materials (ASTM) standards, proximate analysis can give the gross compositions of the biomass without expensive precision instruments. In this work, the dried PP biomass showed the values (in duplicate) of 73.38 ± 2.33 wt% for volatile matter, 4.63 ± 0.25 wt% for ash, and 14.14 wt% for fixed carbon (determined by difference). For this biomass, the ash content had a moderate value, ranging from the low values of woody biomass to the high values of rice residues [40,41]. Using the preliminary analysis of the energy dispersive X-ray spectroscopy (EDS), the elemental contents of the dried PP biomass included carbon (C, 54.07 wt%), oxygen (O, 40.77 wt%), and minor elements, which will be addressed in Section 3.3. Furthermore, the thermogravimetric analysis (TGA) and its derivative thermogravimetry (DTG) curves of the dried PP biomass are shown in Figure 1. The maximal weight loss rate occurred in the pyrolysis temperature range from 200 to 450 °C, which should be attributed to the thermal decomposition of lignocellulosic constituents, especially for hemicellulose [21–23]. Based on the TGA results, the physical activation conditions were operated at 500 °C in the first pyrolysis stage to produce a carbon-rich matrix which was subsequently activated at 800 °C by flowing CO_2 gas.

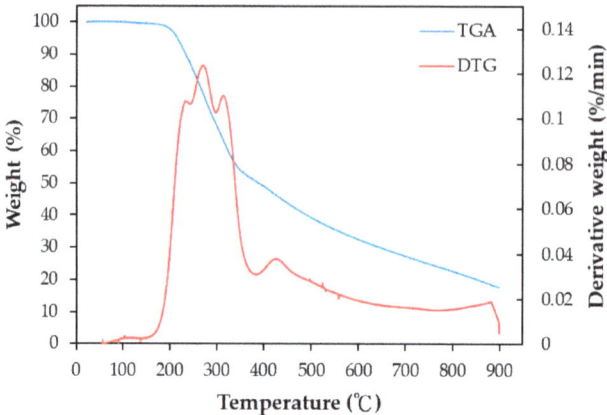

Figure 1. TGA and DTG curves of dried pineapple peel (blue line: TGA; red line: DTG).

3.2. Pore Analysis of Resulting Carbon Materials

In this work, the process conditions were set at a temperature of 800 °C and a holding time of 30 min using N_2/CO_2 gases and post-washing. The mass yields of the resulting carbon materials (i.e., PP-BC and PP-AC) were about 28 wt%. After post-washing with deionized water and dilute acid (i.e., 0.1 M HCl), the residual percentages indicated approximately 84 wt% and 79 wt%, respectively. Concerning the pore properties of the resulting carbon materials (i.e., PP-BC, PP-BC-WW, PP-BC-AW, PP-AC, PP-AC-WW and PP-AC-AW), they are listed in Table 1. Obviously, the values of the resulting biochar products were significantly smaller than those of the resulting activated carbon products. For example, the values of the BET surface area in the PP-BC and PP-AC samples were 100.20 m^2/g and 569.56 m^2/g, respectively. The effect of activation by CO_2 gas on the pore property of the carbon material played a determining role in pore development. On the other hand, the post-washing of the crude carbon products also had an influential effect of enhancing pore property. It could be attributed to the removal of the residual inorganic minerals (or particles) that block the pore entrance, thus leading to more available pores and giving larger pore properties. By comparison, the enhancement in pore property reached about a 50% increase rate, increasing from 569.56 m^2/g for the PP-AC to 843.09 m^2/g for the PP-AC-WW or 799.25 m^2/g for the PP-AC-AW. However, it seemed to show no significant variations on pore property by post-washing methods of DI water and dilute acid. From the viewpoints of economic cost and environmental protection, the post-washing with water was superior to the use of dilute acid in the production of porous carbon materials from pineapple peel.

Furthermore, the data in Table 1 can be derived from the N_2 adsorption/desorption isotherms at −196 °C, mesopore size distributions, and micropore size distributions of the resulting activated carbon products, which are depicted in Figure 2, Figure 3, and Figure 4, respectively. Based on the adsorption–desorption isotherms in Figure 2, the microscale structures of the resulting activated carbon products were mainly microporous. In this regard, the 2D-NLDFT-HS model was adopted to depict their micropore size distributions, which are depicted in Figure 4. These curves were consistent with the data in Table 1. The significant results were further discussed as follows:

1. Based on the classification by the International Union of Pure and Applied Chemistry (IUPAC), these carbon materials are typical of microporous solids, where micropore filling occurs significantly at very low relative pressure (P/P_0). In this regard, they feature the Type I isotherms [42,43]. However, the adsorption–desorption isotherms also possess a hysteresis loop from the relative pressure of about 0.45, which should

    be associated with mesoporous solids. This isotherm shape belongs to the Type IV isotherms, where the capillary condensation occurs [42,43].
2. As compared to PP-based activated carbons produced by chemical activation [22,23,25–29], it was clearly shown that the resulting activated carbons produced by physical activation in this work had slightly lower pore properties (e.g., BET surface area), as shown in Table 2. For example, the BET surface area of PP-based activated carbon produced by KOH activation was 1160 m^2/g [29], which was higher than the optimal value (843 m^2/g) in Table 2.
3. Using the Harrett–Joyner–Halenda (BJH) method and the data on desorption isotherms for the textural characteristics of the mesoporous solids, it was found that the significant peaks of mesopore size distribution occurred at about 3.5 nm. In addition, these peaks were more obvious for the resulting activated carbon materials (i.e., PP-AC-WW and PP-AC-AW) by post-washing in comparison with the crude product (i.e., PP-AC), thus leading to higher pore properties, as listed in Table 1.
4. As mentioned above, the micropore size distributions of the resulting activated carbon materials can be verified in Figure 3, which was obtained by the 2D-NLDFT-HS model. The micropore peak was observed at about 0.6 nm, where it was in the range of less than 2.0 nm. Assuming the cylindrical geometry for all pores, the data on the average pore diameter (or width) were slightly smaller than 2.0 nm (1.6–1.9 nm). Therefore, other significant peaks in the pore size distribution curves were observed at the left side (less than 2.0 nm), indicating micropores are present in all activated carbon products.
5. In order to see the porous textures on the surface of the resulting carbon materials, Figure 5 shows the scanning electron microscopy (SEM) images (i.e., ×300 and ×1000) for the resulting activated carbon samples (i.e., PP-AC, PP-AC-WW and PP-AC-AW). Obviously, the resulting activated carbon products displayed a porous texture on the rigid surface without significant difference. However, post-washing removed the residual impurities and/or particles, thus producing a cleaner surface and greater pore properties, as listed in Table 1.

Table 1. Pore properties of resulting carbon products.

Pore Property	PP-BC	PP-BC-WW	PP-BC-AW	PP-AC	PP-AC-WW	PP-AC-AW
BET surface area [a]	100.20	243.60	269.94	569.56	843.09	799.25
t-plot micropore area [b]	92.15	213.14	237.60	498.59	717.90	689.26
External surface area	8.05	30.46	31.34	70.97	125.19	109.99
Total pore volume [c]	0.042	0.110 [e]	0.121	0.253	0.391	0.371
t-plot micropore area [b]	0.038	0.089	0.097	0.204	0.294	0.283
Average pore width [d]	1.661	1.806	1.802	1.778	1.854	1.856

[a] Based on a relative pressure range of 0.05–0.100 (4–6 points) by the Brunauer–Emmett–Teller (BET) equation. [b] Using the t-plot method. [c] Calculated at a relative pressure of about 0.995. [d] Estimated by the ratio of the total pore volume (V_t) to the BET surface area (S_{BET}) (i.e., average pore width = 4 × V_t/S_{BET}). [e] Analogically estimated by the total pore volume/micropore volume of the PP-BC-AW sample and the micropore volume of the PP-BC-WW sample.

Table 2. Elemental contents of resulting carbon products by EDS spectra.

Elemental Content (wt%)	PP	PP-BC	PP-BC-WW	PP-BC-AW	PP-AC	PP-AC-WW	PP-AC-AW
Carbon (C)	54.072	81.826	88.391	85.916	71.120	78.835	74.921
Oxygen (O)	40.773	14.451	10.186	12.402	20.466	15.819	14.406
Sodium (Na)	0.071	0.018	0.007	0.029	0.000	0.000	0.000

Table 2. Cont.

Elemental Content (wt%)	PP	PP-BC	PP-BC-WW	PP-BC-AW	PP-AC	PP-AC-WW	PP-AC-AW
Magnesium (Mg)	0.229	0.493	0.075	0.309	0.775	1.132	0.207
Aluminum (Al)	4.470	0.123	0.067	0.150	0.006	0.111	2.555
Silicon (Si)	0.227	0.287	0.254	0.283	1.050	1.371	2.530
Phosphorus (P)	0.101	2.421	0.370	0.131	3.719	1.076	0.409
Sulfur (S)	0.058	0.345	0.544	0.460	1.738	0.632	2.913
Calcium (Ca)	0.000	0.035	0.106	0.319	1.126	1.024	2.060

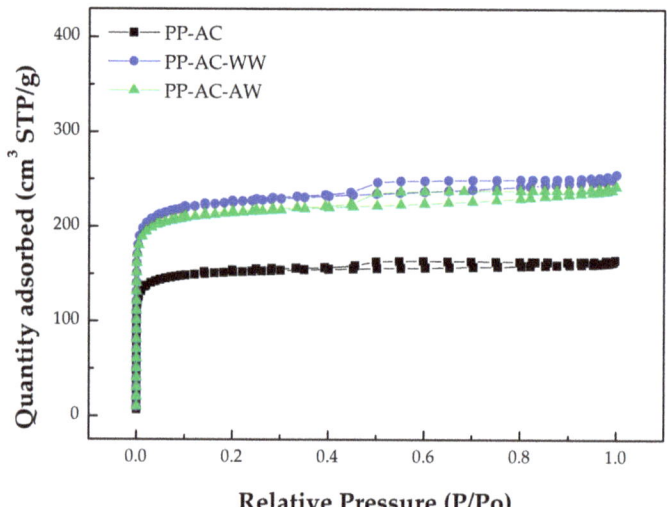

Figure 2. N_2 adsorption–desorption isotherms of the resulting activated carbon materials.

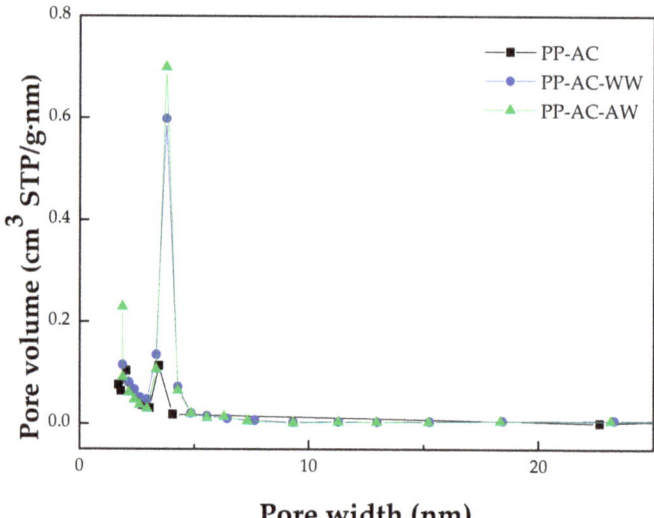

Figure 3. Mesopore size distributions of the resulting activated carbon materials.

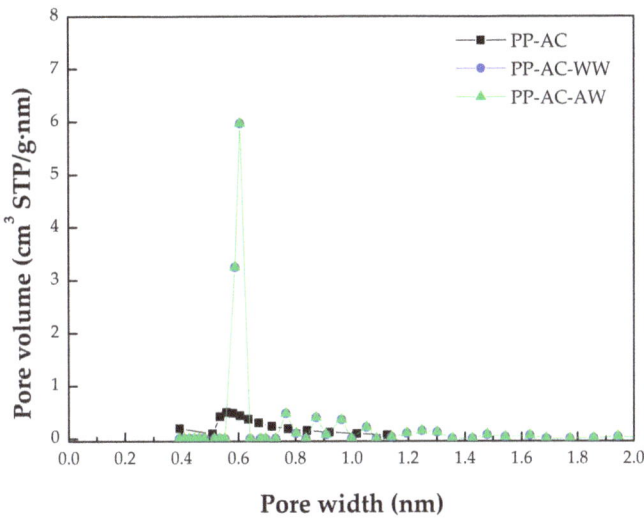

Figure 4. Micropore size distributions of the resulting activated carbon materials.

(a)

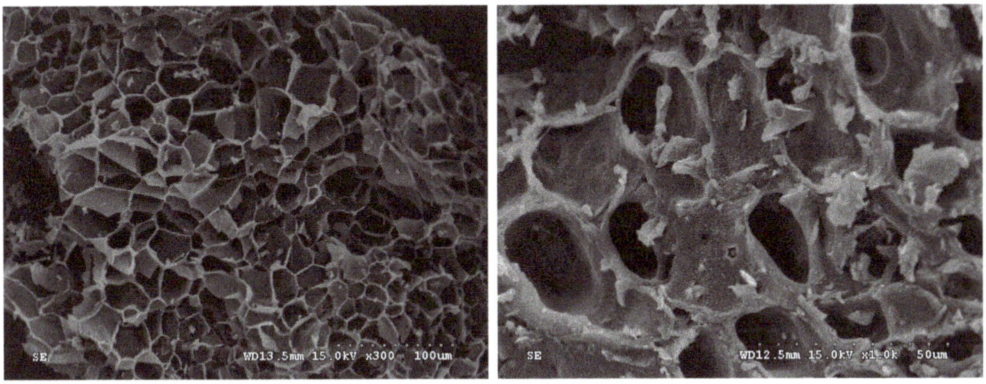

(b)

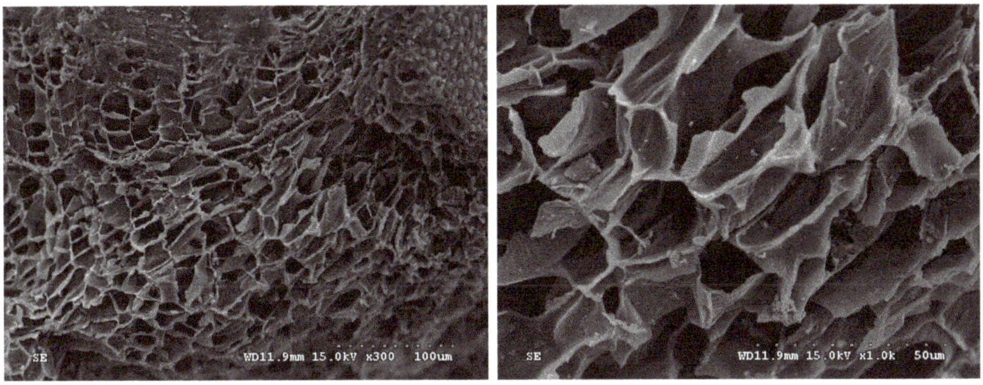

Figure 5. *Cont.*

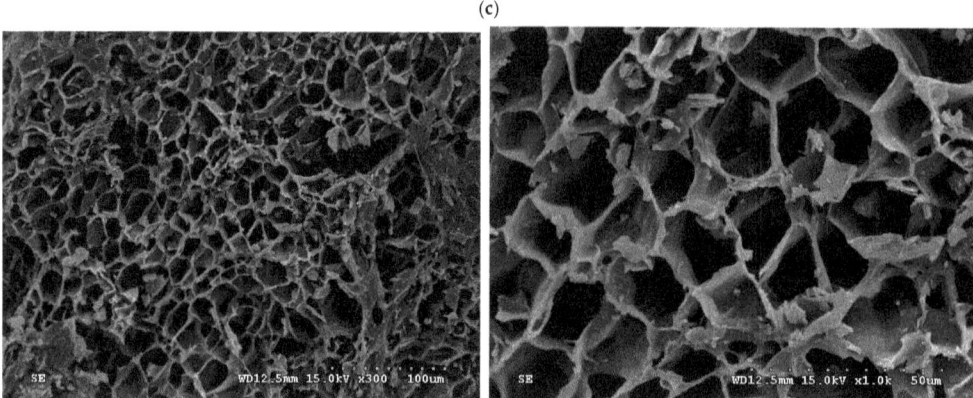

Figure 5. SEM images (left: ×300; right: ×1000) of the resulting activated carbon materials for (**a**) PP-AC, (**b**) PP-AC-WW, and (**c**) PP-AC-AW.

3.3. Chemical Characteristics of Resulting Carbon Materials

The carbon materials also contain to some extent contents of ash, which is derived from the starting feedstock. These inorganic minerals consist mainly of silica, alumina, alkaline/alkaline earth metals, and transition metals in the forms of oxides and carbonates [44]. On the other hand, these carbon materials are generally hydrophilic because of the presence of oxygen-containing complexes on the surface, which should be derived from their lignocellulosic compositions in the starting feedstock. In order to identify the potential for the polar feature, the resulting carbon products were measured by using the Fourier Transform infrared spectroscopy (FTIR) and the dispersive X-ray spectroscopy (EDS). As seen in Figure 6, it depicts the peaks by % transmittance (T) for the corresponding functional groups of the typical activated carbon (PP-AC-AW) in comparison with the starting biomass (PP). It shows seven peaks in Figure 6, indicating the wavenumbers at around 3500, 2924, 2360, 2655, 1550, 1385, 1115, and 737 cm^{-1}, respectively. Based on the FTIR spectra of the corresponding functional groups of carbon materials [45–48], the peak at around 3500 cm^{-1} generally refers to the hydroxyl (O-H) stretching due to the adsorbed or inherent water molecule (H_2O). The peak at about 2924 cm^{-1} could be associated with C-H stretching. The peak at 2360 cm^{-1} may be due to the C≡C stretching in alkyne groups. The peaks (i.e., 1655 and 1550 cm^{-1}) in the range from 1700 to 1550 cm^{-1} are aligned to C=C stretching. The most significant peak at 1385 cm^{-1} can be attributed to oxygen-containing functional groups like C=O and C–O of carboxylic groups, or O–H bending. The peak at 1115 cm^{-1} could correspond to the stretching vibration of the C–O group. The peak at 737 cm^{-1} may be related to the bending vibration of C-H. It also shows a significant difference between the oxygen-containing functional groups of PP-AC-AW and PP. Due to its highly aromatic structure, the spectrum of the resulting activated carbon seems to not have strong peaks, especially in the hydroxyl (O-H) stretching peak at around 3500 cm^{-1}, where it was almost disappeared in the activated carbon spectrum.

Table 2 lists the values of elemental compositions on the surface for all of the resulting carbon materials, which were preliminarily obtained by the energy dispersive X-ray spectroscopy (EDS). Obviously, the carbon contents of the resulting carbon materials (71–89 WT%) were significantly larger than that (54 wt%) of the starting feedstock (i.e., PP), showing that the thermal treatment by carbonization/activation had a determining effect on the textural characteristics. This result was consistent with the variations on the spectra by FTIR analysis. In addition, the carbon contents of the resulting activated carbon materials were smaller than those of the resulting biochar materials. It can be postulated that the oxygen content was reduced because of the reaction of biochar carbon with CO_2 (gasification gas) and the release of CO gas. On the other hand, the oxygen contents of the resulting carbon materials indicated a ramping reduction. This result can be attributed to the devolatilization of non-carbon elements (e.g., oxygen and hydrogen) from the lignocellulosic constituents in the forms of oxygen-containing gases (i.e., CO, CO_2, and H_2O) during the thermochemical process. Concerning the inorganic elements, they could be concentrated on alkaline and alkaline earth metal oxides during the thermal process. These oxygen-containing organic groups and inorganic minerals may increase the hydrophilicity of the resulting PP-based activated carbon materials, which are advantageous for various adsorption treatment processes for effective removal of micropollutants from the aqueous phase. This application can be attributed to the excellent pore structure and electrostatic attraction between the negatively charged surface and cationic targets.

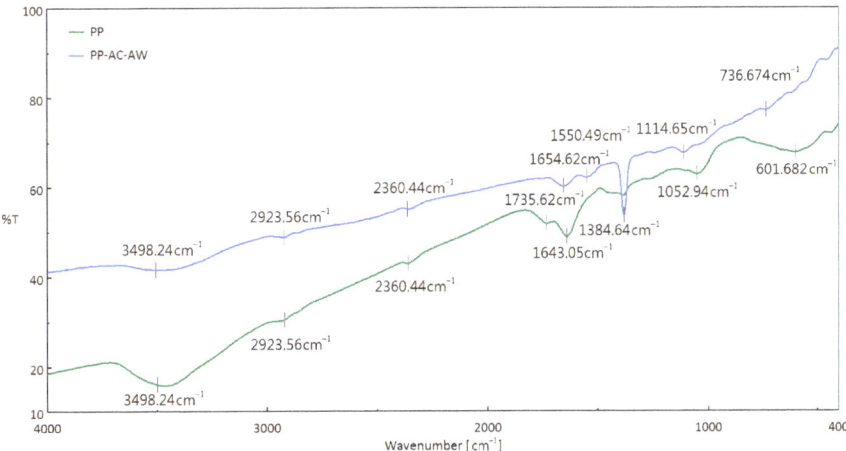

Figure 6. Fourier Transform infrared spectroscopy (FTIR) spectra of pineapple peel (PP) and the resulting activated carbon (PP-AC-AW).

4. Conclusions

The effect of post-washing on the textural characteristics of carbon materials derived from pineapple peel biomass was investigated in this study. These carbon products (i.e., biochar and activated carbon) were produced at an elevated temperature (i.e., 800 °C) holding for 30 min under a combined N_2-pyrolysis and CO_2-activation process. The findings showed that the effect of activation by CO_2 gas on the pore property of the carbon material played a determining role in pore development, with the maximal pore properties for the activated carbon product produced by post-water washing (i.e., BET surface area of 843 m^2/g, and total pore volume of 0.391 cm^3/g). These resulting carbon materials were porous carbon materials with microporous and mesoporous structures. On the other hand, post-washing of the crude carbon products also had an influential effect of enhancing pore property. However, there seemed to be no significant variations in pore property with post-washing methods of water and dilute acid (0.1 M HCl). According to the results of the energy dispersive X-ray spectroscopy (EDS) and the Fourier Transform infrared spectroscopy (FTIR), the chemical characteristics of the resulting carbon materials were rich in carbon (>70 wt%) and also posed hydrophilicity because of the significant oxygen-containing functional groups or complexes on the surface.

Author Contributions: Conceptualization, W.-T.T.; methodology, C.-H.T.; formal analysis, C.-H.T. and L.-A.K.; data curation, C.-H.T.; writing—original draft preparation, W.-T.T.; writing—review and editing, W.-T.T. All authors have read and agreed to the published version of the manuscript.

Funding: This research received no external funding.

Institutional Review Board Statement: Not applicable.

Informed Consent Statement: Not applicable.

Data Availability Statement: Data are contained within the article.

Acknowledgments: Sincere appreciation is expressed to acknowledge the National Pingtung University of Science and Technology for the assistance in the scanning electron microscopy (SEM) analyses.

Conflicts of Interest: The authors declare no conflict of interest.

References

1. Upadhyay, A.; Lama, J.P.; Tawata, S. Utilization of pineapple waste: A review. *J. Food Sci. Technol. Nepal* **2013**, *6*, 10–18. [CrossRef]
2. Cheok, C.Y.; Mohd Adzahan, N.; Abdul Rahman, R.; Zainal Abedin, H.; Hussain, N.; Sulaiman, R.; Chong, G.H. Current trends of tropical fruit waste utilization. *Crit. Rev. Food Sci. Nutr.* **2018**, *58*, 335–361. [CrossRef] [PubMed]
3. Aili Hamzah, A.F.; Hamzah, M.H.; Che Man, H.; Jamali, N.S.; Siajam, S.I.; Ismail, M.H. Recent updates on the conversion of pineapple waste (*Ananas comosus*) to value-added products, future perspectives and challenges. *Agronomy* **2021**, *11*, 2221. [CrossRef]
4. Hikal, W.M.; Mahmoud, A.A.; Ahl, H.A.H.S.-A.; Bratovcic, A.; Tkachenko, K.G.; Kačániová, M.; Rodriguez, R.M. Pineapple (*Ananas comosus* L. Merr.), waste streams, characterisation and valorisation: An overview. *Open J. Ecol.* **2021**, *11*, 610–634. [CrossRef]
5. Fouda-Mbanga, B.G.; Tywabi-Ngeva, Z. Application of pineapple waste to the removal of toxic contaminants: A review. *Toxics* **2022**, *10*, 561. [CrossRef] [PubMed]
6. Tran, T.V.; Nguyen, D.T.C.; Nguyen, T.T.T.; Nguyen, D.H.; Alhassan, M.; Jalil, A.A.; Nabgan, W.; Lee, T. A critical review on pineapple (Ananas comosus) wastes for water treatment, challenges and future prospects towards circular economy. *Sci. Total Environ.* **2023**, *856 Pt 1*, 158817. [CrossRef] [PubMed]
7. Mubarik, S.; Qureshi, N.; Sattar, Z.; Shaheen, A.; Kalsoom, A.; Imran, M.; Hanif, F. Synthetic approach to rice waste-derived carbon-based nanomaterials and their applications. *Nanomanufacturing* **2021**, *1*, 109–159. [CrossRef]
8. Krishni, R.R.; Foo, K.Y.; Hameed, B.H. Food cannery effluent, pineapple peel as an effective low-cost biosorbent for removing cationic dye from aqueous solutions. *Desalin. Water Treat.* **2014**, *52*, 6093–6103. [CrossRef]
9. Ahmad, A.; Khatoon, A.; Mohd-Setapar, S.H.; Kumar, R.; Rafatulah, M. Chemically oxidized pineapple fruit peel for the biosorption of heavy metals from aqueous solutions. *Desalin. Water Treat.* **2016**, *57*, 6432–6442. [CrossRef]
10. Koc, S.N.T.; Kipcak, A.S.; Derun, E.M.; Tugrul, N. Removal of zinc from wastewater using orange, pineapple and pomegranate peels. *Int. J. Environ. Sci. Technol.* **2021**, *18*, 2781–2792.
11. Gómez-Aguilar, D.L.; Rodríguez-Miranda, J.P.; Salcedo-Parra, O.J. Fruit peels as a sustainable waste for the biosorption of heavy metals in wastewater: A review. *Molecules* **2022**, *27*, 2124. [CrossRef]
12. Yılmaz, O.; Tugrul, N. Zinc adsorption from aqueous solution using lemon, orange, watermelon, melon, pineapple, and banana rinds. *Water Pract. Technol.* **2022**, *17*, 318–328. [CrossRef]
13. Marsh, H.; Rodriguez-Reinoso, F. *Activated Carbon*; Elsevier: Amsterdam, The Netherlands, 2006.
14. Basu, P. *Biomass Gasification, Pyrolysis and Torrefaction*, 3rd ed.; Academic Press: San Diego, CA, USA, 2018.
15. Fu, B.; Ge, C.; Yue, L.; Luo, J.; Deng, H.; Yu, H. Characterization of biochar derived from pineapple peel waste and its application for sorption of oxytetracycline from aqueous solution. *BioResources* **2016**, *11*, 9017–9035. [CrossRef]
16. Wang, C.; Gu, L.; Liu, X.; Zhang, X.; Cao, L.; Hu, X. Sorption behavior of Cr(VI) on pineapple-peel-derived biochar and the influence of coexisting pyrene. *Int. Biodeterior. Biodegrad.* **2016**, *111*, 78–84. [CrossRef]
17. Shakya, A.; Agarwal, T. Removal of Cr(VI) from water using pineapple peel derived biochars: Adsorption potential and re-usability assessment. *J. Mol. Liq.* **2019**, *293*, 111497. [CrossRef]
18. Hu, X.; Zhang, X.; Ngo, H.H.; Guo, W.; Wen, H.; Li, C.; Zhang, Y.; Ma, C. Comparison study on the ammonium adsorption of the biochars derived from different kinds of fruit peel. *Sci. Total Environ.* **2020**, *707*, 135544. [CrossRef] [PubMed]
19. Mahmuda, K.N.; Wen, T.H.; Zakaria, Z.A. Activated carbon and biochar from pineapple waste biomass for the removal of methylene blue. *Environ. Toxicol. Manag.* **2021**, *1*, 30–36. [CrossRef]
20. Otieno, A.O.; Home, P.G.; Raude, J.M.; Murunga, S.I.; Ngumba, E.; Ojwang, D.O.; Tuhkanen, T. Pineapple peel biochar and lateritic soil as adsorbents for recovery of ammonium nitrogen from human urine. *J. Environ. Manag.* **2021**, *293*, 112794. [CrossRef]
21. Tsai, W.T.; Ayestas, R.; Tsai, C.H.; Lin, Y.Q. Preparation and characterization of porous materials from pineapple peel at elevated pyrolysis Temperatures. *Materials* **2022**, *15*, 4686. [CrossRef]
22. Foo, K.Y.; Hameed, B.H. Porous structure and adsorptive properties of pineapple peel based activated carbons prepared via microwave assisted KOH and K_2CO_3 activation. *Microporous Mesoporous Mater.* **2012**, *148*, 191–195. [CrossRef]
23. Yacob, A.R.; Azmi, A.; Mustajab, M.K.A.A. Physical and chemical activation effect on activated carbon prepared from local pineapple waste. *Appl. Mech. Mater.* **2014**, *699*, 87–92. [CrossRef]
24. Mahamad, M.N.; Zaini, M.A.A.; Zakaria, Z.A. Preparation and characterization of activated carbon from pineapple waste biomass for dye removal. *Int. Biodeterior. Biodegrad.* **2015**, *102*, 274–280. [CrossRef]
25. Onia, B.A.; Abatana, O.G.; Busarib, A.; Odunlamia, O.; Nweke, C. Production and characterization of activated carbon from pineapple waste for treatment of kitchen wastewater. *Desalin. Water Treat.* **2020**, *183*, 413–424. [CrossRef]

26. Dubey, P.; Shrivastav, V.; Singh, M.; Maheshwari, P.H.; Sundriyal, S.; Dhakate, S.R. Electrolytic study of pineapple peel derived porous carbon for all-solid-state supercapacitors. *ChemistrySelect* **2021**, *6*, 11736–11746. [CrossRef]
27. Veeramalai, S.; Ramlee, N.N.; Mahdi, H.I.; Manas, N.H.A.; Ramli, A.N.M.; Illias, R.M.; Azelee, N.I.W. Development of organic porous material from pineapple waste as a support for enzyme and dye adsorption. *Ind. Crops Prod.* **2022**, *181*, 114823. [CrossRef]
28. Rosli, N.A.; Ahmad, M.A.; Noh, T.U.; Ahmad, N.A. Pineapple peel-derived carbon for adsorptive removal of dyes. *Mater. Chem. Phys.* **2023**, *306*, 128094. [CrossRef]
29. Rosli, N.A.; Ahmad, M.A.; Noh, T.U. Unleashing the potential of pineapple peel-based activated carbon: Response surface methodology optimization and regeneration for methylene blue and methyl red dyes adsorption. *Inorg. Chem. Commun.* **2023**, *155*, 111041. [CrossRef]
30. Gao, Z.; Zhang, Y.; Song, N.; Li, X. Biomass-derived renewable carbon materials for electrochemical energy storage. *Mater. Res. Lett.* **2017**, *5*, 69–88. [CrossRef]
31. Devi, N.S.; Hariram, M.; Vivekanandhan, S. Modification techniques to improve the capacitive performance of biocarbon materials. *J. Energy Storage* **2021**, *33*, 101870. [CrossRef]
32. Mehdi, R.; Khoja, A.H.; Naqvi, S.R.; Gao, N.; Amin, N.A.S. A Review on production and surface modifications of biochar materials via biomass pyrolysis process for supercapacitor applications. *Catalysts* **2022**, *12*, 798. [CrossRef]
33. Tsai, W.T.; Hsu, C.H.; Lin, Y.Q.; Tsai, C.H.; Chen, W.S.; Chang, Y.-T. Enhancing the pore properties and adsorption performance of cocoa pod husk (CPH)-derived biochars via post-acid treatment. *Processes* **2020**, *8*, 144. [CrossRef]
34. Sierra, I.; Iriarte-Velasco, U.; Cepeda, E.A.; Gamero, M.; Aguayo, A.T. Preparation of carbon-based adsorbents from the pyrolysis of sewage sludge with CO_2. Investigation of the acid washing procedure. *Desalin. Water Treat.* **2016**, *57*, 16053–16065. [CrossRef]
35. Liang, Z.; Guo, S.; Dong, H.; Li, Z.; Liu, X.; Li, X.; Kang, H.; Zhang, L.; Yuan, L.; Zhao, L. Modification of activated carbon and-its application in selective hydrogenation of naphthalene. *ACS Omega* **2022**, *7*, 38550–38560. [CrossRef] [PubMed]
36. Wei, X.; Huang, S.; Wu, Y.; Wu, S. Effects of washing pretreatment on properties and pyrolysis biochars of penicillin mycelial residues. *Biomass Bioenergy* **2022**, *161*, 106477. [CrossRef]
37. Tsai, W.T.; Lin, Y.Q.; Tsai, C.H.; Shen, Y.H. Production of mesoporous magnetic carbon materials from oily sludge by combining thermal activation and post-washing. *Materials* **2022**, *15*, 5794. [CrossRef] [PubMed]
38. Tsai, W.T.; Lin, Y.Q. Preparation and characterization of porous carbon composites from oil-containing sludge by a pyrolysis-activation process. *Processes* **2022**, *10*, 834. [CrossRef]
39. Tsai, C.H.; Tsai, W.T. Optimization of physical activation process by CO_2 for activated carbon preparation from Honduras Mahogany pod husk. *Materials* **2023**, *16*, 6558. [CrossRef] [PubMed]
40. Jenkins, B.M.; Baxter, L.L.; Miles, T.R., Jr.; Miles, T.R. Combustion properties of biomass. *Fuel Process. Technol.* **1998**, *54*, 17–46. [CrossRef]
41. Vassilev, S.V.; Baxter, D.; Andersen, L.K.; Vassileva, C.G. An overview of the chemical composition of biomass. *Fuel* **2010**, *89*, 913–933. [CrossRef]
42. Gregg, S.J.; Sing, K.S.W. *Adsorption, Surface Area, and Porosity*; Academic Press: London, UK, 1982.
43. Lowell, S.; Shields, J.E.; Thomas, M.A.; Thommes, M. *Characterization of Porous Solids and Powders: Surface Area, Pore Size and Density*; Springer: Dordrecht, The Netherlands, 2006.
44. Suzuki, M. *Adsorption Engineering*; Elsevier: Amsterdam, The Netherlands, 1990.
45. Li, L.; Yao, X.; Li, H.; Liu, Z.; Ma, W.; Liang, X. Thermal stability of oxygen-containing functional groups on activated carbon surfaces in a thermal oxidative environment. *J. Chem. Eng. Jpn.* **2004**, *47*, 21–27. [CrossRef]
46. Islam, M.S.; Ang, B.C.; Gharekhkhani, S.; Afifi, A.B.M. Adsorption capability of activated carbon synthesized from coconut shell. *Carbon Lett.* **2016**, *20*, 1–9. [CrossRef]
47. Johnston, C.T. Biochar analysis by Fourier-Transform Infra-Red Spectroscopy. In *Biochar: A Guide to Analytical Methods*; Singh, B., Camps-Arbestain, M., Lehmann, J., Eds.; CRC Press: Boca Raton, FL, USA, 2017; pp. 199–228.
48. Qiu, C.; Jiang, L.; Gao, Y.; Sheng, L. Effects of oxygen-containing functional groups on carbon materials in supercapacitors: A review. *Mater. Des.* **2023**, *230*, 111952. [CrossRef]

Disclaimer/Publisher's Note: The statements, opinions and data contained in all publications are solely those of the individual author(s) and contributor(s) and not of MDPI and/or the editor(s). MDPI and/or the editor(s) disclaim responsibility for any injury to people or property resulting from any ideas, methods, instructions or products referred to in the content.

Article

Characterization of Densified Pine Wood and a Zero-Thickness Bio-Based Adhesive for Eco-Friendly Structural Applications

Shahin Jalali [1], Catarina da Silva Pereira Borges [1], Ricardo João Camilo Carbas [1,2,*], Eduardo André de Sousa Marques [2], João Carlos Moura Bordado [3] and Lucas Filipe Martins da Silva [2]

[1] Institute of Science and Innovation in Mechanical and Industrial Engineering (INEGI), Rua Dr. Roberto Frias, 4200-465 Porto, Portugal
[2] Departamento de Engenharia Mecânica, Faculdade de Engenharia, Universidade do Porto, Rua Dr. Roberto Frias, 4200-465 Porto, Portugal; lucas@fe.up.pt (L.F.M.d.S.)
[3] Centro de Recursos Naturais E Ambiennte (CERENA), Instituto Superior Técnico, University of Lisbon, 1049-001 Lisbon, Portugal; jcbordado@tecnico.ulisboa.pt
* Correspondence: rcarbas@fe.up.pt

Abstract: This study investigates a sustainable alternative for composites and adhesives in high-performance industries like civil and automotive. This study pioneers the development and application of a new methodology to characterize a bio-based, zero-thickness adhesive. This method facilitates precise measurements of the adhesive's strength and fracture properties under zero-thickness conditions. The research also encompasses the characterization of densified pine wood, an innovative wood product distinguished by enhanced mechanical properties, which is subsequently compared to natural pine wood. We conducted a comprehensive characterization of wood's strength properties, utilizing dogbone-shaped samples in the fiber direction, and block specimens in the transverse direction. Butt joints were employed for adhesive testing. Mode I fracture properties were determined via compact tension (CT) and double cantilever beam (DCB) tests for wood and adhesive, respectively, while mode II response was assessed through end-loaded split (ELS) tests. The densification procedure, encompassing chemical and mechanical processes, was a focal point of the study. Initially, wood was subjected to acid boiling to remove the wood matrix, followed by the application of pressure to enhance density. As a result, wood density increased by approximately 100 percent, accompanied by substantial improvements in strength and fracture energy along the fiber direction by about 120 percent. However, it is worth noting that due to the delignification nature of the densification method, properties in the transverse direction, mainly reliant on the lignin matrix, exhibited compromises. Also introduced was an innovative technique to evaluate the bio-based adhesive, applied as a zero-thickness layer. The results from this method reveal promising mechanical properties, highlighting the bio-based adhesive's potential as an eco-friendly substitute for synthetic adhesives in the wood industry.

Keywords: bio-based adhesive; densified wood; pine wood; sustainable adhesive joints; zero-thickness adhesive

1. Introduction

In recent years, there has been a growing interest in the development of sustainable alternatives to fossil-fuel-based products. As part of these endeavors, bio-based materials have emerged as a promising solution to mitigate the environmental impact across various industries. Among these applications, the production of load-bearing structures stands out, wherein natural fibers from plants, such as flax, jute, and palm trees, can be integrated into composite materials to create eco-friendlier alternatives. Another approach involves the utilization of natural composite materials directly. Wood, in particular, has been used for various purposes throughout history due to it being a natural and renewable resource. Its durability, environmental resistance, mechanical strength, low weight, flexibility in shaping, abundance in many geographic regions, and reasonable cost make it a desirable material [1–3].

Venkatesan et al. [4] prepared biodegradable composites from poly(butylene adipate-co-terephthalate) (PBAT) and carbon nanoparticles. They characterized these composites in terms of morphology, thermal stability, mechanical properties, and biodegradability. They found that the composites showed improved thermal and mechanical performance compared to pure PBAT, and also exhibited good biodegradability under composting conditions. In another study [5], the authors prepared nanocomposite films from PBAT and zinc oxide nanoparticles, determining the mechanical, thermal, and biological properties of the films. They found that the films had good antimicrobial activity against *E. coli* and *S. aureus*.

Wood is a natural composite material that is susceptible to stress concentrations and notches, making traditional joining methods, such as riveting and bolting, unsuitable. Polymers from liquefied biomass were synthesized and used as wood adhesives. The bio-based polymers exhibited comparable or superior performance to petroleum-based adhesives in terms of bond strength, thermal stability, and water resistance [6].

Adhesive bonding is often seen as a better option as it offers a larger and more uniform bonded area without introducing stress concentrations [7,8]. However, the joint's failure load and mode depend on the combined properties of the adhesive and substrates [9]. When bonding wood substrates, the peel loading can cause delamination between different grain plies, resulting in complete joint failure. To prevent this failure mode, recent techniques, such as densification, substrate toughening, and physical modifications of the adhesive have been proposed [10,11]. Additionally, working with wood substrates can present challenges in terms of mechanical characterization and consistency [12,13], due to wood being a complex and heterogeneous material, with properties that can vary depending on factors such as species, growth conditions, grain slope and size, defects, knots, shakes (cracks and notches of the wood), and age. This can make it difficult to develop standardized testing protocols to ensure a consistent performance in different applications [14,15]. To overcome these problems, wood densification processes can be used to alter the failure mode of this material, improving result consistency. Other methods, like using wood particles or laminates, and wooden composites also can help reduce these design challenges since the material properties can be made to be more uniform. This is particularly important to achieve a predictable behavior in structural applications [16–18].

Additionally, the mechanical properties of hemicellulose and lignin, which are key components of wood, change significantly with service temperature, depending on whether they are above or below their glass transition temperature (T_g). When dry, the T_g of lignin ranges from 134–235 °C, while the T_g of hemicellulose ranges from 167–217 °C. Since the service temperature is mostly below the T_g, these values are generally not a design concern. However, T_g has been observed to significantly decrease as moisture content increases. A decrease in T_g can be problematic as it can lead to reduced mechanical strength, dimensional instability, and increased susceptibility to deformation or failure. It can also affect thermal stability, promoting creep and relaxation at elevated temperatures. Additionally, a lower T_g can accelerate chemical reactions, causing degradation and reducing the material's long-term durability. Moisture acts as a plasticizer that weakens the secondary bond between polymer chains, leading to increased flexibility of the molecules. Proper moisture management is crucial for successful wood densification and preserving structural integrity [19–21]. The sensitivity of wood to moisture leads to a phenomenon called set-recovery, which has implications for both the absorbed energy in calluses and the covalent bonds between polymeric chains.

Set-recovery in wood refers to its ability to partially regain its original shape and dimensions after undergoing deformation or stress. This behavior has an impact on the energy-absorption capacity of calluses (scar tissue) formed within the wood and the strength of the covalent bonds holding the polymer chains together. Sadat Nezhad et al. [22] employed thermo-hydro-mechanical (THM) methods to increase wood density and observed a set-recovery of approximately 44% after three wet–dry cycles. Additionally,

Laine et al. [20] reported a set-recovery of around 60% after saturating densified wood samples with water, which was later reduced to nearly zero through the application of thermal modification post-compression [23].

To address the problem of set-recovery while achieving greater levels of densification, chemical pre-treatments have been utilized [18]. These treatments are based on the delignification process used in the paper industry, which can be adapted to modify wood properties. By removing lignin and hemicellulose, the resulting densified wood exhibits changes in mechanical properties. While the passage does not provide specific details, it suggests that the removal of lignin decreases the stiffness of the wood in the transverse direction. However, the overall effect on stiffness and strength depends on factors such as wood species, processing conditions, and the degree of densification. Additionally, the physical interlocking of cellulose fibers within the densified wood structure enhances its mechanical strength by providing structural support and resistance to deformation. The specific mechanical properties achieved through this treatment process will vary depending on the intended application and desired characteristics [22–24].

Recently, the drive for more sustainable bonded structures has led to the development of various structural and non-structural bio-based adhesives and the adaptation and characterization of many bio-based polymers, which are natural, renewable, and non-petroleum-based, for use in diverse applications. However, the number of materials that can fulfil this role is limited. Tannin, lignin, carbohydrates, unsaturated oils, proteins, and protein hydrolysates are some of the natural materials that have been used as adhesives with good results. In addition, dissolved wood and wood welding with self-adhesion have also been presented as potential alternatives to bonding [25–28].

The current study focuses specifically on natural-oil-based polyurethane adhesive, a bio-based adhesive known for its advantageous properties. Oil-based polyurethane adhesives have shown great potential in providing strong and durable bonds, while also possessing eco-friendly characteristics.

Depending on the polar urethane group employed, a wide range of adhesive behaviors, ranging from rubber-like elasticity to brittle–hard characteristics, can be achieved [29]. These adhesives can be categorized as one-component or two-component systems. Two-component polyurethane adhesives consist of separate isocyanate and polyol components that are mixed prior to application, offering faster curing rates and unlimited depth of cure. In contrast, one-component polyurethane adhesives are prepolymers containing isocyanate groups that react with moisture in the air or on the substrate to cure, eliminating the need for mixing equipment but having limitations in depth of cure [27]. Several recent studies have examined the influence of various factors on the properties of moisture-cured polyurethane (PU) adhesives, shedding light on their potential applications.

This study develops and applies a new methodology to characterize a bio-based zero-thickness polyurethane adhesive that uses 70% natural resources as raw materials. This method allows for the accurate measurement of the strength and fracture properties of the adhesive under zero-thickness conditions. The study also characterizes the densified pine wood, a novel wood product with enhanced mechanical properties, and compares it with natural pine wood. The main contributions of this work are the advancement of the knowledge on bio-based adhesives and densified wood products, and the demonstration of their superior performance over conventional materials. To understand the wood's mechanical properties, dogbone-shaped samples were used for the fiber direction, while block specimens were employed for the transverse direction. Fracture properties were determined through testing compact tension (CT) and end-loaded split (ELS) specimens. For the bio-based adhesive, tensile properties were obtained using butt joints with a wooden substrate, and fracture properties were measured using double cantilever beam (DCB) and ELS joints. This study aims to assess the potential of densified pine wood and the zero-thickness bio-based adhesive as sustainable alternatives by comprehensively characterizing their mechanical properties. Given the focus on structural applications, established processes and testing procedures commonly associated with structural adhesives were employed.

These techniques transcend those typically used for low-strength or wood adhesives. In this context, the tests yield precise material properties, underpinned by rigorous finite element analysis. This analytical approach enables accurate modeling of the tests, ensuring that the obtained properties align with the loads acting on the adhesive layer.

2. Materials

2.1. Wood

2.1.1. Natural Pine Wood

In this study, Pinus pinaster wood (pine wood) sourced from the Alentejo region in South Portugal was used as the main material. Pine wood was chosen as the main material for various reasons, which include its wide availability, low cost, and good mechanical qualities, such as durability, stiffness, and strength. The wood samples were extracted from trees that were 15 years old and located in the coastal area of the region. The age and location of the wood samples are important factors that affect the quality and characteristics of the wood, as they influence its density, moisture content, mechanical properties, and durability. The precise origin of the wood samples was considered in this study, as it can have a significant impact on the performance of the wood-based products. The wood samples were selected based on the criteria proposed by Moura et al. [15], who studied the properties of Pinus pinaster wood from the same regions of Portugal.

The geometries and dimensions of the pine wood blocks, used for both pine wood characterization and the densification process, are shown in Figure 1. As represented, the wood had the rings as parallel as possible to one of the sides of the timber. The initial length of the timber was 1 m, and smaller pieces were cut to the required dimensions for the tests. Pine wood was chosen for its availability, affordability, and favorable mechanical properties, including strength, stiffness, and durability. The elastic constants and strength properties have to be determined in the longitudinal (L), radial (R), and tangential (T) directions. Previous work by Oliviera et al. [30] fully mechanically characterized this type of natural pine wood, and the summarized results are presented in Table 1.

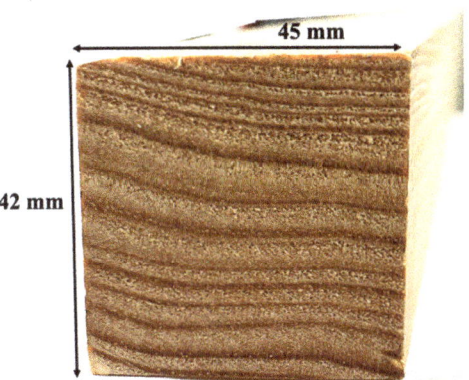

Figure 1. Wood timber dimensions.

Table 1. Elastic properties of pine wood determined by Olivera et al. [30].

E_L [GPa]	E_R [GPa]	E_T [GPa]	ν_{LT}	ν_{LR}	ν_{TR}	G_{LR} [GPa]	G_{LT} [GPa]	G_{TR} [GPa]
12.0	1.9	1.0	0.5	0.4	0.3	1.1	1.0	0.3

2.1.2. Densified Pine Wood

The blocks used for wood densification were cut from natural pine wood into pieces measuring 45 × 40 mm, with an average length of 240 mm.

The densification process was based on the method developed by Song et al. [31] and involved two main steps, as in Figure 2. During the first step, wood blocks were boiled in a chemical bath containing a solution of 2.5 M NaOH and 0.4 M Na_2SO_3 for seven hours, allowing the chemical catalyst to penetrate the cell walls and increase cell volume. This resulted in the destruction of hemicellulose and lignin matrices. Subsequently, the blocks were boiled in deionized water for an hour to remove the catalyst, with the reaction continuing until complete elimination. To ensure thorough chemical removal, the deionized water was changed every 30 min. Water absorption during this step further increased cell volume, creating empty spaces between the cells filled with water. The second step of the densification process involved a thermo-mechanical procedure. The wood blocks were placed in a hot-press for 24 h under a pressure of 3 MPa and 100 °C using a steel mold developed in a previous study [32], to compress and deform the cell walls. This caused the collapse of the cell walls without damaging the fibers, resulting in increased density and strength. Maintaining precise humidity levels was paramount throughout the production of both pine wood and densified pine wood. Typically, wood undergoes conditioning to achieve a moisture content ranging between 12% and 20%. This is done by exposing the wood to controlled humidity and temperature conditions for a certain period of time. For densified pine wood, maintaining moisture content within this specified range was of utmost importance. This ensures that the material preserves its intended strength, flexibility, and dimensional stability. Deviations from these critical moisture levels could cause issues such as warping, cracking, or a decline in mechanical properties, all of which could significantly affect the quality and performance of the final product. The moisture content range of 12% to 20% was chosen based on the expected service conditions of the densified pine wood products, as well as the recommendations from previous studies on wood densification. To further curtail moisture content of densified wood and align it with that of pine wood, a meticulous process was employed. Therefore, densified wood blocks were kept in silica gel (with a diameter of 2–5 mm with moisture indicator changing the color), maintained at a temperature of 70 °C for a duration of 48 h. This method harmonized the moisture content, enhancing both durability and resistance to chemical reactions.

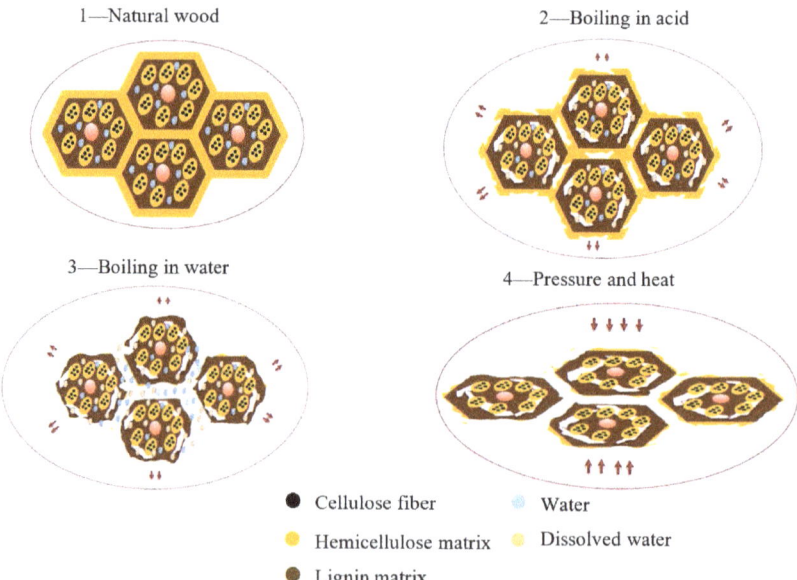

Figure 2. Simplified scheme illustrating the modification of wood's cellular structure through a delignification process involving a chemical reaction, followed by a compression stage.

2.2. Bio-Based Adhesive

A polyurethane bio-based adhesive, derived from 70% of renewable biomass sources, such as vegetable oils according to the ASTM D6866 standard, was characterized. This adhesive was designed for strong adhesion to wood which is a prototype product, developed by the team of Professor João Bordado at Instituto Superior Técnico. It is not yet commercially available, but it shows potential as a sustainable alternative to synthetic adhesives. It is produced in an irreversible reaction, without humidity, in a reactor under a nitrogen atmosphere, and heating is done with a thermal oil coil. It uses an aliphatic isocyanate as a basis, which contains 70% plant matter, which are more easily biodegradable. Manufacturing the bio-adhesive is estimated to consume 15 to 20% less energy than those derived from petroleum. The bio-adhesive contains pentamethylene diisocyanate and polyisocyanate, which react with the hydroxyl (OH) groups in the wood substrate. This reaction creates strong bonds, and, therefore, increasing the humidity of the substrates could speed up the curing process. To ensure uniform curing of the bio-adhesive in the joints, it was important to keep consistent moisture levels across all samples. The synergistic interplay between the bio-adhesive and the wood's OH groups, along with the influence of humidity, brings about multiple benefits. Not only does it enhance mechanical interlocking, bolstering the physical and mechanical bonds within the joints, but it also facilitates superior chemical bonding. High-strength oak wood was used as the substrate for reliable testing. Curing bulk samples was challenging due to the zero-thickness bond requirement. Also, the absence of adhesive thickness was confirmed through the bonding of substrates directly to each other by applying pressure without using any spacer, thereby ensuring a 'zero-thickness' condition. The adhesive undergoes an initial curing phase at 100 °C for 8 h, followed by a recommended 48 h curing period at room temperature, as suggested by the developer. This approach ensures that the curing conditions are in line with the recommendations of the data sheet.

3. Experimental Details

The mechanical tests described in this section were conducted under quasi-static conditions, using a uniaxial universal testing machine (Instron 3367, united states of America based) with a load cell capacity of 30 kN and a displacement rate of 1 mm/min. For each condition, at least three specimens were tested, and also the dimensions of all tested specimens were controlled using a caliper with an accuracy of 0.1 mm.

3.1. Characterization of the Natural Pine Wood and Densified Pine Wood
3.1.1. Density Measurement

To evaluate the effect of the wood densification procedure on the actual densification of wood, the volume of each wood block when natural and densified was measured, as well as its mass. The dimensions were measured using a caliper with an uncertainty of 0.05 mm and the mass using a digital scale with an uncertainty of 0.01 g. The density was then determined by calculating the quotient between the mass and the volume of the block.

3.1.2. Strength Tests

Bulk tensile test

To assess the strength along the fiber direction, dogbone-shaped samples were manufactured for both natural pine wood and densified wood, as shown in Figure 3a. Reduced scale specimens were used due to the geometric restrictions of the densified wood block. These specimens were validated against standard specimens of pine wood, comparing with the standard specimen results of Moura et al. [15]. To prevent sample failure at the grips, 1 mm thick steel tabs were bonded to both ends of the samples. Adhesive fillets were also applied to ensure a more uniform stress transfer to the gage length of the specimens; as depicted in Figure 3b, five specimens were tested.

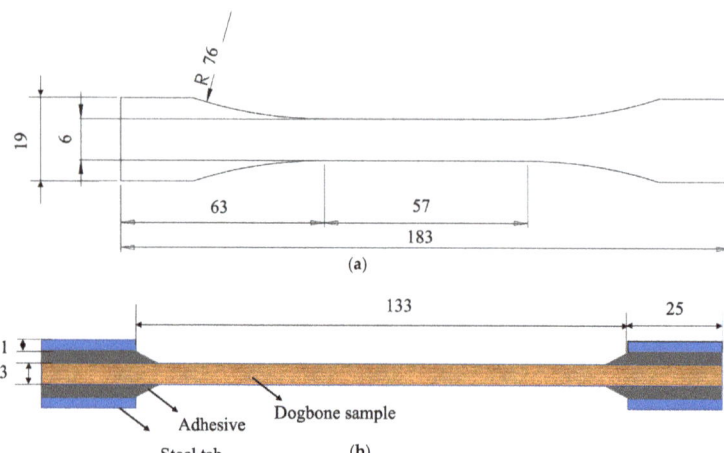

Figure 3. Geometry of wooden dogbone specimen (**a**), and attachment of tabs for testing (**b**). (Dimensions in mm).

Block specimen test

To determine the strength of the wood in the direction parallel to the fibers, blocks of wood and densified wood were cut to dimensions of 20 × 25 × 20 mm. In order to mount the wood in the testing machine, steel blocks (shown in Figure 4) were bonded to the wood using Araldite AV138, an epoxy adhesive. The adhesive was cured for 24 h at room temperature. Once cured, the adhesive fillets were carefully cleaned using sandpaper to ensure accurate measurements. To minimize any potential influence of the adhesive and steel specimens on the measurements, the strain field of all samples was obtained using digital image correlation (DIC); four specimens were tested. This helped in accurately assessing the properties of the wood without interference from the bonding materials.

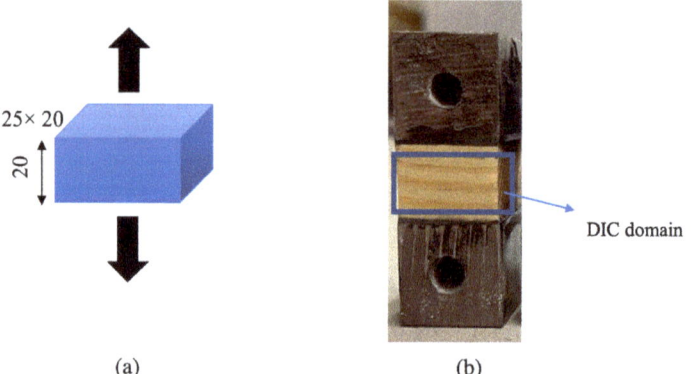

Figure 4. Block specimen geometry and dimensions (**a**), and digital image correlation (DIC) domain (**b**). (Dimensions in mm).

3.1.3. Fracture Tests

Compact tension test (CT)

CT specimens were manufactured to determine the mode I fracture energy of wood. The CT test is commonly used to determine the fracture toughness of brittle materials such as wood. These CT specimens were designed with a centrally located crack in both the fiber and transverse directions, which was loaded in tension to create a pure mode I loading.

The dimensions of the CT specimens are presented in Figure 5a. Six specimens were tested to ensure the repeatability of the results.

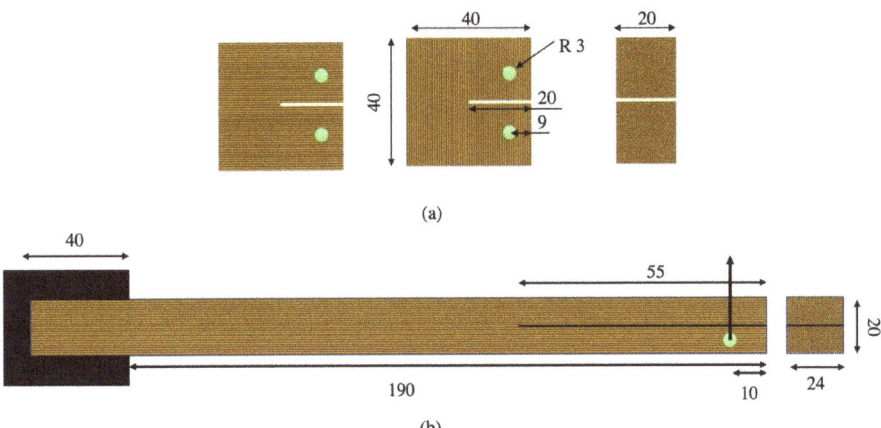

Figure 5. Representation of the test specimens used to conduct compact tension (CT) tests (a), and end-loaded split (ELS) tests (b). (Dimensions in mm).

End-loaded split test (ELS)

ELS specimens were produced to determine the mode II fracture energy of the wood. Similar to the CT specimens, the ELS specimens were designed with a centrally located crack, which was loaded in a transverse direction to achieve a pure mode II condition. Due to the geometrical constraints of the densified wood block, the length of ELS specimens in the fiber direction was limited to 230 mm. Three specimens were tested to ensure the repeatability of the results. The dimensions of the ELS specimens are shown in Figure 5b. Mode II fracture energy is the energy required to propagate a crack perpendicular to the direction of the applied load. To calculate the fracture energy from the load displacement curves, the compliance-based beam method (CBBM) [33] was chosen as the preferred data-reduction approach. CBBM [33] was used to determine the fracture energy without the need for measuring crack propagation during testing. Additionally, CBBM takes into account the fracture process zone (FPZ) formed ahead of the crack tip allowing for the calculation of a corrected or equivalent crack length (a_{eq}). In this study, load-displacement data obtained from the universal tensile test machine were used to compute the fracture energy using CBBM.

3.2. Characterization of Bio-Based Adhesive

This study focused on characterizing a prototype adhesive that relies on the moisture in wood substrates for curing, necessitating a zero-thickness bond. To ensure that failure occurred only in the adhesive and not the wood, stronger oak wood was used. Nonetheless, surface preparation was critical, involving polishing with 400-grade sandpaper for a smooth, uniform surface. Compressed air was used to remove dust particles that could hinder effective bonding, and acetone was applied for thorough cleaning, removing any contaminants.

3.2.1. Strength Tests

Butt-joint test

To measure the strength properties of the bio-based adhesive, wood butt joints were used. To prepare these joints, oak wood with an area of 20 × 25 mm was cut with the thickness of 10 mm (Figure 6a). After the surface preparation described above, the bio-based adhesive was applied on the surfaces of both the substrates. The substrates were then bonded to each other carefully, to avoid any misalignments, and pressure was applied to ensure even contact between the wood substrates, using a clamp. The adhesive was cured and steel blocks to allow for testing were bonded as described for the wood block specimens.

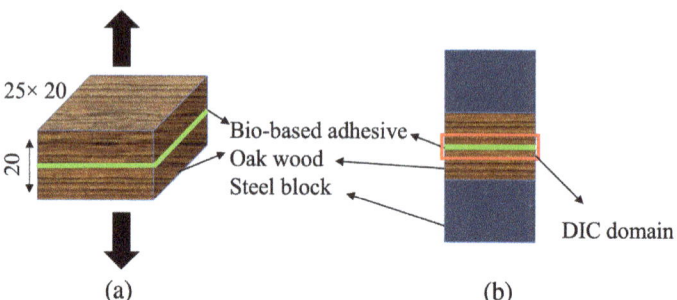

Figure 6. Butt-joint geometry and dimensions (**a**); considered digital image correlation (DIC) domain (**b**). (Dimensions in mm).

DIC was used to measure strain near the bondline (Figure 6), through a speckle pattern introduced to the bonded area. This approach eliminated additional elongation caused by the wood, resulting in a precise determination of the bio-based-adhesive strain.

Thick adherend shear test (TAST)

The TAST was used to assess the shear properties of the adhesive. The testing setup consisted of two different joint configurations, designed to explore variations in substrate geometry and thickness. The two joints are shown in Figure 7a,b. The first joint, shown in Figure 7a, featured a thicker substrate loaded perpendicular to the grain direction, while the second joint, shown in Figure 7b, employed a slightly thinner substrate, loaded in the grain direction.

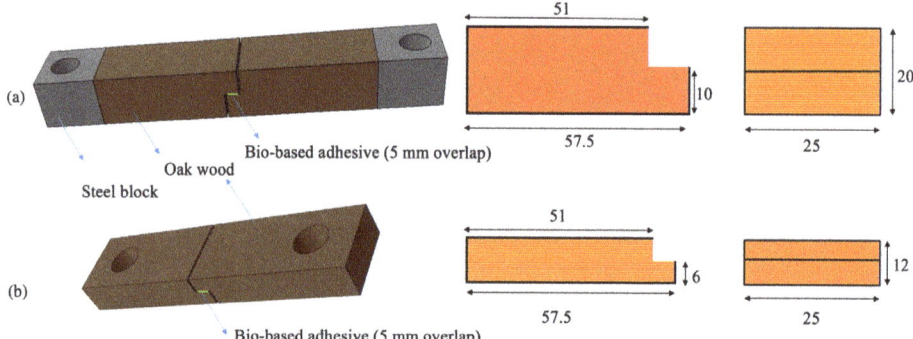

Figure 7. Thick adherend shear test (TAST) specimen geometries with the wood loaded in the transverse (**a**) and fiber direction (**b**). (Dimensions in mm).

3.2.2. Fracture Tests

Double cantilever beam test (DCB)

To determine the mode I fracture energy of the bio-based adhesive, DCB specimens were used. Following the surface-preparation procedures, the bio-based adhesive was applied to both sides of the joint. In order to introduce a pre-crack of 45 mm, a 0.1 mm thick Teflon film was placed between the two substrates. To ensure uniformity in the joint-preparation process, a total of four clamps were employed, applying controlled pressure over the entire bondline. The testing process incorporated specimens with specific dimensions and geometries, as depicted in Figure 8a. To calculate the fracture energy of the bio-based adhesive under quasi-static conditions, CBBM [33] was utilized.

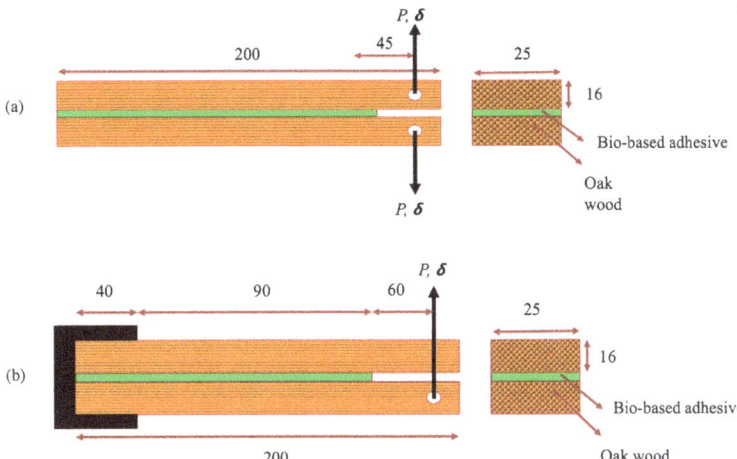

Figure 8. Representation of the test specimens used to conduct double cantilever beam test (DCB) tests (a) and end-loaded split (ELS) tests (b). (Dimension in mm).

End-loaded split test (ELS)

To investigate the mode II fracture energy of the bio-based adhesive, ELS specimens were employed in the experimental procedure. The manufacturing process followed the same set of procedures used for the DCB, since it has the same joint geometry. For the ELS joints, a pre-determined pre-crack length of 60 mm was established. The specimen geometry and the testing procedure are shown in Figure 8b.

4. Results and Discussion

4.1. Characterization of the Wood and Densified Wood

4.1.1. Density Measurement

The pine wood displayed a substantial increase in its average density, changing from 0.56 ± 0.03 g/cm^3 to 1.23 ± 0.12 g/cm^3 (Figure 9). These findings demonstrate a successful enhancement of the wood's density through the densification process.

Figure 9. Wood beam before (left) and after (right) densification.

4.1.2. Strength Tests

Figure 10 illustrates a representative stress–strain curve obtained from the manufactured samples of both pine wood and densified pine wood for strength properties. The determined values of Young's modulus and strength in the fiber and transverse direction can be seen in Table 2.

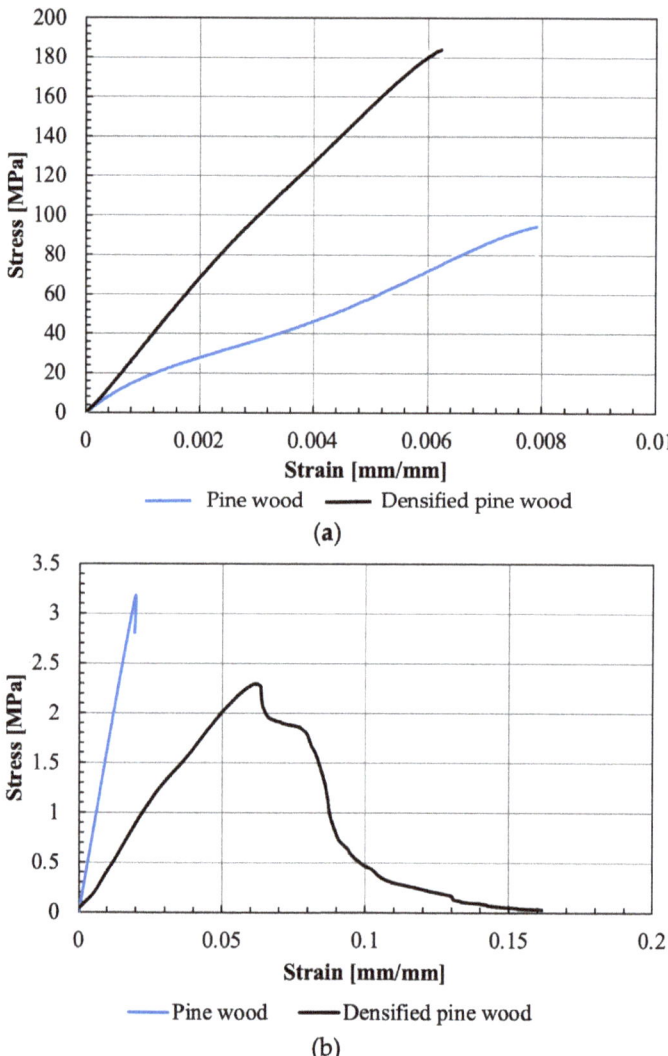

Figure 10. Representative tensile stress–strain curve of wood and densified wood in the tensile (**a**) and transverse (**b**) directions.

In the case of pine wood, the results for natural pine wood were in line with the study by Moura et al. [15], that reported a Young's modulus and strength of 12 GPa and 97.5 MPa, respectively. This validates the reduced-scale dogbone specimens used. Regarding transverse properties, Moura et al. [15] registered a strength of 4.2 MPa, slightly higher than the values obtained with the block specimens used in this study.

Table 2. Strength properties of wood (W) and densified wood (D).

	$E_{\text{Fiber direction}}$ [GPa]	$\sigma_{\text{Fiber direction}}$ [MPa]	$E_{\text{Transverse direction}}$ [MPa]	$\sigma_{\text{Transverse direction}}$ [MPa]
W	12 ± 1	97.3 ± 8.3	155 ± 22	2.9 ± 0.2
D	31 ± 1	180.1 ± 12.9	43 ± 3	2.1 ± 0.1

Densified pine wood displayed a significant increase in tensile strength and stiffness, due to the significant increase in the volume fraction of the fibers in the wood. However, unlike what was found for the properties in the fiber direction, the densification process resulted in a decrease in the transverse properties of pine wood, as the wood matrix (responsible for the transverse strength) was degraded during this process.

4.1.3. Fracture Tests

The fracture toughness for natural pine wood and densified pine wood obtained are presented in Table 3. For the CT tests, it was seen that, both in the fiber direction and the transverse direction, the crack propagated between the grains of wood, as shown in Figure 11. This demonstrates that the densification process significantly enhances the fracture properties of wood.

Table 3. Fracture properties of wood (W) and densified wood (D), along fiber direction (LR) and perpendicular to the fiber (RL).

	W_{LR}	D_{LR}	W_{RL}	D_{RL}
K_{IC} [MPa/m]	17.7 ± 0.3	39.4 ± 1.5	24.0 ± 4.1	128.1 ± 15.1
G_{IC} [N/mm]	0.20 ± 0.01	0.76 ± 0.09	0.33 ± 0.04	0.75 ± 0.10

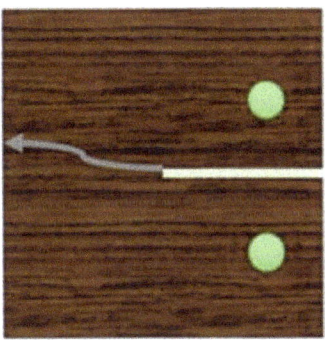
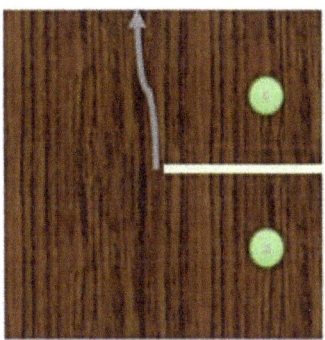

LR　　　　　　　　　RL

Figure 11. Crack propagating along the grains perpendicular (LR) and parallel to the loading direction (RL).

Mode II fracture energy of densified wood was measured by testing ELS specimens, with CBBM being used to generate an R-curve from a P-δ curve. The average mode II fracture energy of pine wood and densified pine wood was about 0.9 ± 0.1 N/mm, and 1.7 ± 0.2 N/mm, respectively. Typical behavior of ELS specimens is shown in the P-δ curve ELS in Figure 12a and the obtained R-curve is presented in Figure 12b.

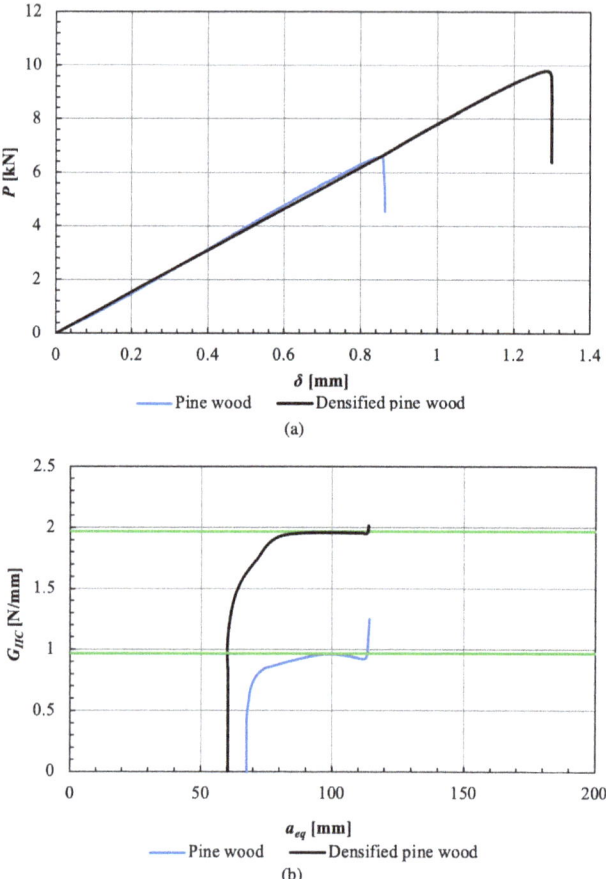

Figure 12. ELS test results for the woods: P-δ curve (**a**) and R-curve (**b**). Green lines presented the obtained fracture energy.

The densification process increases wood density, leading to improved stiffness, strength, and dimensional stability. The alignment of fibers during densification further enhances mechanical properties, making it suitable for structural applications. In the fiber direction, densified wood showed a 90% increase in Young's modulus and an 85% increase in strength, while transverse properties decreased due to lignin- and wood-matrix destruction. However, overall, densified wood remains still appear to be a valid option for structural applications and a sustainable alternative to traditional materials. Finally, the fracture properties of densified wood exhibited higher resistance to failure, with significant improvements in fracture toughness and energy.

4.2. Characterization of Bio-Based Adhesive

4.2.1. Strength Tests

Figure 13a presents stress–strain curves for the bio-based adhesive using DIC to obtain the described. The Young's modulus obtained was 1.32 ± 0.05 GPa, and the average tensile strength of the adhesive was 16.45 ± 0.55 MPa, values comparable to petroleum-based structural adhesives. To ensure the testing process' accuracy, the fracture surfaces were meticulously analyzed to verify cohesive failure in the adhesive. This was confirmed by the presence of adhesive on top of both substrates. All samples exhibited visibly cohesive failure without any delamination, which indicates that the bio-based adhesive exhibited an appropriate bonding strength to the oak-wood substrates (see Figure 13b).

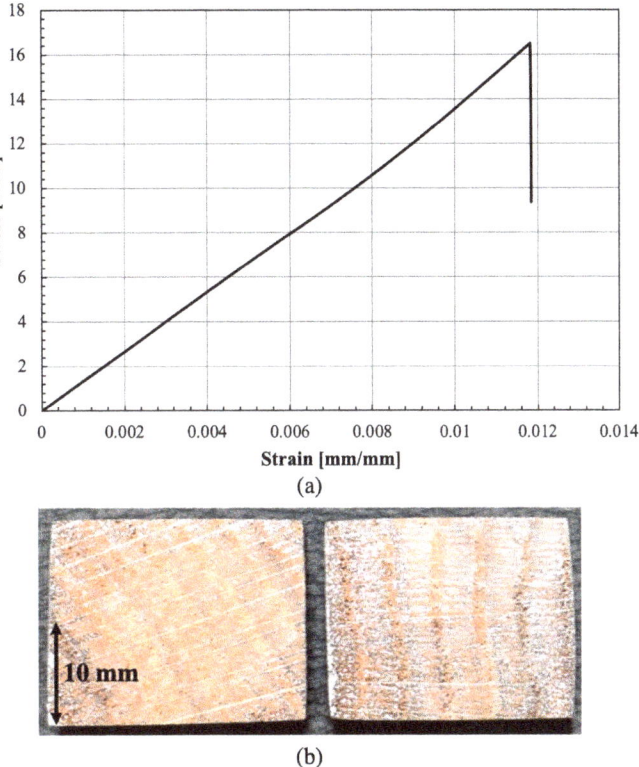

Figure 13. Representative tensile stress–strain curve of the bio-based adhesive (a), and fracture surface (b).

During the testing process aimed at determining the shear strength and modulus of the adhesive, consistent failure occurred across the wood substrate, resulting in either the wood breaking or delamination (Figure 14). Consequently, it became evident that accurately measuring the shear properties of the adhesive itself was not feasible through this method. The main challenge arose from the nature of the adhesive's curing process, which solely takes place on wood surfaces. This leads to interpenetration between the adhesive and the wood substrate, contributing to the overall joint strength. As a result, it was not possible to isolate and accurately quantify the shear properties of the adhesive alone.

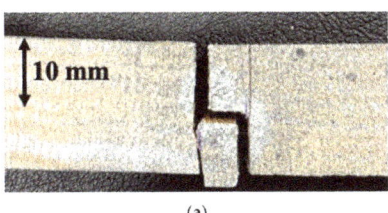 

(a) (b)

Figure 14. Fracture surface of TAST specimen: substrate loaded in transverse direction (a); substrate loaded in fiber direction (b).

4.2.2. Fracture Tests

The fracture toughness of the adhesive in mode I was obtained from the P-δ curve of the DCB test, shown in Figure 15a, and through applying CBBM and generating the R-curve, shown in Figure 15b. The average mode I fracture energy of the bio-based adhesive was 0.33 ± 0.03 N/mm. The samples exhibited cohesive failure, thus returning a fracture toughness representative of that of the adhesive layer, shown in Figure 16.

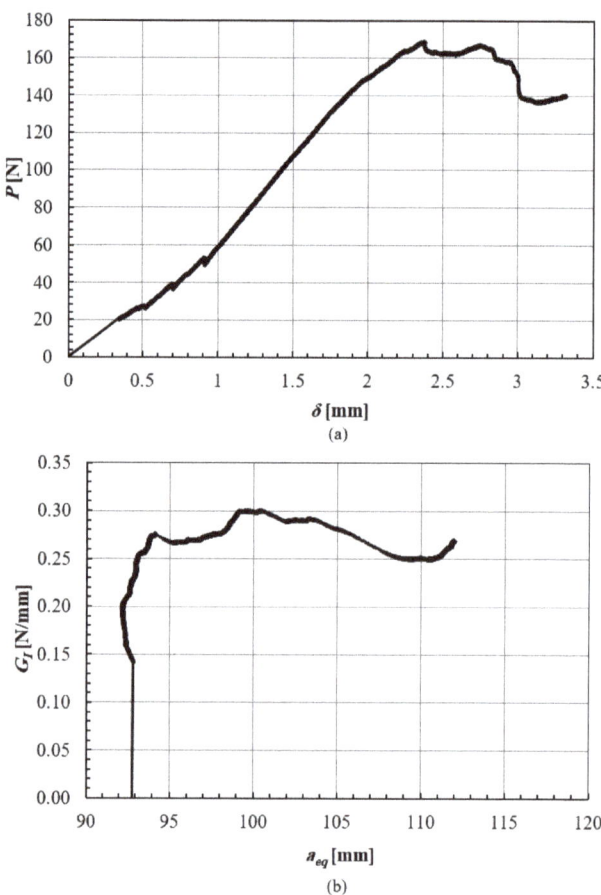

Figure 15. DCB test results for the adhesive: P-δ curve (**a**) and R-curve (**b**).

Figure 16. DCB specimens' fracture surfaces.

The same procedure was conducted for mode II from the P-δ curve of the ELS test, shown in Figure 17a, and through applying CBBM and obtaining the R-curve, shown in Figure 17b. The average mode II fracture energy was determined to be 1.2 ± 0.2 N/mm.

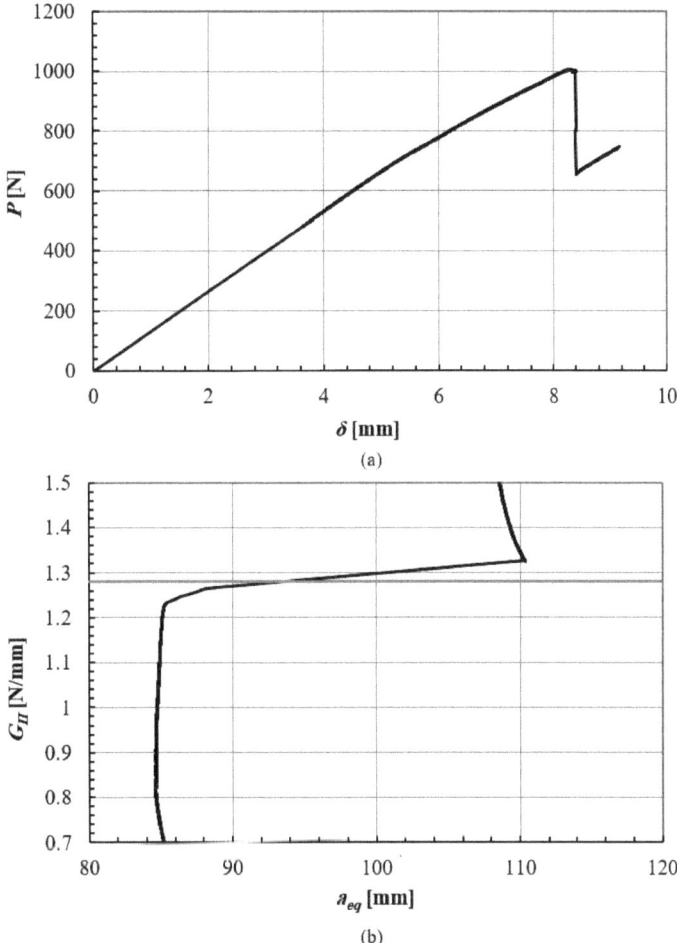

Figure 17. ELS test results for the adhesive: P-δ curve (**a**) and R-curve (**b**). Blue line presented the obtained fracture energy.

Typically, the average fracture energy for mode I fracture in commercial synthetic-urea formaldehyde adhesives falls within the range of 0.1 to 0.2 N/mm. For mode II fracture, this value extends between 0.2 to 0.4 N/mm [34,35]. In contrast, the bio-based adhesive showcased substantially elevated fracture energy values when compared to these conventional synthetic-urea–formaldehyde adhesives indicating its potential to enhance the durability and dependability of wooden products. In conclusion, the study highlights the promising potential of the bio-based adhesive as a sustainable alternative for the wood industry, offering superior fracture resistance and reliability for various fracture modes.

The cohesive properties of the bio adhesive are reviewed in Table 4.

Table 4. Bio-based-adhesive properties.

E [MPa]	σ [MPa]	G_{IC} [N/mm]	G_{IIC} [N/mm]
197.09 ± 9.76	3.27 ± 0.14	0.30 ± 0.03	1.27 ± 0.10

The study's results reveal the large potential of the studied bio-based polyurethane adhesive, which has demonstrated good strength and fracture energy for a bio-derived product. This performance positions it as a compelling choice for high-performance applications across multiple industries where sustainability is mandatory. This research also highlights the adhesive's role as an eco-conscious alternative to traditional adhesives, driven by its outstanding bonding capabilities, environmentally friendly composition, and reduced emissions of harmful compounds. It is essential to note that the adhesive's advantages extend beyond the study's scope, encouraging future research to explore a wider array of applications and delve deeper into the molecular structure and formulation for further enhancements.

5. Conclusions

The study examined natural pine wood and its densified counterpart, which showed significant increases in density, stiffness, and strength (120%, 44%, and 85%, respectively). However, densification negatively impacted transverse properties due to lignin- and wood-matrix destruction. Nevertheless, densified wood remains promising for structural applications given the very large increases in strength and stiffness attained in the fiber direction.

The bio-based-adhesive characterization process revealed a significant tensile strength and cohesive failure, indicating strong adhesion to wood surfaces. Compared to synthetic adhesives, the bio-based adhesive exhibited high fracture energy values, offering a reliable and sustainable alternative. Overall, the combination of densified pine wood and the bio-based adhesive enhances mechanical properties and adhesion capabilities, with potential implications for various applications, improving performance, durability, and sustainability of wood-related products.

Author Contributions: Conceptualization, E.A.d.S.M., J.C.M.B. and L.F.M.d.S.; Methodology, S.J., C.d.S.P.B. and R.J.C.C.; Validation, C.d.S.P.B., R.J.C.C., E.A.d.S.M., J.C.M.B. and L.F.M.d.S.; Formal analysis, S.J., C.d.S.P.B., R.J.C.C., J.C.M.B. and L.F.M.d.S.; Investigation, S.J. and C.d.S.P.B.; Resources, C.d.S.P.B. and E.A.d.S.M.; Data curation, S.J.; Writing—original draft, S.J.; Writing—review & editing, C.d.S.P.B., R.J.C.C., E.A.d.S.M., J.C.M.B. and L.F.M.d.S.; Supervision, R.J.C.C. and L.F.M.d.S.; Project administration, R.J.C.C.; Funding acquisition, L.F.M.d.S. All authors have read and agreed to the published version of the manuscript.

Funding: This research was funded by the Project No. PTDC/EME-EME/6442/2020 "A smart and eco-friendly adhesively bonded structure for the next generation mobility platforms", and the individual grants 2022.12426.BD and CEECIND/03276/2018, funded by national funds through the Portuguese Foundation for Science and Technology (FCT).

Institutional Review Board Statement: Not applicable.

Informed Consent Statement: Not applicable.

Data Availability Statement: Data are contained within the article.

Conflicts of Interest: The authors declare no conflict of interest.

References

1. Yang, C.Z.; Yaniger, S.I.; Jordan, V.C.; Klein, D.J.; Bittner, G.D. Most plastic products release estrogenic chemicals: A potential health problem that can be solved. *Environ. Health Perspect.* **2011**, *119*, 989–996. [CrossRef]
2. Hahladakis, J.N.; Velis, C.A.; Weber, R.; Iacovidou, E.; Purnell, P. An overview of chemical additives present in plastics: Migration, release, fate and environmental impact during their use, disposal and recycling. *J. Hazard. Mater.* **2018**, *344*, 179–199. [CrossRef]
3. Borges, C.S.P.; Akhavan-Safar, A.; Tsokanas, P.; Carbas, R.J.C.; Marques, E.A.S.; da Silva, L.F.M. From fundamental concepts to recent developments in the adhesive bonding technology: A general view. *Discov. Mech. Eng.* **2023**, *2*, 8. [CrossRef]

4. Venkatesan, R.; Surya, S.; Suganthi, S.; Muthuramamoorthy, M.; Pandiaraj, S.; Kim, S.-C. Biodegradable composites from poly(butylene adipate-co-terephthalate) with carbon nanoparticles: Preparation, characterization and performances. *Polym. Adv. Technol.* **2020**, *31*, 2267–2279. [CrossRef]
5. Venkatesan, R.; Rajeswari, N. ZnO/PBAT nanocomposite films: Investigation on the mechanical and biological activity for food packaging. *Polym. Adv. Technol.* **2016**, *27*, 1632–1640. [CrossRef]
6. He, Z. (Ed.) *Bio-Based Wood Adhesives: Preparation, Characterization, and Testing*; CRC Press: Boca Raton, FL, USA, 2017.
7. Gajula, S.; Antonyraj, C.A.; Odaneth, A.A.; Srinivasan, K. A consolidated road map for economically gainful efficient utilization of agro-wastes for eco-friendly products. *Biofuels Bioprod. Biorefining* **2019**, *13*, 899–911. [CrossRef]
8. Zhang, Y.; Duan, C.; Bokka, S.K.; He, Z.; Ni, Y. Molded fiber and pulp products as green and sustainable alternatives to plastics: A mini review. *J. Bioresour. Bioprod.* **2022**, *7*, 14–25. [CrossRef]
9. Jalali, S.; Ayatollahi, M.R.; Akhavan-Safar, A.; da Silva, L.F.M. Effects of impact fatigue on residual static strength of adhesively bonded joints. *Proc. Inst. Mech. Eng. Part L J. Mater. Des. Appl.* **2021**, *235*, 1519–1523. [CrossRef]
10. Oliveira, P.R.; May, M.; Panzera, T.H.; Scarpa, F.; Hiermaier, S. Reinforced biobased adhesive for eco-friendly sandwich panels. *Int. J. Adhes. Adhes.* **2020**, *98*, 102550. [CrossRef]
11. Budhe, S.; Banea, M.D.; De Barros, S.; Da Silva, L.F.M. An updated review of adhesively bonded joints in composite materials. *Int. J. Adhes. Adhes.* **2017**, *72*, 30–42. [CrossRef]
12. Yildizhan, Ş.; Çalik, A.; Özcanli, M.; Serin, H. Bio-composite materials: A short review of recent trends, mechanical and chemical properties, and applications. *Eur. Mech. Sci.* **2018**, *2*, 83–91. [CrossRef]
13. Gurr, J.; Barbu, M.C.; Frühwald, A.; Chaowana, P. The bond strength development of coconut wood in relation to its density variations. *J. Adhes.* **2022**, *98*, 1520–1533. [CrossRef]
14. Souza, A.M.; Nascimento, M.F.; Almeida, D.H.; Silva, D.A.L.; Almeida, T.H.; Christoforo, A.L.; Lahr, F.A.R. Wood-based composite made of wood waste and epoxy based ink-waste as adhesive: A cleaner production alternative. *J. Clean. Prod.* **2018**, *193*, 549–562. [CrossRef]
15. De Moura, M.F.S.F.; Dourado, N. *Wood Fracture Characterisation*; CRC Press: Boca Raton, FL, USA, 2018.
16. Yang, H.; Gao, M.; Wang, J.; Mu, H.; Qi, D. Fast Preparation of high-performance wood materials assisted by ultrasonic and vacuum impregnation. *Forests* **2021**, *12*, 567. [CrossRef]
17. Grunwald, C.; Fecht, S.; Vallée, T.; Tannert, T. Adhesively bonded timber joints–Do defects matter? *Int. J. Adhes. Adhes.* **2014**, *55*, 12–17. [CrossRef]
18. Modenbach, A.A.; Nokes, S.E. Effects of sodium hydroxide pretreatment on structural components of biomass. *Trans. ASABE* **2014**, *57*, 1187–1198.
19. Morsing, N. Densification of Wood. The Influence of Hygrothermal Treatment on Compression of Beech Perpendicular to Gain. Ph.D. Thesis, Technical University of Denmark, Kongens Lyngby, Denmark, 2000.
20. Laine, K.; Belt, T.; Rautkari, L.; Ramsay, J.; Hill, C.A.S.; Hughes, M. Measuring the thickness swelling and set-recovery of densified and thermally modified Scots pine solid wood. *J. Mater. Sci.* **2013**, *48*, 8530–8538. [CrossRef]
21. Schwarzkopf, M. Densified wood impregnated with phenol resin for reduced set-recovery. *Wood Mater. Sci. Eng.* **2021**, *16*, 35–41. [CrossRef]
22. Sadatnezhad, S.H.; Khazaeian, A.; Sandberg, D.; Tabarsa, T. Continuous surface densification of wood: A new concept for large-scale industrial processing. *BioResources* **2017**, *12*, 3122–3132. [CrossRef]
23. Laine, K.; Segerholm, K.; Wålinder, M.; Rautkari, L.; Hughes, M. Wood densification and thermal modification. Hardness, set-recovery and micromorphology. *Wood Sci. Technol.* **2016**, *50*, 883–894. [CrossRef]
24. Follrich, J.; Höra, M.; Müller, U.; Teischinger, A.; Gindl, W. Adhesive bond strength of end grain joints in balsa wood with different density. *Wood Res.* **2010**, *55*, 21–32.
25. Pizzi, A. *Advanced Wood Adhesives Technology*; CRC Press: Boca Raton, FL, USA, 1994.
26. Pizzi, A. Recent developments in eco-efficient bio-based adhesives for wood bonding: Opportunities and issues. *J. Adhes. Sci. Technol.* **2006**, *20*, 829–846. [CrossRef]
27. Pizzi, A.; Mittal, K.L. *Handbook of Adhesive Technology*; CRC Press: Boca Raton, FL, USA, 2017.
28. Pizzi, A.; Mittal, K.L. *Wood Adhesives*; VSP: Rancho Cordova, CA, USA, 2010.
29. Sahoo, S.; Mohanty, S.; Nayak, S.K. Biobased polyurethane adhesive over petroleum based adhesive: Use of renewable resource. *J. Macromol. Sci. Part A* **2018**, *55*, 36–48. [CrossRef]
30. Oliveira, J.M.Q.; De Moura, M.F.S.F.; Silva, M.A.L.; Morais, J.J.L. Numerical analysis of the MMB test for mixed-mode I/II wood fracture. *Compos. Sci. Technol.* **2007**, *67*, 1764–1771. [CrossRef]
31. Song, J.; Chen, C.; Zhu, S.; Zhu, M.; Dai, J.; Ray, U.; Li, Y.; Kuang, Y.; Li, Y.; Quispe, N.; et al. Processing bulk natural wood into a high-performance structural material. *Nature* **2018**, *554*, 224–228. [CrossRef]
32. Corte-Real, L.M.; Jalali, S.; Borges, C.S.; Marques, E.A.; Carbas, R.J.; da Silva, L.F. Development and Characterisation of Joints with Novel Densified and Wood/Cork Composite Substrates. *Materials* **2022**, *15*, 7163. [CrossRef]
33. De Moura, M.; Campilho, R.; Gonçalves, J.P.M. Pure mode II fracture characterization of composite bonded joints. *Int. J. Solids Struct.* **2009**, *46*, 1589–1595. [CrossRef]

34. Hu, M.; Duan, Z.; Zhou, X.; Du, G.; Li, T. Effects of surface characteristics of wood on bonding performance of low-molar ratio urea–formaldehyde resin. *J. Adhes.* **2023**, *99*, 803–816. [CrossRef]
35. Xu, G.; Liang, J.; Zhang, B.; Wu, Z.; Lei, H.; Du, G. Performance and structures of urea-formaldehyde resins prepared with different formaldehyde solutions. *Wood Sci. Technol.* **2021**, *55*, 1419–1437. [CrossRef]

Disclaimer/Publisher's Note: The statements, opinions and data contained in all publications are solely those of the individual author(s) and contributor(s) and not of MDPI and/or the editor(s). MDPI and/or the editor(s) disclaim responsibility for any injury to people or property resulting from any ideas, methods, instructions or products referred to in the content.

Article

Effect of the Sintering Mechanism on the Crystallization Kinetics of Geopolymer-Based Ceramics

Nur Bahijah Mustapa [1], Romisuhani Ahmad [1,2,*], Mohd Mustafa Al Bakri Abdullah [2,3,*], Wan Mastura Wan Ibrahim [1,2], Andrei Victor Sandu [4,5,6], Ovidiu Nemes [7,*], Petrica Vizureanu [4,8], Christina W. Kartikowati [9] and Puput Risdanareni [10]

[1] Faculty of Mechanical Engineering & Technology, Universiti Malaysia Perlis (UniMAP), Arau 02600, Perlis, Malaysia; bahijahmustapa@gmail.com (N.B.M.); wanmastura@unimap.edu.my (W.M.W.I.)
[2] Centre of Excellence Geopolymer and Green Technology (CEGeoGTech), Universiti Malaysia Perlis (UniMAP), Arau 02600, Perlis, Malaysia
[3] Faculty of Chemical Engineering & Technology, Universiti Malaysia Perlis (UniMAP), Arau 02600, Perlis, Malaysia
[4] Faculty of Material Science and Engineering, Gheorghe Asachi Technical University of Iasi, 700050 Iasi, Romania; sav@tuiasi.ro (A.V.S.) peviz2002@yahoo.com (P.V.)
[5] Romanian Inventors Forum, Str. Sf. P. Movila 3, 700089 Iasi, Romania
[6] Academy of Romanian Scientists, 54 Splaiul Independentei St., Sect. 5, 050094 Bucharest, Romania
[7] Department of Environmental Engineering and Sustainable Development Entrepreneurship, Faculty of Materials and Environmental Engineering, Technical University of Cluj-Napoca, B-dul Muncii 103-105, 400641 Cluj-Napoca, Romania
[8] Technical Sciences Academy of Romania, Dacia Blvd 26, 030167 Bucharest, Romania
[9] Department of Chemical Engineering, Universitas Brawijaya, Malang 65145, Indonesia; christinawahyu@ub.ac.id
[10] Department of Civil Engineering, Faculty of Engineering, Universitas Negeri Malang, Malang 65145, Indonesia; puput.risdanareni.ft@um.ac.id
* Correspondence: romisuhani@unimap.edu.my (R.A.); mustafa_albakri@unimap.edu.my (M.M.A.B.A.); ovidiu.nemes@sim.utcluj.ro (O.N.)

Abstract: This research aims to study the effects of the sintering mechanism on the crystallization kinetics when the geopolymer is sintered at different temperatures: 200 °C, 400 °C, 600 °C, 800 °C, 1000 °C, and 1200 °C for a 3 h soaking time with a heating rate of 5 °C/min. The geopolymer is made up of kaolin and sodium silicate as the precursor and an alkali activator, respectively. Characterization of the nepheline produced was carried out using XRF to observe the chemical composition of the geopolymer ceramics. The microstructures and the phase characterization were determined by using SEM and XRD, respectively. The SEM micrograph showed the microstructural development of the geopolymer ceramics as well as identifying reacted/unreacted regions, porosity, and cracks. The maximum flexural strength of 78.92 MPa was achieved by geopolymer sintered at 1200 °C while the minimum was at 200 °C; 7.18 MPa. The result indicates that the flexural strength increased alongside the increment in the sintering temperature of the geopolymer ceramics. This result is supported by the data from the SEM micrograph, where at the temperature of 1000 °C, the matrix structure of geopolymer-based ceramics starts to become dense with the appearance of pores.

Keywords: geopolymer; geopolymer-based ceramics; ceramics; sintering mechanism; crystallization kinetics

1. Introduction

The rising demand for ceramics in the industrial manufacturing, metallurgical, energy production, and biomedical sectors has attracted worldwide interest. The manufacturing of ceramic products entails utilizing abundant natural resources that contain a significant proportion of clay minerals. This involves a series of steps, including dehydration and subjecting the materials to high sintering temperatures of up to 1600 °C [1,2]. However, the conventional method of fabricating ceramics demands an elevated temperature, which reaches up to 1600 °C, and a lengthy heating period, and it also has issues with agglomeration, irregular grain growth, and furnace contamination. Moreover, the primary natural resources required, such as limestone, coal, clay, and others, are being depleted at a rapid rate. To overcome these challenges, a dedicated effort is being made by scientists, researchers, engineers, and industrial workers to explore and develop new, sustainable, and innovative construction materials, as well as alternative binders [3]. Therefore, geopolymerization is used as a substitute method to produce ceramic materials with excellent mechanical properties, low production costs, short fabrication times, and, with the increasing threat to the environment, to develop applications for geopolymer technology. The imperative to adopt sustainable, rational, and ecologically sound construction methods propels the pursuit of innovative alternatives like geopolymerization and alkali activation. These approaches are garnering growing attention in the construction industry to address these needs [4].

Geopolymers are a class of inorganic, non-metallic materials that are produced by the reaction of aluminosilicate materials and alkaline activators under highly alkaline conditions. In the 1970s, geopolymers were first developed as an alternative to traditional cement-based materials, such as ordinary Portland cement (OPC) [5]. Nowadays, geopolymers have become an interesting subject of extensive research and development. This is due to the appealing properties that they offer, including improved mechanical properties, higher thermal stability, and reduced environmental impact [6]. This is associated with the structure of geopolymers, which consist of a three-dimensional and cross-linked network of aluminosilicate bonds [7], which contribute to the uniqueness of their properties. The microstructure of the geopolymer material is strongly influenced by factors such as the selection of raw materials, curing conditions, and sintering temperature [8]. By carefully controlling these factors, researchers can tailor the properties of geopolymer materials to suit specific applications.

In this study, kaolin is used as the aluminosilicate source, while sodium hydroxide (NaOH) and sodium silicate ($NaSiO_3$) are mixed to produce the alkali activator which provides the necessary alkalinity to initiate the geopolymerization reaction [9]. Kaolin is an inorganic material that has been identified as geopolymer-compatible with excellent performance. Kaolin is composed mainly of the mineral kaolinite, which has a layered structure consisting of alternating layers of silica tetrahedra (SiO_4) and alumina octahedra (AlO_6) [10]. The layered structure of kaolinite allows the formation of pores and a high surface area, which can enhance the reactivity of the materials. When dissolved in an alkaline solution during the geopolymerization process, it will trigger the dissolution and reorganisation of the tetrahedral and octahedral elements to form a three-dimensional network of linked units.

The geopolymerization process converts the aluminosilicate material, for example kaolin, into geopolymer materials with desirable properties by chemically reacting with an alkaline activator solution [11–13]. Past research had concluded that the geopolymerization process typically involves dissolution, polycondensation, and curing stages. The dissolution stage occurs when kaolin is mixed with an alkaline activator forming a slurry. The alkaline activator initiates the dissolution of the precursor, leading to the release of silica and alumina ions. Then, the dissolved silica and alumina ions undergo polycondensation reactions [14], which involve the formation of covalent bonds between the tetrahedral and octahedral units. The resulting product is a three-dimensional network of linked tetrahedral and octahedral units, forming the geopolymer materials. In the curing stage, the geopolymer is set to cure or harden. During this time, the geopolymer material undergoes further chemical and physical changes, such as the formation of additional covalent bonds and the development of its final mechanical properties. The specific processing conditions, including the type of aluminosilicate source materials [15], the concentration of the alkaline activator [16], and the curing conditions [17], can significantly influence the properties of the geopolymer. By carefully controlling these factors, it is possible to tailor the properties of geopolymer materials to meet specific application requirements, such as mechanical strength, thermal stability, and chemical resistance.

From the idea of geopolymer production, geopolymer-based ceramics are introduced as an alternative in the field of ceramic production due to their ability to offer enhanced thermal stability, chemical resistance, and mechanical strength [18], and most importantly, they required low sintering temperatures compared to conventional ceramic fabrication, where the sintering temperature goes up to 1600 °C. Several other problems are also found in using the conventional method, such as prolonged heating time, irregular grain growth, and furnace contamination. Therefore, the geopolymerization method has been adapted to the fabrication of geopolymer-based ceramics to progress the application of the geopolymer technology. Apart from geopolymerization, the sintering temperature also affects the properties of a geopolymer when heat is applied to geopolymer bodies. Therefore, in this study, the effects of the sintering mechanism are investigated to study how it affects the properties and crystallization of the geopolymer ceramics produced.

2. Experimental Method

2.1. Materials

Kaolin is a clay mineral mainly containing a chemical composition of $Al_2Si_2O_5(OH)_4$. In this study, kaolin was used as a starting material for geopolymerization. The kaolin used was supplied by Associated Kaolin Industries Malaysia as Si-Al source materials, where the large compounds found are SiO_2 and Al_2O_3. Table 1 shows the chemical composition of kaolin obtained by X-ray fluorescence (XRF).

Table 1. Chemical composition of kaolin (wt.%) obtained by X-ray fluorescence.

Element	SiO_2	Al_2O_3	K_2O	Fe_2O_3	TiO_2	MnO_2	ZrO_2	LOI
Wt. (%)	54.0	31.7	6.05	4.89	1.14	0.11	0.10	1.74

A mixture of 12 M sodium hydroxide (NaOH) and sodium silicate (Na_2SiO_3) as an alkaline activator was used to activate the source material. The ratio of Na_2SiO_3/NaOH was set at 0.24. The sodium hydroxide caustic soda micro-pearls with a purity of 99% were supplied by Formosoda-P from Taiwan. Meanwhile, sodium silicate (Na_2SiO_3) was provided by South Pacific Chemicals Industries Sdn. Bhd. (SPCI), Pahang, Malaysia with the chemical composition of H_2O (60.5%), SiO_2 (31.1%), and Na_2O (9.4%).

2.2. Sample Preparation

Figure 1 shows the overall process of synthesizing geopolymer ceramics. The influence of sintering temperature on the green bodies in the production of kaolin-based geopolymer ceramics was investigated by systematically varying the sintering temperature within the range of 200 °C to 1200 °C. The aluminosilicate source, kaolin was mixed with an alkali activator at a solid-to-liquid ratio of 1.0 to activate the source material. The ratio of the alkali activator and molarity of NaOH was fixed at 0.24 and 12 M, respectively. The solution needed to be prepared 24 h before it was used to obtain a homogeneous solution. The mixture was then cured at 80 °C in an oven for 24 h. To obtain the fine powder, the kaolin geopolymer was crushed using a mechanical crusher and sieved by using a 150 μm sieve. Subsequently, the compacted geopolymer (86 MPa, 2 min) underwent a sintering process at various temperatures (200 °C, 400 °C, 600 °C, 800 °C, 1000 °C, and 1200 °C), with a soaking duration of 180 min and a heating rate of 5 °C/min. Previously, Ahmad et al. [19] had studied the sintering profile of geopolymer-based ceramics in the temperature range of 900 °C to 1200 °C, and obtained 1200 °C as the maximum sintering temperature. Therefore, in this research, a thorough procedure is prepared from low sintering temperature of 200 °C to study the evolution of the properties of geopolymer-based ceramics. Figure 2 shows the sintering profile for the fabrication of geopolymer ceramics.

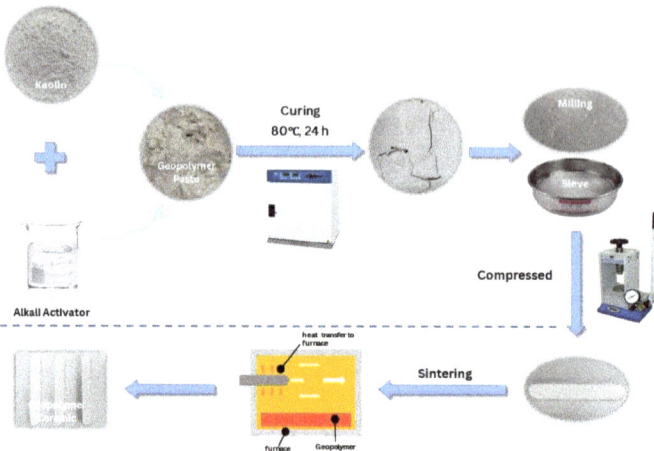

Figure 1. The overall process of synthesizing geopolymer ceramics.

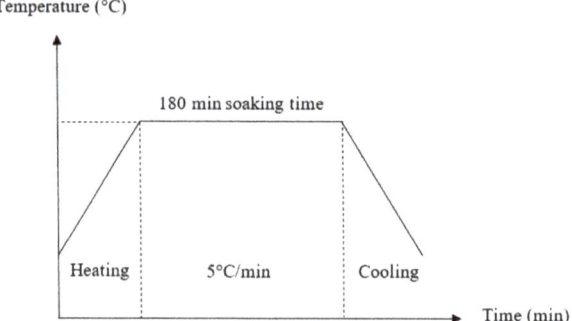

Figure 2. Sintering profile for fabrication of geopolymer ceramics.

To examine the influence of temperature on the properties of geopolymer ceramics, an extensive characterization of all prepared samples was conducted. The mechanical properties, specifically the flexural strength of the geopolymer ceramics, were evaluated. The three-point bending method, adhering to the guidelines of ASTM C-1163b, was implemented with a support span length of 30 mm and a crosshead speed of 0.3 mm/min. The morphology of the geopolymer bodies was studied by using a JEOL JSM-6460LA scanning electron microscopy (SEM). The microstructural development of the geopolymer and geopolymer ceramics as well as identified reacted/unreacted regions, porosity, and cracks were examined to study the effect of the sintering mechanism. To create a conductive layer, the sample was coated with gold using a JEOL JFC 1600 model auto fine coater. This data is supported by the results of the Synchrotron radiation X-ray tomographic (SXTM), used to study the porosity of the geopolymer ceramics. The test was carried out at the Synchrotron Light Research Institute (SLRI), Thailand.

The phase composition of the samples was determined by an XRD 6000, SHIMADZU diffractometer. The sample was ground into a fine powder by using a ring mill small enough to ensure that the X-ray can penetrate the sample and generate a diffraction pattern. The sample was then mounted onto a sample holder and evenly distributed. To ensure the diffraction pattern is accurate and reproducible, the sample was aligned so that the beam was perpendicular to the surface of the sample. The data were collected over a range of 10° to 65°.

3. Results and Discussions
3.1. Mechanical Properties of Geopolymer-Based Ceramics

The effect of thermal treatment on the mechanical properties of geopolymer-based ceramics is shown in Figures 3–6, based on the mean of the result of the flexural strength, density, shrinkage, and water absorption of the sintered geopolymer ceramics, respectively. The error bars in the graph represent the standard deviations of the data. Higher sintering temperatures typically lead to denser ceramic materials due to increased grain diffusion and densification. As a result, higher sintering temperatures generally lead to higher flexural strength as well, as the denser material is better able to resist bending and deformation under mechanical stress. Upon increasing the temperature, the flexural strength increases from 7.18 MPa for unsintered geopolymers, caused by crystal growth, which strengthens the materials and reduces their susceptibility to deformation and cracking. When geopolymer-based ceramics are exposed to 1000 °C, a high flexural strength of 53.5 MPa is recorded. When further sintered at 1200 °C, the strength achieves its maximum flexural strength of 78.9 MPa, which is caused by the densified and crystallized matrix and the enhanced fiber/matrix interface bonding as the sintering mechanism occurs [20].

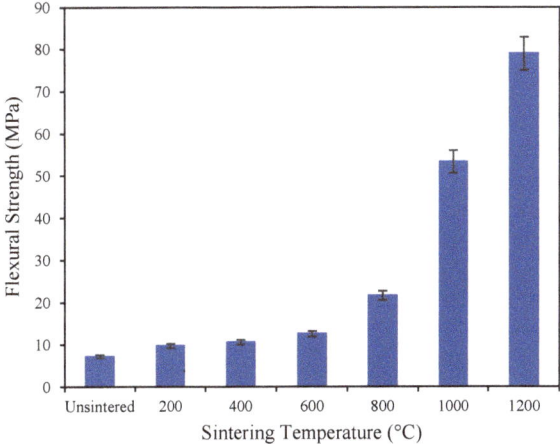

Figure 3. Flexural strength of unsintered and sintered geopolymer ceramics.

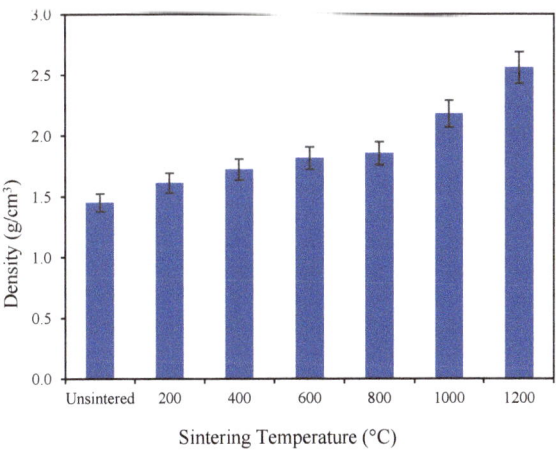

Figure 4. Density of unsintered and sintered geopolymer ceramics.

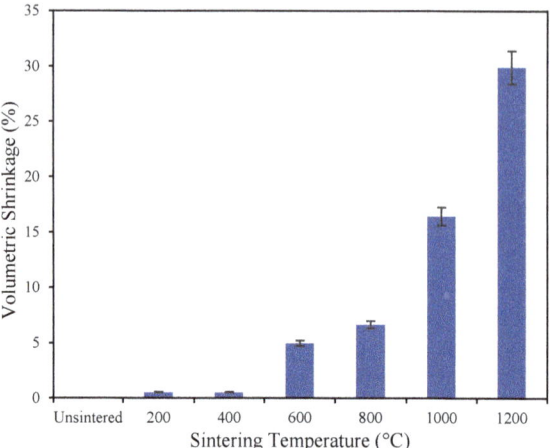

Figure 5. The shrinkage of geopolymer-based ceramics after the sintering process.

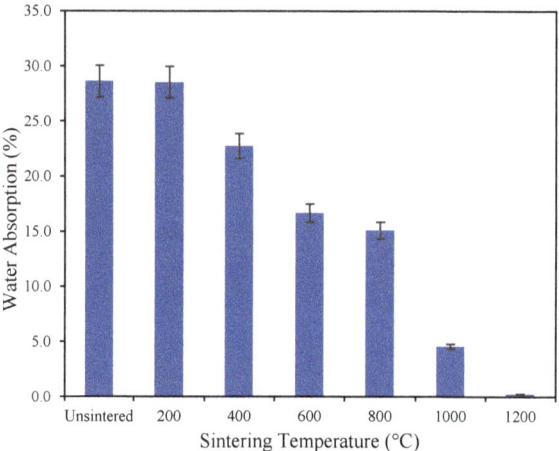

Figure 6. The water absorption of geopolymer-based ceramics.

Figure 4 presents the density of the unsintered and sintered geopolymer at various sintering temperatures of 200 °C, 400 °C, 600 °C, 800 °C, 1000 °C, and 1200 °C. As displayed, along with the increments of the sintering temperature, the density of the geopolymer-based ceramics also changes in increments. At 1200 °C, the density recorded the highest value of 2.56 g/cm^3, compared to a geopolymer sintered at 200 °C (1.61 g/cm^3) and an unsintered geopolymer (1.45 g/cm^3). The increase in the density of the geopolymer-based ceramics is related to the greater grain boundary diffusion which promotes densification of the ceramics, thus increasing the strength [21]. High sintering temperature crystallize the inner structure of the geopolymer-based ceramics, thus enhancing the mechanical performance of the ceramics. The density of the ceramics is dependent on the sintering temperature. The sintering temperature plays a significant role in the densification process. Exposure to a high temperature eliminates the pores between particles and also facilitates fast grain growth during the sintering mechanism [22].

When sintered at 200 °C, the occurrence of dehydration stages leads to shrinkage and deformation. Linear shrinkage was defined as the reduction percentage in the thickness of the central part of the powder compacts. This can be attributed to capillary contraction induced by the shape of water within micro and nanopore solutions. Generally, linear shrinkage increases as the sintering temperature increases. Increasing the sintering temperature led to greater shrinkage in the materials, as the higher temperatures cause more particle rearrangement and compaction. The body of the ceramics achieved its maximum shrinkage of 29.91% when exposed to 1200 °C. A study by Jaya et al. [23] stated that sintering above 1200 °C is not suitable as the geopolymer palette starts to melt as the temperature rises to 1250 °C, turning the color from milky white to brown. This happens due to the exceeding of the melting point of the geopolymer ceramics.

As the geopolymer undergoes sintering at elevated temperatures, the structure of the material is modified. These modifications encompass the elimination of water molecules and the formation of a new crystalline phase embedded within the geopolymer matrix. The surface energy of incoherent particles decreases during the sintering mechanism, leading to a decrease in the overall surface area. The application of high temperatures during the sintering process induces structural rearrangement and the development of crystalline phases, including nepheline ($NaAlSiO_4$), kalsilite ($KAlSiO_4$), and mullite ($3Al_2O_3 \cdot 2SiO_2$) [24]. This sintering mechanism leads to enhanced material density and facilitates the growth of grains through diffusion. As the liquid flows between particles, shrinkage takes place, exerting greater pressure at the contact areas and prompting the material to relocate, resulting in closer proximity between particle centers.

As for the water absorption (WA), the decreasing trend is possibly due to the densification process of the geopolymer matrix in the ceramic materials as the sintering temperature increases. When sintered above 1200 °C, the geopolymer-based ceramics absorbed 0.23% of water in 24 h compared to unsintered and a geopolymer sintered at 200 °C which absorbed 28.63% and 28.55% of water, respectively. The fluctuation in water absorption may be due to the elevated temperature, resulting in the closing of some open pores [25], thus reducing the amount of available pore space for water to be absorbed. This phenomenon causes the rate of water absorption to decrease while increasing the pore strength. The low water absorption exhibits a better property for geopolymer-based ceramics as it contributes to an increase in strength and creates barriers to the formation of cracks and voids [26].

3.2. Morphology Analysis and Porosity of Geopolymer-Based Ceramics

The SEM micrograph provides a visual representation of the unsintered geopolymer, illustrating the scanning image obtained during the activation of kaolin with the alkali activator. The plate-like morphology of unreacted kaolin in geopolymer ceramics can be observed in Figure 7. It has a unique morphological structure characterized by its elongated, plate-like particles, which function to reinforce the geopolymer matrix, thus increasing the strength and toughness of the resulting geopolymer-based ceramics.

Figure 8 shows the SEM micrograph of unsintered and sintered kaolin-based geopolymer ceramics at various sintering temperatures of 200, 400, 600, 800, 1000, and 1200 °C. All of the samples have been imaged at a fracture section of the geopolymer-based ceramic. Only a few micro-level cracks were visible on the sample surface when exposed to a higher temperature. Following the heat treatment, as an effect of the grain coarsening phenomenon, the small pores within the matrix gradually disappear, and the larger pores form. There is no significant change in the microstructure of geopolymer when exposed to temperatures of 200 °C, 400 °C, and 600 °C. During heating from room temperature (RT) to 200 °C, the water content in the geopolymer evaporates causing weight loss and minimal shrinkage. The shrinkage and deformation could be attributed to the capillary contraction induced by the escape of water from the pores [27]. However, as the geopolymer sintered at 800 °C, it can be observed that the geopolymer matrix starts to become more dense. This phenomenon can be attributed to the dehydroxylation stages, during which condensation and polymerization between T-OH groups leads to the escape of water and subsequent shrinkage at high sintering temperatures. Sintering above 800 °C causes the flexural strength to increase [28]. The thermal analysis revealed the occurrence of the sintering process, which was evident from the observed shrinkage resulting from crystal coarsening. In accordance with Li et al. [29], with an increase in sintering temperature, there is a promotion in the growth of sintering necks and sintering densification. This, in turn, leads to an enhancement in the flexural strength of the geopolymer-based ceramics. Subsequent sintering to 1200 °C induces the formation of a molten amorphous glass phase, promoting further sintering and densification. This process significantly contributes to the maximum shrinkage observed. Small pores form from an amorphous geopolymer into nepheline geopolymer-based ceramics.

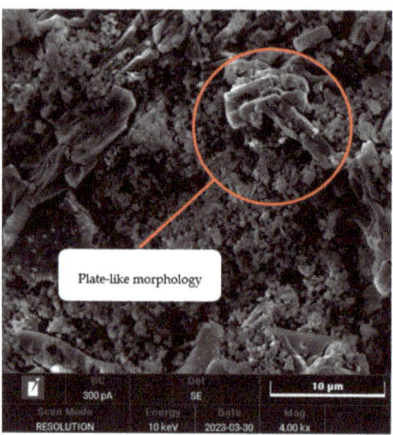

Figure 7. SEM micrograph of unreacted plate-like kaolinite at 4000× magnification.

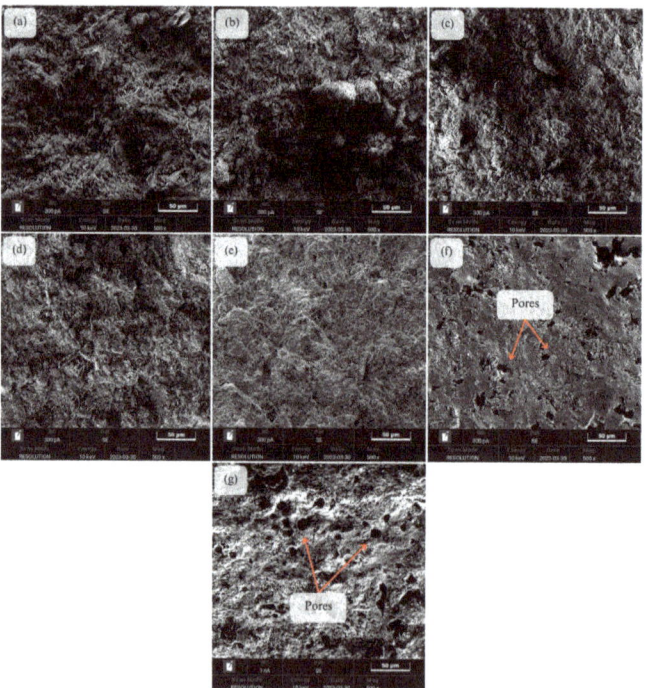

Figure 8. SEM micrograph of ceramic samples for (**a**) unsintered and sintered samples at (**b**) 200 °C, (**c**) 400 °C, (**d**) 600 °C, (**e**) 800 °C, (**f**) 1000 °C, and (**g**) 1200 °C.

The sintering process above 1000 °C resulted in the development of a glassy morphology, signifying the closure of the majority of the small pores and the attainment of ceramic densification [30]. Figure 9 shows the proposed scheme on the grain growth of the geopolymer-based ceramics during the sintering mechanism which leads to the densification of the geopolymer matrix. During the initial stage of sintering, the coalescence and orientation of particles reduces the coordination number around pores and alters the balance of surface tension around pore surfaces, which causes the closure of pores. As the sintering temperature increases, the grain grows and leads to the formation of densified geopolymer-based ceramics.

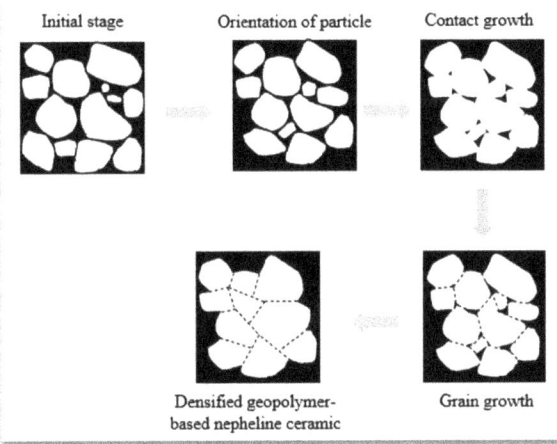

Figure 9. Proposed scheme of the grain growth during the sintering mechanism of geopolymer-based ceramics.

At a high temperature, the ceramic structure undergoes significant structural changes, where the porosity and mechanical properties are affected. As the geopolymer is exposed to high sintering temperatures, the geopolymer structure begins to collapse and form cracks. This may create new pores and increase the overall porosity of the material. While increasing porosity is expected to decrease mechanical strength, there are cases where the mechanical strength could be increased as the porosity increases. This is due to the function of the pores that can act as stress concentrators, which helps in dissipating stress and preventing the initiation and propagation of cracks [31].

The porosity of geopolymer-based ceramics can be observed in Figure 10 using data from the XTM technique, where it is shown that the total number of pores is highest at 6.61% when sintered at 1200 °C, at which point they mainly consist of closed pores. This is due to the sintering mechanism where, as the sintering temperature increases, the small pores merge forming a larger pore due to moisture hydration. Even though the total number of pores is highest at this temperature, the closed pores present contribute to the high density, thus improving the strength of the geopolymer-based ceramics. Contrary to sintered ceramics, the unsintered geopolymer happens to have the lowest number of pores with 0.2%. This result is thus related to the SEM micrograph obtained.

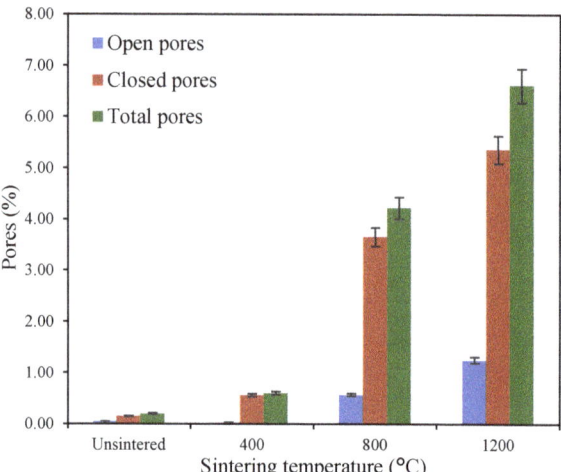

Figure 10. The porosity of unsintered and sintered geopolymer-based ceramics from XTM.

3.3. Phase Analysis of Geopolymer-Based Ceramics

Figure 11 shows the XRD pattern of unsintered kaolin-based geopolymer. The presence of zeolite (Z), kaolinite (K), and quartz (Q) were detected in the kaolin-based geopolymer. The kaolinite ($Al_2SiO_5(OH)_4$) and quartz (SiO_2) represent inherent mineral constituents of kaolin, whereas zeolite typically crystallizes through the activation of kaolin with the alkali activator, originating from the transformation of amorphous aluminosilicate gel. Geopolymerization is initiated by the combination of NaOH and Na_2SiO_3 solutions, leading to the dissolution of aluminosilicate minerals in an alkali activator. The dissolved components undergo a series of processes involving nucleation, growth, and polymerization before ultimately solidifying through polycondensation. The alteration in crystallographic composition was evidenced by the disappearance of the amorphous phase hump upon heat treatment, indicative of the transition from an amorphous to a crystalline state.

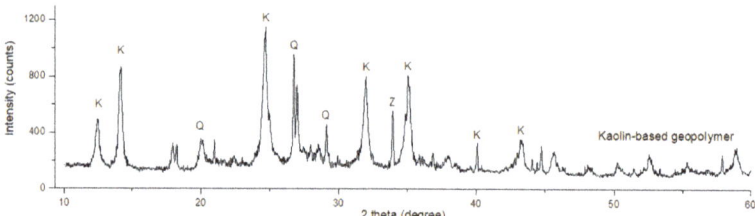

Figure 11. XRD patterns of unsintered kaolin-based geopolymer (K = kaolinite, Q = quartz, Z = zeolite).

Figure 12 provides a characterization of geopolymer-based nepheline ceramics, encompassing sintered samples exposed to different temperatures of 200 °C, 400 °C, 600 °C, 800 °C, 1000, and 1200 °C. At the temperatures of 200 °C, 400 °C, and 600 °C, amorphous humps and reflections from quartz present between the range of a 20° to 40° diffraction angle (2θ) were introduced by the raw kaolin. Sharp peaks in the reflection of nepheline ($NaAlSiO_4$) start to appear when the geopolymer-based ceramics are sintered at a temperature of 800 °C. The absence of muscovite ($KAl_2(AlSi_3O_{10})(OH)$) could be attributed to the exposure of the kaolin-based geopolymer to high temperatures during the sintering process. As the sintering temperature increased to 800 °C, the intensity of the peaks increased, and the amorphous phase diminished, indicating the initiation of geopolymer crystallization [32].

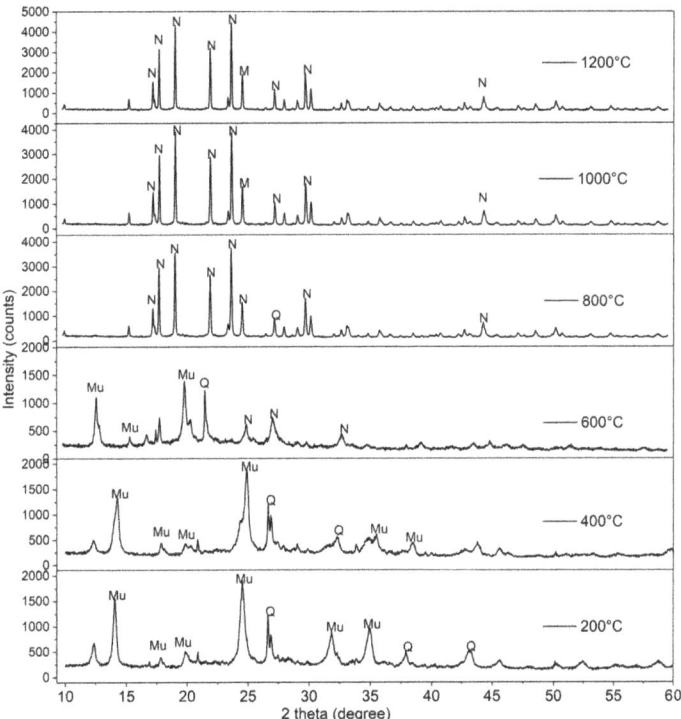

Figure 12. XRD patterns of sintered kaolin-based geopolymer at 200 °C, 400 °C, 600 °C, 800 °C, 1000 °C, and 1200 °C (Mu = muscovite ICDD# 05-0652, Q = quartz ICDD# 46-1045, N = nepheline ICDD# 35-0424).

Sintering at 1000 °C and 1200 °C resulted in the formation of mullite ($3Al_2O_3 \cdot 2SiO_2$), which exhibits excellent thermochemical stability. The XRD pattern recorded for 800 °C, 1000 °C, and 1200 °C showed an incremental rise in the intensity of the nepheline peaks as the sintering temperature increased. Nepheline contributes to the strength and stability of ceramics. It also helps in improving their resistance to heat and erosion. Furthermore, nepheline can impact the physical and mechanical properties of ceramics, such as their density, hardness, and thermal conductivity [33]. Overall, incorporating nepheline into a kaolin-based geopolymer can improve its performance and make it more suitable for a range of applications.

4. Conclusions

Geopolymer-based ceramic materials have numerous potential uses, such as building materials, refractories, high-temperature coatings, and even as an alternative to conventional cement-based materials. The spectrum of potential uses for geopolymer materials is projected to broaden with continued study and development, making them a more crucial component of contemporary engineering and building.

From the obtained results, it can be inferred that the sintering process has an impact on the characteristics and properties of geopolymer-based ceramics. The phase analysis conducted by using XRD indicates that the results align with the compressive strength and the morphology analysis. By detecting the major crystalline components present in the geopolymer-based ceramic systems, phase analysis aids in explaining the properties of the geopolymer. As the geopolymer is sintered at high temperatures above 800 °C, the XRD pattern shows an incremental increase in the intensity. Additionally, from the SEM micrograph of the fracture section of the geopolymer-based ceramics, the morphology analysis of geopolymer-based ceramics sintered at 1200 °C reveals a denser surface appearance compared to the unsintered geopolymer. Furthermore, sintering the ceramics at a temperature of 1200 °C enhances their mechanical performance as they achieved a maximum flexural strength of 78.9 MPa, a peak density of 2.56 g/cm^3, and their lowest water absorption at 0.23%.

Author Contributions: Writing—original draft, N.B.M.; conceptualization and writing—review and editing, R.A. and O.N; visualization and supervision, M.M.A.B.A.; investigation and methodology, W.M.W.I.; validation, visualization, and corrections, A.V.S., O.N., P.V., C.W.K. and P.R. All authors have read and agreed to the published version of the manuscript.

Funding: This research was funded by the Fundamental Research Grant Scheme (FRGS) under grant number FRGS/1/2021/TK0/UNIMAP/02/17 from the Ministry of Higher Education, Malaysia. This research was also partially supported by project 38 PFE in the frame of the PDI-PFE-CDI 2021 program.

Institutional Review Board Statement: Not applicable.

Informed Consent Statement: Not applicable.

Data Availability Statement: Not applicable.

Acknowledgments: The authors of the present work wish to dedicate their great thanks to the Centre of Excellence Geopolymer and Green Technology, (CeGeoGTech), UniMAP and the Faculty of Mechanical Engineering Technology, UniMAP.

Conflicts of Interest: The authors declare no conflict of interest.

References

1. Ren, Y.; Ren, Q.; Wu, X.; Zheng, J.; Hai, O. Mechanism of low temperature sintered high-strength ferric-rich ceramics using bauxite tailings. *Mater. Chem. Phys.* **2019**, *238*, 121929. [CrossRef]
2. Hofer, A.-K.; Kocjan, A.; Bermejo, R. High-strength lithography-based additive manufacturing of ceramic components with rapid sintering. *Addit. Manuf.* **2022**, *59*, 103141. [CrossRef]
3. Allaoui, D.; Nadi, M.; Hattani, F.; Majdoubi, H.; Haddaji, Y.; Mansouri, S.; Oumam, M.; Hannache, H.; Manoun, B. Eco-friendly geopolymer concrete based on metakaolin and ceramics sanitaryware wastes. *Ceram. Int.* **2022**, *48*, 34793–34802. [CrossRef]
4. Burduhos Nergis, D.D.; Vizureanu, P.; Lupescu, S.; Burduhos Nergis, D.P.; Perju, M.C.; Sandu, A.V. Microstructural Analysis of Ambient Cured Phosphate Based-Geopolymers with Coal-Ash as Precursor. *Arch. Metall. Mater.* **2022**, *67*, 595–600. [CrossRef]
5. Davidovits, J. Geopolymers—Inorganic polymeric new materials. *J. Therm. Anal.* **1991**, *37*, 1633–1656. [CrossRef]
6. Almutairi, A.L.; Tayeh, B.A.; Adesina, A.; Isleem, H.F.; Zeyad, A.M. Potential applications of geopolymer concrete in construction: A review. *Case Stud. Constr. Mater.* **2021**, *15*, e00733. [CrossRef]
7. Kovářík, T.; Hájek, J.; Pola, M.; Rieger, D.; Svoboda, M.; Beneš, J.; Šutta, P.; Deshmukh, K.; Jandová, V. Cellular ceramic foam derived from potassium-based geopolymer composite: Thermal, mechanical and structural properties. *Mater. Des.* **2021**, *198*, 109355. [CrossRef]
8. Bykov, Y.V.; Egorov, S.; Eremeev, A.; Holoptsev, V.; Plotnikov, I.; Rybakov, K.; Semenov, V.; Sorokin, A. Temperature profile optimization for microwave sintering of bulk Ni-Al2O3 functionally graded materials. *J. Mater. Process Technol.* **2014**, *214*, 210–216. [CrossRef]
9. Yuan, J.; Li, L.; He, P.; Chen, Z.; Lao, C.; Jia, D.; Zhou, Y. Effects of kinds of alkali-activated ions on geopolymerization process of geopolymer cement pastes. *Constr. Build. Mater.* **2021**, *293*, 123536. [CrossRef]
10. Lecomte, I.; Liégeois, M.; Rulmont, A.; Cloots, R.; Maseri, F. Synthesis and characterization of new inorganic polymeric composites based on kaolin or white clay and on ground-granulated blast furnace slag. *J. Mater. Res.* **2003**, *18*, 2571–2579. [CrossRef]
11. Tchadjie, L.N.; Ekolu, S.O. Enhancing the reactivity of aluminosilicate materials toward geopolymer synthesis. *J. Mater. Sci.* **2017**, *53*, 4709–4733. [CrossRef]
12. Taki, K.; Mukherjee, S.; Patel, A.K.; Kumar, M. Reappraisal review on geopolymer: A new era of aluminosilicate binder for metal immobilization. *Environ. Nanotechnol. Monit. Manag.* **2020**, *14*, 100345. [CrossRef]
13. Hamdane, H.; Tamraoui, Y.; Mansouri, S.; Oumam, M.; Bouih, A.; El Ghailassi, T.; Boulif, R.; Manoun, B.; Hannache, H. Effect of alkali-mixed content and thermally untreated phosphate sludge dosages on some properties of metakaolin based geopolymer material. *Mater. Chem. Phys.* **2020**, *248*, 122938. [CrossRef]

14. Aziz, I.H.; Abdullah, M.M.A.B.; Salleh, M.A.A.M.; Yoriya, S.; Razak, R.A.; Mohamed, R.; Baltatu, M.S. The investigation of ground granulated blast furnace slag geopolymer at high temperature by using electron backscatter diffraction analysis. *Arch. Metall. Mater.* **2022**, *67*, 227–231. [CrossRef]
15. Kohout, J.; Koutník, P.; Hájková, P.; Kohoutová, E.; Soukup, A. Effect of Different Types of Aluminosilicates on the Thermo-Mechanical Properties of Metakaolinite-Based Geopolymer Composites. *Polymers* **2022**, *14*, 4838. [CrossRef] [PubMed]
16. Tuyan, M.; Andiç-Çakir, Ö.; Ramyar, K. Effect of alkali activator concentration and curing condition on strength and microstructure of waste clay brick powder-based geopolymer. *Compos. B Eng.* **2018**, *135*, 242–252. [CrossRef]
17. Jiao, Z.; Li, X.; Yu, Q. Effect of curing conditions on freeze-thaw resistance of geopolymer mortars containing various calcium resources. *Constr. Build. Mater.* **2021**, *313*, 125507. [CrossRef]
18. Luukkonen, T.; Olsen, E.; Turkki, A.; Muurinen, E. Ceramic-like membranes without sintering via alkali activation of metakaolin, blast furnace slag, or their mixture: Characterization and cation-exchange properties. *Ceram. Int.* **2023**, *49*, 10645–10651. [CrossRef]
19. Ahmad, R.; Abdullah, M.M.A.B.; Ibrahim, W.M.W.; Hussin, K.; Zaidi, F.H.A.; Chaiprapa, J.; Wysłocki, J.J.; Błoch, K.; Nabiałek, M. Role of sintering temperature in production of nepheline ceramics-based geopolymer with addition of ultra-high molecular weight polyethylene. *Materials* **2021**, *14*, 1077. [CrossRef] [PubMed]
20. He, P.; Jia, D. Interface evolution of the Cf/leucite composites derived from Cf/geopolymer composites. *Ceram. Int.* **2013**, *39*, 1203–1208. [CrossRef]
21. Mashhadi, M.; Taheri-Nassaj, E.; Sglavo, V.M. Pressureless sintering of boron carbide. *Ceram. Int.* **2010**, *36*, 151–159. [CrossRef]
22. Khattab, R.M.; Wahsh, M.M.S.; Khalil, N.M. Preparation and characterization of porous alumina ceramics through starch consolidation casting technique. *Ceram. Int.* **2012**, *38*, 4723–4728. [CrossRef]
23. Jaya, N.A.; Abdullah, M.M.A.B.; Ghazali, C.M.R.; Hussain, M.; Hussin, K.; Ahmad, R. Characterization and Microstructure of Kaolin-Based Ceramic Using Geopolymerization. *Key Eng. Mater.* **2016**, *700*, 3–11.
24. Ahmad, R.; Ibrahim, W.M.W.; Abdullah, M.M.A.B.; Pakawanit, P.; Vizureanu, P.; Abdullah, A.S.; Sandu, A.V.; Zaidi, F.H.A. Geopolymer-Based Nepheline Ceramics: Effect of Sintering Profile on Morphological Characteristics and Flexural Strength. *Crystals* **2022**, *12*, 1313.
25. Ramli, M.I.I.; Salleh, M.A.A.M.; Abdullah, M.M.A.B.; Aziz, I.H.; Ying, T.C.; Shahedan, N.F.; Kockelmann, W.; Fedrigo, A.; Sandu, A.V.; Vizureanu, P.; et al. The Influence of Sintering Temperature on the Pore Structure of an Alkali-Activated Kaolin-Based Geopolymer Ceramic. *Materials* **2022**, *15*, 2667. [CrossRef]
26. Azevedo, A.R.G.; Vieira, C.M.F.; Ferreira, W.M.; Faria, K.C.P.; Pedroti, L.G.; Mendes, B.C. Potential use of ceramic waste as precursor in the geopolymerization reaction for the production of ceramic roof tiles. *J. Build. Eng.* **2020**, *29*, 101156.
27. He, P.; Jia, D.; Wang, M.; Zhou, Y. Effect of cesium substitution on the thermal evolution and ceramics formation of potassium-based geopolymer. *Ceram. Int.* **2010**, *36*, 2395–2400. [CrossRef]
28. Zhu, Z.; Xiao, J.; He, W.; Wang, T.; Wei, Z.; Dong, Y. A phase-inversion casting process for preparation of tubular porous alumina ceramic membranes. *J. Eur. Ceram. Soc.* **2015**, *35*, 3187–3194. [CrossRef]
29. Li, H.; Liu, Y.; Liu, Y.; Zeng, Q.; Hu, K.; Lu, Z.; Liang, J. Effect of sintering temperature in argon atmosphere on microstructure and properties of 3D printed alumina ceramic cores. *J. Adv. Ceram.* **2020**, *9*, 220–231. [CrossRef]
30. Chen, W.; Garofalo, A.C.; Geng, H.; Liu, Y.; Wang, D.; Li, Q. Effect of high temperature heating on the microstructure and performance of cesium-based geopolymer reinforced by cordierite. *Cem. Concr. Compos.* **2022**, *129*, 104474. [CrossRef]
31. Arriagada, C.; Navarrete, I.; Lopez, M. Understanding the effect of porosity on the mechanical and thermal performance of glass foam lightweight aggregates and the influence of production factors. *Constr. Build. Mater.* **2019**, *228*, 116746. [CrossRef]
32. Payakaniti, P.; Chuewangkam, N.; Yensano, R.; Pinitsoontorn, S.; Chindaprasirt, P. Changes in compressive strength, microstructure and magnetic properties of a high-calcium fly ash geopolymer subjected to high temperatures. *Constr. Build. Mater.* **2020**, *265*, 120650. [CrossRef]
33. Bih, N.L.; Mahamat, A.A.; Hounkpè, J.B.; Onwualu, P.A.; Boakye, E.E. The Effect of Polymer Waste Addition on the Compressive Strength and Water Absorption of Geopolymer Ceramics. *Appl. Sci.* **2021**, *11*, 3540. [CrossRef]

Disclaimer/Publisher's Note: The statements, opinions and data contained in all publications are solely those of the individual author(s) and contributor(s) and not of MDPI and/or the editor(s). MDPI and/or the editor(s) disclaim responsibility for any injury to people or property resulting from any ideas, methods, instructions or products referred to in the content.

Article

Research on Thermal Insulation Performance and Impact on Indoor Air Quality of Cellulose-Based Thermal Insulation Materials

Cristian Petcu [1], Andreea Hegyi [2,3,*], Vlad Stoian [4,*], Claudiu Sorin Dragomir [1,5], Adrian Alexandru Ciobanu [6], Adrian-Victor Lăzărescu [2] and Carmen Florean [2,3]

1. National Institute for Research & Development URBAN-INCERC Bucharest Branch, 266 Soseaua Pantelimon, 021652 Bucharest, Romania; cristian.petcu@yahoo.com (C.P.); dragomirclaudiusorin@yahoo.com (C.S.D.)
2. National Institute for Research & Development URBAN-INCERC Cluj-Napoca Branch, 117 Calea Floresti, 400524 Cluj-Napoca, Romania; adrian.lazarescu@incerc-cluj.ro (A.-V.L.); carmen.florean@incerc-cluj.ro (C.F.)
3. Faculty of Materials and Environmental Engineering, Technical University of Cluj-Napoca, 103-105 Muncii Boulevard, 400641 Cluj-Napoca, Romania
4. Department of Microbiology, Faculty of Agriculture, University of Agricultural Sciences and Veterinary Medicine of Cluj-Napoca, 3-5 Calea Mănăştur, 400372 Cluj-Napoca, Romania
5. Faculty of Land Reclamation and Environmental Engineering, University of Agronomic Sciences and Veterinary Medicine of Bucharest, 59 Mărăşti Boulevard, 011464 Bucharest, Romania
6. National Institute for Research & Development URBAN-INCERC Iasi Branch, 6 Anton Sesan Street, 700048 Iasi, Romania; adrian.ciobanu@incd.ro
* Correspondence: andreea.hegyi@incerc-cluj.ro (A.H.); vlad.stoian@usamvcluj.ro (V.S.)

Citation: Petcu, C.; Hegyi, A.; Stoian, V.; Dragomir, C.S.; Ciobanu, A.A.; Lăzărescu, A.-V.; Florean, C. Research on Thermal Insulation Performance and Impact on Indoor Air Quality of Cellulose-Based Thermal Insulation Materials. *Materials* 2023, 16, 5458. https://doi.org/10.3390/ma16155458

Academic Editor: Gianpiero Colangelo

Received: 10 July 2023
Revised: 27 July 2023
Accepted: 2 August 2023
Published: 3 August 2023

Copyright: © 2023 by the authors. Licensee MDPI, Basel, Switzerland. This article is an open access article distributed under the terms and conditions of the Creative Commons Attribution (CC BY) license (https://creativecommons.org/licenses/by/4.0/).

Abstract: Worldwide, the need for thermal insulation materials used to increase the energy performance of buildings and ensure indoor thermal comfort is constantly growing. There are several traditional, well-known and frequently used thermal insulation materials on the building materials market, but there is a growing trend towards innovative materials based on agro-industrial waste. This paper analyses the performance of 10 such innovative thermal insulation materials obtained by recycling cellulosic and/or animal waste, using standardised testing methods. More precisely, thermal insulation materials based on the following raw materials were analysed: cellulose acetate, cigarette filter manufacturing waste; cellulose acetate, cigarette filter manufacturing waste and cigarette paper waste; cellulose acetate, waste from cigarette filter manufacturing, waste cigarette paper and waste aluminised paper; cellulose from waste paper (two types made by two independent manufacturers); wood fibres; cellulose from cardboard waste; cellulose from waste cardboard, poor processing, inhomogeneous product; rice husk waste and composite based on sheep wool, recycled PET fibres and cellulosic fibres for the textile industry. The analysis followed the performance in terms of thermal insulating quality, evidenced by the thermal conductivity coefficient (used as a measurable indicator) determined for both dry and conditioned material at 50% RH, in several density variants, simulating the subsidence under its own weight or under various possible stresses arising in use. The results showed in most cases that an increase in material density has beneficial effects by reducing the coefficient of thermal conductivity, but exceptions were also reported. In conjunction with this parameter, the analysis of the 10 types of materials also looked at their moisture sorption/desorption capacity (using as a measurable indicator the amount of water stored by the material), concluding that, although they have a capacity to regulate the humidity of the indoor air, under low RH conditions the water loss is not complete, leaving a residual quantity of material that could favour the development of mould. Therefore, the impact on indoor air quality was also analysed by assessing the risk of mould growth (using as a measurable indicator the class and performance category of the material in terms of nutrient content conducive to the growth of microorganisms) under high humidity conditions but also the resistance to the action of two commonly encountered moulds, *Aspergillus niger* and *Penicillium notatum*. The results showed a relative resistance to the action of microbiological factors, indicating however the need for intensified biocidal treatment.

Keywords: bio-based insulation material; heat transfer; resistance to micro-organisms; air quality

1. Introduction

Today, worldwide, but especially in the highly and medium-developed countries, there is a change in the lifestyle of the population. On one hand, more and more activities are taking place inside buildings and, on the other hand, there is a growing awareness of the need for sustainable use of natural resources, reduced energy consumption and reduced environmental impact [1]. In this context, there is a strong orientation towards identifying new possibilities and developing new, more efficient, user-friendly and eco-friendly building materials. Various development possibilities are also being explored in the area of materials for thermal insulation of buildings, especially as the disadvantages of one of the most widely used thermal insulation materials, expanded or extruded polystyrene, are now well-known. Although it is relatively cheap, easily accessible and performs well in terms of thermal insulation, it is characterised by very low biodegradability; dangerous behaviour in the event of fire, forming burning droplets and releasing a lot of smoke; and a negative impact on indoor air quality by reducing the air and water vapour permeability of walls. In other words, it no longer allows the walls to "breathe" and leads to a "sealing" of the indoor environment, a reduction in the degree of ventilation, thus contributing to the creation of favourable conditions for the appearance and growth of mould, algae, lichens or other films of micro-organisms on the surface of the building [2,3]. This leads to a reduction in the quality of the air in the inhabited space and thus to health consequences for the population.

The literature highlights the overall negative impact on the health of users through the term "sick building syndrome (SBS)", which is the negative manifestation of the population working, partially or totally, inside buildings affected by microorganism deposits, due to the degradation of indoor air quality through contamination with spores and toxins [1,4–6]. The most common mycotoxins identified in indoor air and the bodies of the population living in the contaminated environment are produced by moulds such as *Cladosporium, Acremonium, Alternaria, Periconia, Curvularia, Rhizopus, Mucor, Streptomyces, Penicillium, Aspergillus, Stachybotrys, Fusarium,* and *Myrothecium* [7–10], known to be genotoxic, immunotoxic, hepatotoxic, mutagenic; and potentially carcinogenic mycotoxins are ochratoxin [7] (OCT), aflatoxin B1 [9] and trichothecene [6–19].

The thermo-physical characteristics of thermal insulation materials used in the building industry include properties such as thermal conductivity, specific heat capacity, density and thermal diffusivity. These characteristics specify the efficiency of a material in terms of its ability to absorb, transfer and retain heat. Collectively, these properties contribute to a building's energy efficiency and ability to maintain a comfortable indoor temperature, thereby reducing the need for artificial heating or cooling. The literature points to the possibility of developing materials for the thermal insulation of buildings which, in addition to their specific performance, offer some advantages in terms of opening up new opportunities for implementing the concept of the Circular Economy. However, these "alternative" insulating materials, as of 2017, only accounted for 13% of the market, and mostly comprise non-woven, fibre mattresses made from recycled plastics (the most common being polyethylene terephthalate (PET), polypropylene (PP) and polyvinyl chloride (PVC)); fibres from recycled industrial textile waste, animal or vegetable fibres (sheep wool, flax and hemp fibres, cotton wool); and/or other waste, including cellulosic waste from agricultural or industrial waste [20–24]. Of these, insulation materials made from agricultural raw materials, also called "bio-based insulation material", accounted for only 6% (2012), 8% (2017) and 10% (2020) of the insulation materials market, with expectations encouraged by EU policies of 13% in 2030, of which 40–50% should be based on wood and other cellulosic materials [24]. At present, most reports in the literature focus on the methods of making such thermal insulation materials and their physical, mechanical and thermal efficiency performance, fewer on durability and very few on the impact on indoor air quality [24].

Depending on their nature, insulating materials are characterised by a thermal conductivity coefficient, λ, with values ranging from 0.024–0.07 W/mK [20]. The thermal performance of insulation made of homogeneous materials, simple or combined, is usually evaluated by the following parameters: thermal conductivity, thermal transmittance, thermal diffusivity and specific heat [25]. For the most commonly used thermal insulation materials, the current thermal conductivity coefficient range is between 0.030–0.040 W/mK for expanded polystyrene and mineral wool, 0.020–0.030 W/mK for polyurethane-based thermal insulation and 0.033–0.044 W/mK for glass fibre-based insulation [26].

Hadded et al. [23] studied recycled textiles in terms of thermo-physical characteristics (thermal conductivity and thermal diffusivity). Danihelová et al. [27] conducted a study on the performance of recycled technical textiles showing that, in line with other reports [28], mattresses made of recycled waste fibres of a vegetable or animal nature can be good thermal insulators, characterised by a thermal conductivity coefficient around 0.033 W/mK. The results of their research have shown that recycled textiles have competitive thermal properties and can be used as an alternative to the "classic" building insulation materials (extruded polystyrene or mineral wool). Thus, the thermal conductivity of these insulating materials increases with increasing temperature, identifying, for example, a thermal conductivity coefficient, λ, for recycled denim insulation that varies in the range $0.032 \div 0.036$ W/mK in the temperature range $10\ °C \div -30\ °C$ and decreases with increasing density. In the same context, Valverde et al. [29] analysed the influence of the density of the thermal insulation product made by recycling textile waste on the thermal conductivity coefficient, indicating a non-linear variation, with the highlighting of a density range for which the thermal insulation performance is superior. Patnaik and Mvubua [30] created panels from layers of unspun wool, (coring wool (CW)—15 mm thick and 66.66 kg/m^3 and dorper wool (DW)—17 mm thick and 58.82 km/m^3), reinforced by interlacing. From the point of view of thermal conductivity, the recorded values indicate that an increase in temperature leads to an increase in thermal conductivity from 0.030 W/mK (at $-5\ °C$) to 0.034 W/mK (at $35\ °C$)—for the CW sample and from 0.031 W/mK (at $-5\ °C$) to 0.034 W/mK (at $35\ °C$)—for the DW sample. Zeinab et al. [28] analysed heat transfer through different types of non-woven fabrics. They studied the dependence of thermal conductivity on the thickness and density of polyester and polypropylene fibre insulation boards and concluded that, based on the measured value of thermal conductivity (approx. 0.033 W/mK), the non-woven materials analysed were suitable for use as thermal insulation material. A collective at the Brno University of Technology, Czech Republic [31] analysed the behaviour of thermal insulation boards made of recycled polypropylene R-PP and 5–20% bi-component polyvinyl chloride PVC fibres. It was concluded that in this case the thermal conductivity coefficient, λ, increases with increasing test temperature, temperature difference and density, and a product density of min. 150 kg/m^3 provides sufficient physico-mechanical performance to allow in-situ vertical handling and positioning of the thermal insulation boards. Patnaik et al. [30] developed and analysed a proposal for a non-woven thermal insulation material based on 50% wool and 50% recycled R-PP fibres, which showed good thermal insulation performance and biological resistance.

Over time, a number of criteria have been established to assess the quality of one insulating material against another. In addition to thermal insulation performance, aspects such as impact on human health from production to end-of-life, dust or fibre emissions, biopersistence, operational safety, environmental impact, fire performance, fire toxicity, affordability in terms of price and purchase, durability and use are now being analysed. Although the development of innovative thermal insulation materials based on recycled waste would apparently solve many problems, the most important of which is the further implementation of the concept of the Circular Economy, while at the same time reducing energy consumption for indoor comfort and reducing environmental impact, a number of other challenges and difficulties arise. Thus, most of these materials, especially those developed by recovering agro-industrial wastes or by-products, are highly sensitive to water and water vapour. At the same time, because of their structure—they are often made in the form of non-woven or loose material (which requires supporting structures when put into operation)—they have a low stability of shape and dimensions. This type of material frequently weighs under its own weight, mechanical strengths are low; therefore, it is also sensitive in terms of thermal insulation performance (this is also influenced by dimensional and density aspects) [32]. Consequently, they are often conditioned by their location inside buildings, unlike the most commonly used thermal insulation materials such as expanded/extruded polystyrene or mineral wool.

Therefore, with both advantages and disadvantages, innovative thermal insulation materials, developed by recycling waste or industrial by-products, represent an area of interest with potential for exploitation, but which requires further research.

In terms of the possibilities for recycling cellulosic waste into thermal insulation materials, the advantage of these materials is that, as they are often in bulk, they can be used to insulate areas that are difficult to access for the application of other forms of insulation material (boards, panels, etc.) [32]. However, these cellulosic waste insulations, in addition to their sensitivity to water, have a very low resistance to fire, which makes it necessary to identify methods of improvement, some of which use the properties of aerogel-based composites [26]. Other studies have shown that the thermal conductivity of cellulose thermal insulation is influenced by moisture content during use, with thermal conductivity increasing with increasing moisture content. The percentage increase in thermal conductivity is higher than the increase in humidity [32]. Vejelis et al. estimate that a 1% increase in the adsorbed hygroscopic moisture content induces a 1.25–1.5%, or even 2%, increase in the λ [33,34]. As cellulose fibres are dried, their strength increases and porosity decreases, which also influences thermal insulation performance [35–37].

Research by Talukdar et al. [38] showed that the thermal conductivity coefficient, measured for a temperature range between 10 °C and 30 °C at an average temperature value of 22.5 °C, varies according to a polynomial function with respect to moisture content ($\varnothing$), as shown in Equation (1):

$$\lambda = (a + b \cdot \varnothing + c \cdot \varnothing^{1.5} + d \cdot \exp(-\varnothing)) \tag{1}$$

where a, b, c and d are coefficients determined experimentally with the following values: a = 0.092482655, b = 0.15480621, c = 0.066517733, d = 0.1296168.

In the same trend, Sandberg [39] analysed the variation in the thermal conductivity coefficient as a function of the water absorption of the thermal insulation material, identifying a linear equation of the form:

$$\lambda = 0.037 + 0.0002 \cdot w \; (W/mK) \tag{2}$$

where w is the amount of water absorbed per unit volume of cellulose, kg/m^3.

In cellulosic material, water can exist in three different ways: non-freezing bound water, present in the large pores and between the fibres; non-freezing bound water, present in the mycopores of the fibre; and bound water in the hemicellulose. Experimental research has shown that capillary water tends to be lost faster than absorbed water, which induces the advantage that such materials contribute to the regulation of indoor air humidity, i.e., in low humidity conditions, they can release water; and in high humidity conditions, they can retain it [40]. However, there is a risk that under conditions of high humidity for a relatively long period, especially in cold climates, the cellulosic material may form an environment favourable to the growth of mould, which makes antifungal treatment necessary [34]. The most common antifungal treatments, which also have a role in increasing fire resistance, were those based on borax, boric acid, aluminium sulphate or ammonium sulphate, most commonly applied by wet spraying [41,42]. However, it is now known that these treatments have limited durability, may impact on human health and are not environmentally friendly; therefore, more effective methods are being sought that are resistant to accidental water infiltration and have low environmental impact and that comply with the EU Registration, Evaluation, Authorization and Restriction of Chemicals (REACH) regulation to avoid threatening human health and the environment.

From the information presented, we can identify, on one hand, the interest that the potential for revalorization of waste and industrial by-products holds in the investigation of possibilities for creating eco-innovative insulation materials. On the other hand, a series of difficulties arise due to the high degree of diversity in the type and quality of the raw material, leading to notable variations in the performance of the eco-innovative product intended for thermal insulation.

This study aimed to analyse a set of 10 types of "niche" thermal-insulation materials, produced using recycled cellulosic or agro-industrial wastes, available on the building materials market. The comparative analysis was carried out from the point of view of thermal insulation performance, simultaneously with water vapour sorption/desorption capacity and resistance to the action of moulds, all of which have implications in terms of indoor air quality. The study achieves several objectives, as follows:

- It provides a comparative analysis of the performance of a variety of thermal insulation materials that are available in the national and European construction materials market;
- The study highlights concrete possibilities for integrating waste;
- It contributes to establishing a positive environment for interdisciplinary research. This is done by simultaneously highlighting the characteristics of these materials from the viewpoint of their application field (thermal insulation of buildings), as well as the potential impact of their use on indoor air quality, and, consequently, the long-term effects on public health. This includes aspects such as resistance to fungi and other microorganism activity. This approach promotes a deeper understanding and emphasises the necessity to evaluate the performance of construction materials—in this case, thermal insulation materials—not only from the perspective of the response they provide to the requirements of their application field but also through a broader analysis. This wider analysis considers environmental impacts (such as opportunities for recycling waste, the inclusion of agricultural by-products), durability, and effects on the hygiene, safety and security conditions of the population. Historically, such analyses were mainly focused on compatibility with the field of use. However, today, in line with sustainable development strategies founded on the three core pillars (economic, social and environmental) endorsed at both European and global levels, all these requirements form an integral part of the evaluation of all materials designated for use in construction;
- Last but not least, the study seeks to improve the supportive theoretical framework. This is beneficial especially for the practical implementation of technological transfer from applied research to the production of thermal insulation materials. These materials have a high potential for recycling waste or agro-industrial by-products and are optimised from a thermal efficiency point of view and for the necessary treatments to ensure safe and hygienic use.

2. Materials and Methods

2.1. Material Characterisation

The experimental investigations carried out for the comparative analysis of the performance of "niche" thermal insulation materials were conducted under laboratory conditions. From the range of materials available on the building materials market, 10 types were selected (Figure 1). Their characteristics are presented in Table 1.

While some of the selected materials might seem similar, their selection aimed to evaluate comparatively how the degree of raw material sorting and subsequent processing can influence the performances analysed. Thus, although materials P1, P2 and P3, which primarily consist of waste from cigarette filters, appear similar, it is evident that the raw material for P1 had a higher degree of sorting, specifically, just waste from these filters. In contrast, for P2 and P3, this degree of raw material sorting was reduced, with cigarette filter waste mixed with cigarette paper waste (for P2) and even with waste aluminised paper (for P3). The experimental results demonstrate that their thermophysical characteristics are quite distinctive, also. Comparing these three situations could be particularly beneficial for cigarette manufacturers seeking to optimise their waste recycling process, and independent production units, which could impose certain conditions on the raw material they acquire for producing such thermal insulation materials. For similar reasons, materials P4 and P8 were selected. Although they seem to use raw materials from the same waste category (waste paper), they are produced by two different manufacturers, and specific elements of the technological process could influence the final product. Conversely, materials P6 and P7, manufactured by the same entity and derived from similar waste materials (waste cardboard), are subject to varying degrees of technological processing. Therefore, the experimental results for P6 and P7 are particularly beneficial for manufacturers of construction materials. They increase awareness regarding performance and show a high level of product quality is required to ensure the desired energy efficiency and hygiene in the occupied spaces. Despite their similarities or differences, all these selected materials have a common denominator: the cellulose component in various forms.

Figure 1. The appearance of thermal insulation materials.

Table 1. Coding and characterisation of thermal insulation materials.

Sample Code	Density in Natural State (kg/m^3)	Marketing Method	Installing	Raw Material for Production
P1	21.2	bulk	inside the building, loose between rigid panels	cellulose acetate, cigarette filter manufacturing waste
P2	31.7			cellulose acetate, cigarette filter manufacturing waste and cigarette paper waste
P3	38.9			cellulose acetate, waste cigarette filter manufacture, waste cigarette paper and waste aluminised paper
P4	35.1			cellulose from waste paper, producer 1
P5	23.4			wood fibres
P6	55.8			cellulose from cardboard waste
P7	125.5			cellulose from waste cardboard, poor processing, inhomogeneous product
P8	48.0			cellulose from waste paper, producer 2
P9	98.2			rice husk waste
P10	20.7	non-woven mattress	inside the building, on wooden frames	composite based on sheep wool, recycled PET fibres and cellulosic fibres for the textile industry

2.2. Assessing the Impact of Density and Humidity Content on Thermal Conductivity

The thermal insulation efficiency of the products was evaluated using the thermal conductivity coefficient λ (W/m·K). Thermal conductivity was measured with a specialised instrument, the λ-Meter EP500e guarded hot plate equipment (Lambda-Messtechnick GmbH, Dresden, Germany). The entire process was carried out following the guidelines of the SR EN 12667 [43] standard. This approach allowed for a complete characterisation of the products based on their thermal insulation efficiency.

For our experimental investigation, we selected test temperatures of 10 °C and 23 °C and subjected the materials under test to specific humidity conditions (dry material and material conditioned at 50% relative humidity). These choices were made in compliance with the SR EN ISO 10456 standard [44], which explicitly outlines this set of conditions for reporting the values derived for the thermal conductivity coefficient.

Testing of bulk materials and loose insulation typically involves the use of incompressible frames [45–47], often made of polyurethane (PUR) or extruded polystyrene (XPS). These materials are not only more durable for laboratory tasks, but they also prevent mass transfer from the external environment due to their sealed pore structure. For this study, XPS frames with predetermined volume were used. Each frame measures 500 × 500 × 50 mm on the exterior to fit λ-Meter EP500e plates and features a rectangular cavity measuring 200 × 200 × 50 mm where the materials were tested, as is shown in Figure 2. This approach ensures that the testing area, which on λ-Meter EP500e measures 150 × 150 mm [48], exclusively comprises the material under test, eliminating any possibility of inadvertent overlap with the frame. The frame is positioned in the guarded area of the equipment, ensuring that the sample is tested at the specified density. The sample is shielded from external factors such as humidity, and there is no interference with the heat flow generated by the equipment [48]. This heat flow is used to determine the sample's equivalent conductivity using an absolute measurement technique applicable to samples of a rectangular shape [49].

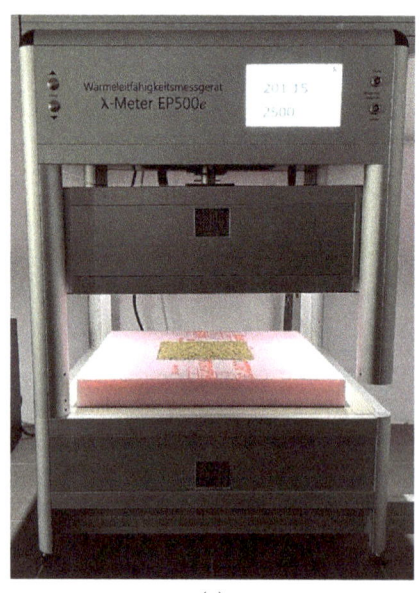

(a)   (b)

Figure 2. Testing the P6 material: (**a**) The setup used for determining thermal conductivity at the specified density; (**b**) The product, after testing it at maximum density and following extraction from the XPS frame.

While adherence to certain standards is not necessarily mandatory from a research perspective, complying with these established guidelines ensures our results are reproducible and can be independently validated. Moreover, the results could be useful during potential revisions of the standards. Therefore, tested material was conditioned following standard SR EN 15101-1+A1: 2019 [50] in an oven at a temperature of (70 ± 2) °C, using air from the laboratory at (23 ± 2) °C and (50 ± 5)% relative humidity. The testing procedure started with identifying the natural density of the product, defined as the lowest density at which the material could sustain its form under its own weight. Every test is done for three average temperatures of the sample: (10 ± 1) °C, (23 ± 1) °C and (40 ± 1) °C. Following the initial conductivity test with the dry material, it was conditioned using a climate chamber Angelantoni Challenge CH250 (Angelantoni Industrie Srl, Massa Martana, Italy), at a temperature of 23 °C $\pm$ $(0.25 \div 0.3)$ °C and a relative air humidity (RH) of (50 ± 1)%. The conditioning time was specific to each type of material analysed, in order to achieve constant mass. Constant mass was considered reached when, between two successive weighings, there was no difference greater than 0.1%. This way of testing was considered necessary because it is known that humidity influences thermal conductivity performance [44,51]. After constant mass was achieved, the product was retested, using the same protocol, at (10 ± 1) °C, (23 ± 1) °C and (40 ± 1) °C.

Another stage in the assessing of the thermal efficiency of the materials involved studying the change in the thermal conductivity coefficient relative to density. To achieve this, after the sample was examined in both dried and normal conditions, extra material was incrementally added to the frame until the necessary density for the next test was reached. The extra material was introduced in uniformly distributed layers. Each layer was compacted before the addition of a new layer, in order to ensure a consistent sample density throughout. This approach to analysis was considered important for two primary reasons: firstly, due to the significant impact a material's density can have on its thermal insulation properties, and secondly, due to the high likelihood of the material compacting under its own weight. The final stage of evaluating the materials' thermal performance analysed how the thermal conductivity coefficient varied in relation to both density and moisture content. To do this, the test samples, once dried to a constant mass and tested for a given density, were placed in a climate-controlled chamber at a temperature of 23 °C $\pm$ $(0.25 \div 0.3)$ °C and a relative humidity of (50 ± 1)% until they achieved a constant mass, after which the thermal conductivity coefficient was determined for the sample conditioned in this way. Due to existing compaction, it was observed that the moisture permeates the material more slowly. As a result, the conditioning period required for the samples to attain a constant mass was longer than it was for samples at their natural density.

2.3. Characterisation of Materials in Terms of Hygroscopicity Performance

The hygroscopicity characteristics were evaluated by analysing the sorption/desorption curves according to the methodology indicated in European standard SR EN ISO 12571:2021 [52]. The sorption/desorption capacity of water vapour was quantified by plotting the characteristic curves of the variation of the specimen mass as a function of the relative air humidity, $w(\%) = f(RH(\%))$. For testing, specimens were exposed to the ambient environment in a closed room at a constant temperature (23 ± 0.5) °C, varying only the humidity parameter (RH) with an accuracy of ± 1%. The climatic chamber used was of the type FDM F.lii Della Marca C700BXPRO RT100 (F.lli Della Marca s.r.l., Rome, Italy), with a temperature accuracy of ± 0.5 °C and a humidity accuracy of ± 1%. Initially, for sorption curve plotting, the relative humidity was increasing in 5 steps, 30%, 45%, 60%, 75% and 90%, respectively. Subsequently, for the analysis of the moisture desorption capacity, the air humidity variation plot was plotted in reverse, also in 5 steps, 90%, 75%, 60%, 45% and 30% respectively. For stabilisation of the monitored parameters, the specimens were kept at each RH value for 7 days, after which they were weighed to an accuracy of 0.0001 g. To ensure repeatability conditions, 5 determinations were performed for each test, presenting the mean value of the results.

2.4. Characterisation of Materials in Terms of Resistance to the Action of Micro-Organisms

The resistance to the action of micro-organisms was carried out in two test variants: in the case of exposure of thermal insulation materials under conditions of high humidity and under conditions of an environment contaminated with micro-organisms, respectively.

2.4.1. Analysis of the Risk of Mould Growth When Thermal Insulation Materials Are Exposed to High Humidity Conditions

In the first testing variant, in accordance with the working methodology indicated in the European guide on the elaboration of technical agreements for the marketing of construction products in accordance with European regulations, EAD 040138-01-1201 [53] and EAD 040005-00-1201 [54], and the European standard SR EN ISO 846:2019 [55], the test specimens were exposed in an environment with high humidity achieved as a result of water evaporation under conditions of normal pressure and constant temperature (23 ± 1) °C, in a closed enclosure. This method is specific to the evaluation of the biological resistance of thermal insulation materials or acoustic insulators made of animal fibres, manufactured in the factory or in-situ. During the 28 days of exposure, the specimens did not come into contact with liquid water. At regular intervals (5, 10, 15, 20 and 25 days) and at the end of the 28 days of testing, the test specimens were examined visually without optical magnification and microscopically using a LEICA SAPO microscope (Leica Microsystems, Wetzlar, Germany) to identify signs of mould growth. The quantification of resistance to the action of microorganisms was evaluated by classifying the fungal growth in rating classes and categories indicating the performance of the product in terms of nutrient content conducive to the growth of microorganisms, according to the SR EN ISO 846:2019 [55] reference, summarised in Table 2. With the character "+", it was evidenced that the fungal load increased significantly from the evaluation performed at exposure time t to the evaluation performed at exposure time t + 1, without however being a sufficient change in order to affect the fungal growth rating class.

Table 2. Evaluation of fungal growth and product performance.

Class	Evaluation of Fungal Growth	Category	Product Performance Estimation
Class 0	No sign of growth on microscopic examination	0	The material is not a nutrient medium for microorganisms (it is inert or fungistatic).
Class 1	Growth invisible to the naked eye but clearly visible under a microscope	1	The material contains few nutrients or is so little contaminated that it allows very little growth.
Class 2	Increase visible to the naked eye, covering up to 25% of the test area	2–3	The material is not resistant to attack by micro-organisms; it contains nutrients that allow them to grow.
Class 3	Increase visible to the naked eye, covering up to 50% of the test area.		
Class 4	Increase visible to the naked eye, covering more than 50% of the test area.		
Class 5	Strong growth covering the entire test surface.		

2.4.2. Evaluation of the Resistance of Thermal Insulation Materials in Conditions of a Micro-Organism Contaminated Environment

In the second variant of testing the resistance of thermal insulation materials to the action of moulds, systems of exposure of specimens in an environment contaminated with two species of moulds, most encountered in daily activity, were carried out. The microbial laboratory procedure was performed on potato dextrose agar (PDA) culture media in the presence of two fungal species—*Penicillium notatum* and *Aspergillus niger*. Each sample of insulation was placed in the middle of a Petri dish, and the two fungal species were inoculated on media in a cross-system. Both species were collected from the indoor walls of buildings and were chosen as fungal contaminants due to their presence in inhabited areas.

Inoculation was performed with a sterile 10 μL loop and the fungal-insulation culture was incubated at 25 °C for 5 days. A secondary culture test was performed, with insulation incubated on culture media. The test was designed to assess the fungal component present on insulation. The readings were performed each 24 h with the recording of results on the 6th level of the proposed scale.

3. Results and Discussion

The results of the experimental research on thermal insulation performance including water vapour sorption/desorption capacity and resistance to mould growth, all of which have implications for indoor air quality, showed similarities and differences between the 10 types of materials analysed. These specific behaviours of each material type are believed to have been influenced by several factors, of which the characteristics of the raw material used to make the material is the main one.

3.1. Characterisation of Materials in Terms of Thermal Insulation Efficiency

Heat transfer through a solid sample of material primarily occurs through conduction and convection mechanisms, as is already established [56–59]. Convection is affected by the arrangement and volume of air gaps between particles [60,61], while conduction is influenced by the contact area and number of contact points between solid particles [62]. These mechanisms collectively determine how heat is transferred. As the density of a material increases due to subsidence, the size of air voids decreases and the number of contact points between particles increases. Therefore, when compressing a uniformly air-filled material, the convective component of the heat transfer is reduced, while the conduction is increased due to improved connectivity between the solid parts [63,64]. For thermal insulators that incorporate recycled textile fibres [25,29] or natural fibrous insulation material, byproducts from the agricultural industry [60,64], polynomial correlations between thermal conductivity and density have been already presented. The test results from the present research indicate that compression of a low-density product with uniformly distributed air gaps leads to a significant decrease in thermal conductivity, as seen for the products P1, P2, P5, and P10 in Figure 3. Compression produces a sharp decline in convective heat flow in the initial segment of the density range, specifically between the natural density of the material and approximatively 40 kg/m^3. For this interval, the reduction in convective heat flow surpasses the increase in the conductive part of heat transfer. However, after a certain density threshold is reached, the rate of increase in heat transfer through conduction surpasses the decrease in convection; therefore, the value of thermal conductivity rises. For easier comparison of the evolution of the thermal conductivity coefficient depending on density, for the dry situation (a), versus conditioned at 23 °C ± (0.25 ÷ 0.3) °C and RH (50 ± 1)% (b), the graphical representation for each of the analysed thermal insulation materials was made while maintaining the same scale (on the X and Y axes).

Increasing material density leads to different variations in thermal conductivity, depending on the type of material. Thus, in the case of samples P1, P2, P10, the increase in density initially causes a reduction in the thermal conductivity coefficient, with a tendency to linearise, possibly with the recording of a minimum value on the graph λ = f(density) marked by an inflection point. A similar trend, with the existence of an inflection point on the λ = f(density) graph, but without a linearisation tendency, i.e., with the existence of an inflection point, is also recorded for the materials P4, P5, P6, P9. It should be noted, however, that in these cases the zone of decrease and existence of the inflection point is more evident, followed by a zone of increase in the thermal conductivity coefficient with increasing density. This trend is particularly evident in dry samples. Testing these materials after they have been conditioned at 50% RH, the λ = f(density) variation graph undergoes changes, generally emphasising the existence of the inflection (minimum) point. Therefore, it is considered that, in the case of these materials, increasing the density to a material-specific value may have beneficial effects by reducing the thermal conductivity coefficient. Above this optimum density value, the effect on the coefficient of thermal conductivity is either not significant (λ increases by less than 10%, e.g., sample P1) or even negative (λ increases by more than 10%, e.g., samples P2, P4, P6).

(a)

(b)

P1—Raw material: cellulose acetate, cigarette filter manufacturing waste

(a)

(b)

P2—Raw material: cellulose acetate, cigarette filter manufacturing waste and cigarette paper waste

(a)

(b)

P3—Raw material: cellulose acetate, waste cigarette filter manufacture, waste cigarette paper and waste aluminised paper

(a)

(b)

P4—Raw material: cellulose from waste paper (producer 1)

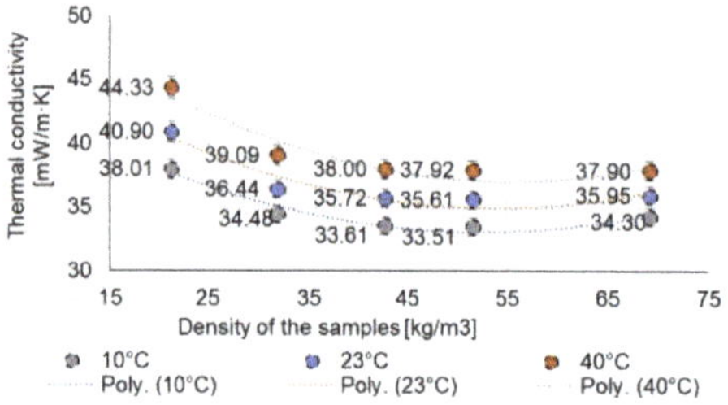

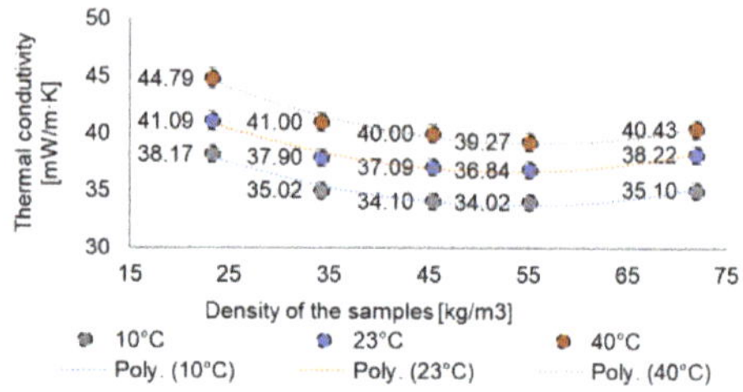

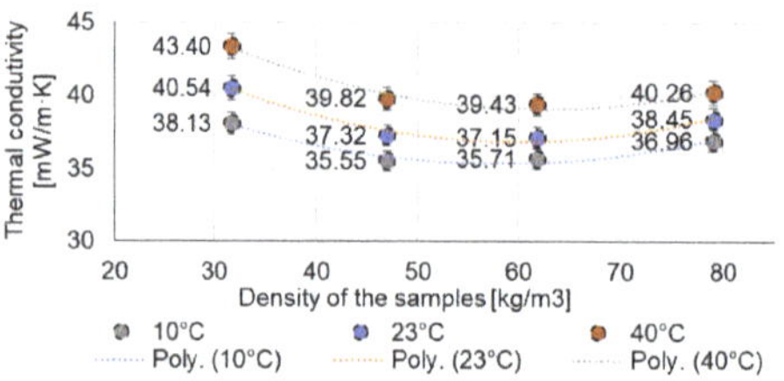

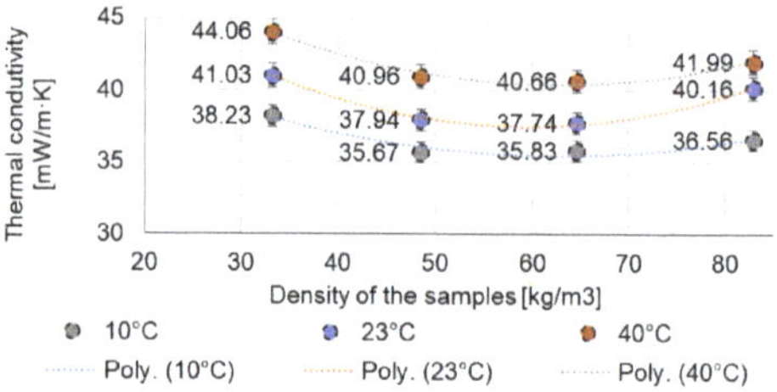

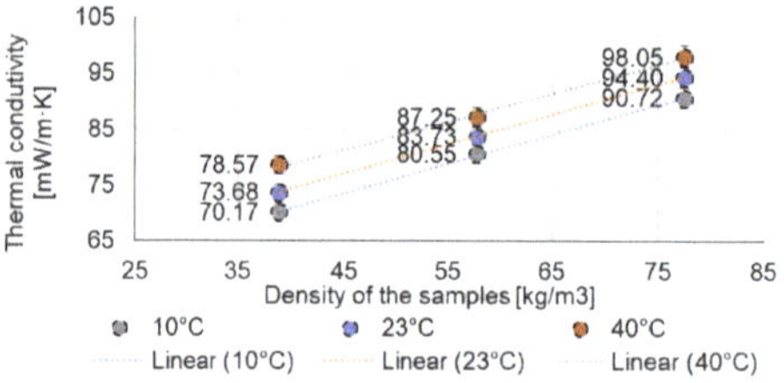

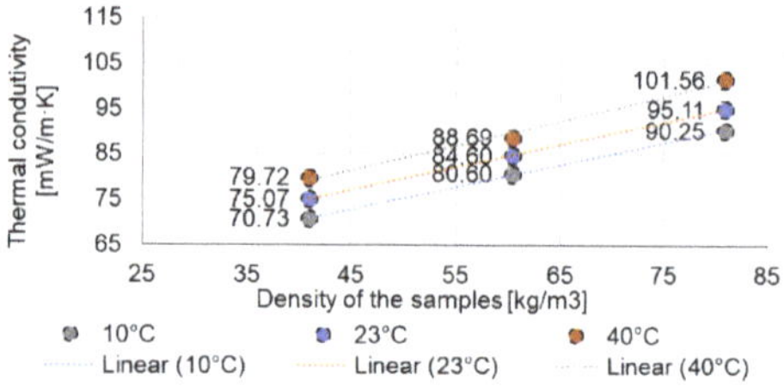

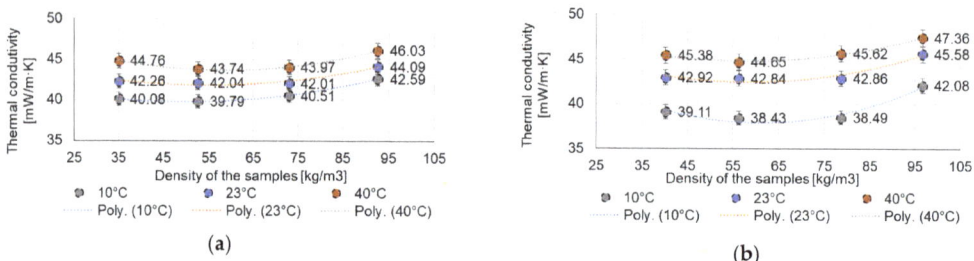

Figure 3. *Cont.*

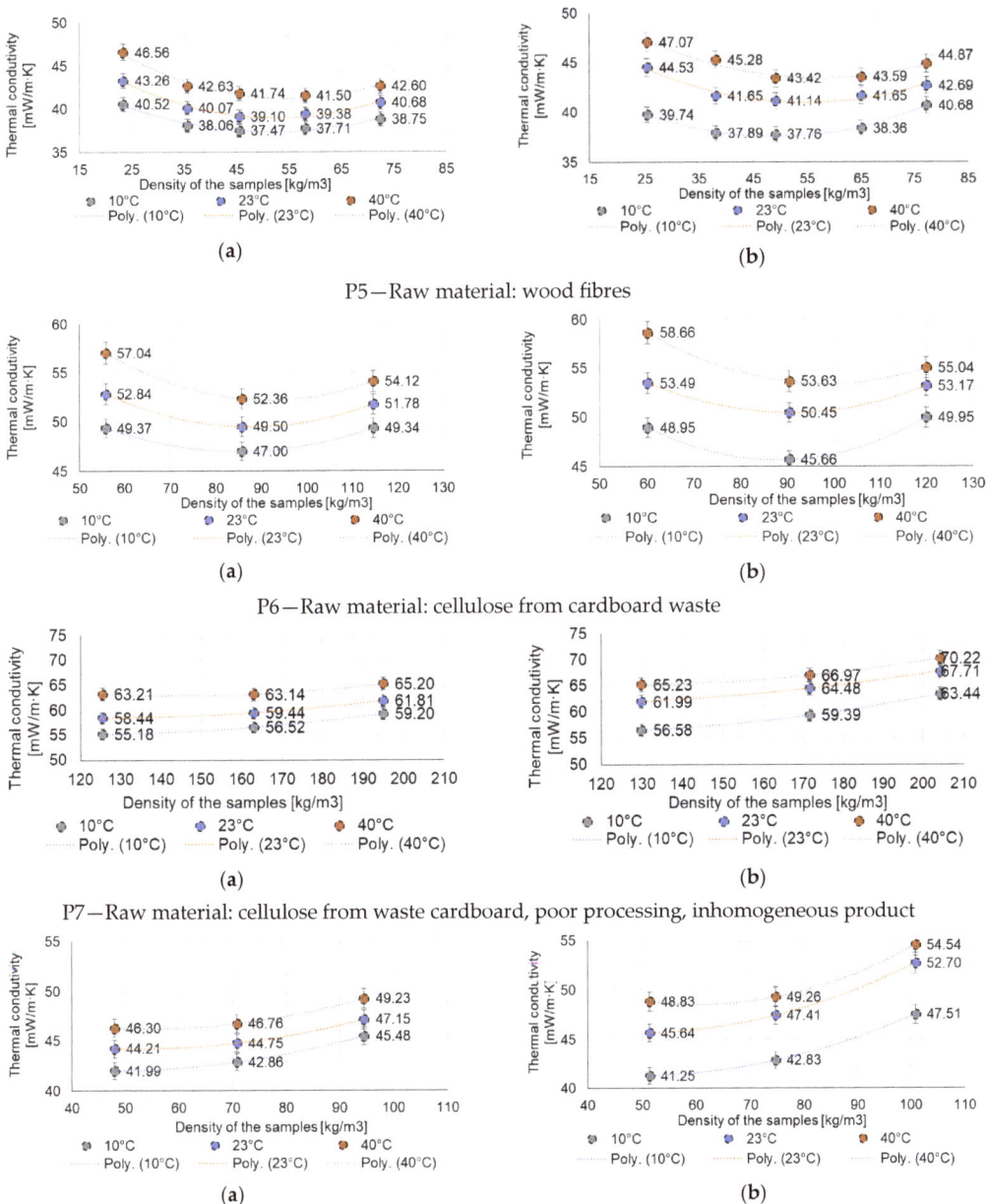

Figure 3. Cont.

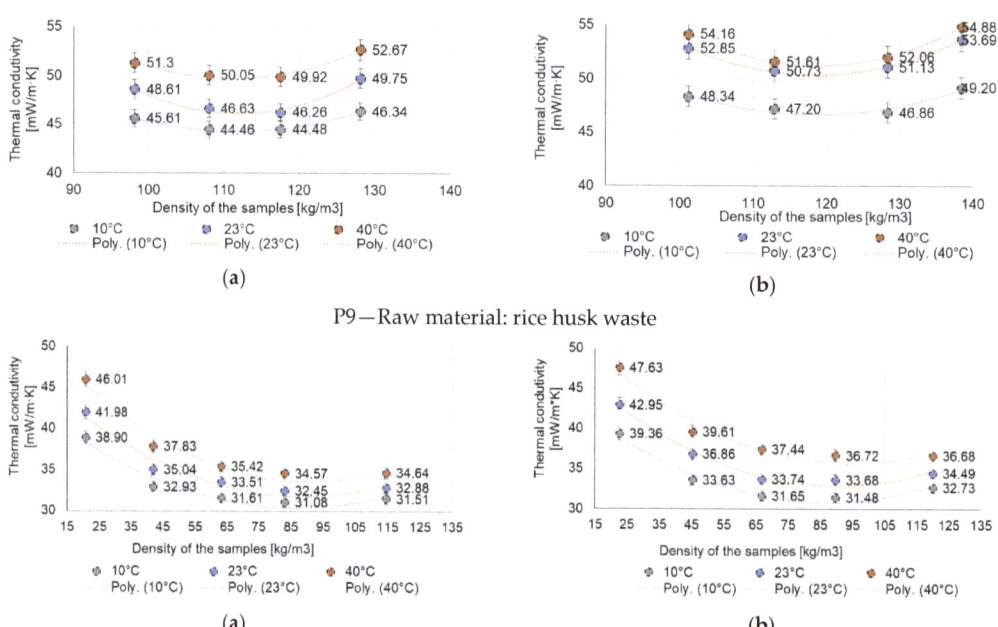

Figure 3. Influence of material density: dry (**a**), respectively, conditioned at 23 °C ± (0.25 ÷ 0.3) °C and RH (50 ± 1)% (**b**), on the thermal conductivity coefficient determined at sample average temperature of 10 °C, 25 °C or 40 °C.

An entirely different aspect is presented by the λ = f(density) curves for samples P3, P7 and P8. In the case of these materials, an increase in the thermal conductivity coefficient is clearly identified as the density increases, both for tests on dry samples and material conditioned at 50% RH. Therefore, in these cases a densification of the material, either as a result of compaction under its own weight or by the way it is laid, induces disadvantages in terms of thermal insulation performance, all the more evident as the temperature difference, warm zone-cold zone, is greater (more obviously identified on the graph of sample P8 conditioned at 50% RH where the distance between the λ = f(density) curve recorded at the 10 °C test temperature is far removed from the λ = f(density) curve recorded at 23 °C).

A comparative analysis of the values recorded for each dry tested material, respectively, after conditioning at 50% RH, shows that, in general, the existence of moisture content in the material leads to slight increases in the thermal efficiency indicator. Exceptions are recorded in the case of testing at 10 °C for samples P3, P4, P5, P8.

Increasing the test temperature of the material (10 °C, 23 °C or 40 °C) generally results, for each material analysed, both dry and conditioned at 50% RH, in a similar appearance and parallel positioning of the λ = f(density) graphs; the higher the test temperature, the higher the thermal conductivity coefficient. Therefore, it is estimated that the benefit obtained by using such materials is greater the lower the temperature, which contributes to obtaining a benefit on the quality of indoor comfort and energy consumption, especially in cold climates.

By analysing the thermal performance indicator, λ, as a function of the macroscopic structure of the material, some similarities in behaviour can be identified. For example, samples P1, P2 and P5, cellulose thermal insulation materials with a fibre-bound appearance and even P10, which is a fibre composite, show λ = f(density) graphs with a similar appearance. Therefore, it is considered that the type, raw material, structure, density and moisture content are important factors influencing the thermal insulation performance, being in correlation with previous studies [38].

The functions showing the variation of the thermal conductivity coefficient with respect to density are shown in Table 3, and those showing the variation of the coefficient of thermal conductivity with respect to the mean temperature of the samples during the test are shown in Table 4. It should be noted that, in order to achieve a correct degree of appreciation of the variation function, the condition was imposed that the correlation factor $R^2 > 0.9$. On the basis of these functions, at value 0 of the derivative of the function, it is possible to determine the density value for which the thermal conductivity coefficient is minimum, as shown in Table 5. Thus, it can be said that an increase of 149.1% (P1), 83.9% (P2), 41.6% (P4), 116.0% (P5), 54.3% (P6), 5.8% (P9), 325.1% (P10), respectively, will lead to a reduction of the thermal conductivity coefficient recorded on dry material, at 10 °C, $\lambda_{10,ct.}$, by 13.2% (P1), 7.4% (P2), 0.7% (P4), 8.2% (P5), 5.2% (P6), 3.5% (P9), 22.4% (P10), representing the highest benefit that can be obtained in this way. As can be seen, in some cases, this effort to increase the material density is substantially beneficial and justified; in other cases, it is less significant in terms of improving thermal performance. An exception is recorded for the P3 material for which the identified function is linear, continuously increasing. In this case, no benefit of reducing thermal conductivity can be obtained by increasing the material density; the $\lambda_{10,ct.}$ value recorded experimentally at natural density being the lowest of the set of experimental values. Similarly, the P8 material is identified as an exception whose density, theoretically, should be reduced by 4.4% in order to achieve an optimal conductivity coefficient $\lambda_{10,ct.}$. Although these theoretical conclusions exist, reducing the natural density of the materials (case P3 and P8) is hardly possible from a practical, implementation point of view, possibly requiring interventions in the manufacturing technology. Another exception was the P7 material, for which, although apparently polynomial functions were identified, it turned out that the evolution of the $\lambda = f(density)$ curve is increasing and with the increase in the density of the material, the thermal conductivity coefficient also increases, the subsidence even under its own weight having a negative effect. In fact, of all the materials analysed, P7 is considered to have the lowest quality, being inhomogeneous, with frequent agglomerations of material, which could be one of the explanations for this behaviour: not only the type of source material but also the quality of its processing, the homogeneity of the final product, have a great influence on the thermal insulation performance.

By comparing the experimentally obtained thermal conductivity coefficient of the materials with general values characteristic of commonly used thermal insulation materials or cellulose-based thermal insulation materials reported in the literature [32] (Table 6), it is judged that in terms of efficiency as an insulating material, P1–P10 materials correspond to the intended field of use.

Table 3. Dependence of thermal conductivity coefficient on dry material density.

Dry Material	Sample Temperature 10 °C		Sample Temperature 25 °C		Sample Temperature 40 °C	
	λ = f(Density)	R^2	λ = f(Density)	R^2	λ = f(Density)	R^2
P1	y = 0.0047x² − 0.4963x + 46.103	0.95	y = 0.0051x² − 0.5519x + 49.817	0.92	y = 0.0058x² − 0.642x + 54.771	0.93
P2	y = 0.0038x² − 0.4432x + 48.245	0.96	y = 0.0045x² − 0.5431x + 53.096	0.98	y = 0.0045x² − 0.5567x + 56.44	0.98
P3	y = 0.5337x + 49.509	0.99	y = 0.5382x + 52.702	1.00	y = 0.5063x + 58.577	0.99
P4	y = 0.0016x² − 0.1591x + 43.732	0.99	y = 0.0016x² − 0.1727x + 46.485	0.95	y = 0.0021x² − 0.2479x + 50.885	0.99
P5	y = 0.0037x² − 0.3886x + 47.437	0.97	y = 0.0049x² − 0.516x + 52.508	0.98	y = 0.0052x² − 0.5732x + 56.907	0.98
P6	y = 0.0027x² − 0.4649x + 66.824	1.00	y = 0.0032x² − 0.5704x + 74.574	1.00	y = 0.0037x² − 0.6797x + 83.456	1.00
P7	y = 0.0007x² − 0.1633x + 64.8	1.00	y = 0.0007x² − 0.1699x + 69.025	1.00	y = 0.001x² − 0.2772x + 82.96	1.00
P8	y = 0.0016x² − 0.1468x + 45.449	1.00	y = 0.0017x² − 0.1745x + 48.743	1.00	y = 0.0018x² − 0.1945x + 51.474	1.00
P9	y = 0.0074x² − 1.6584x + 136.88	0.99	y = 0.0135x² − 3.0344x + 216.08	0.98	y = 0.0099x² − 2.1941x + 171.64	0.98
P10	y = 0.0018x² − 0.3168x + 44.139	0.95	y = 0.0021x² − 0.3684x + 48.13	0.96	y = 0.0025x² − 0.4435x + 53.464	0.96

Table 4. Dependence of the thermal conductivity coefficient on the density of the conditioned material at 23 °C ± (0.25 ÷ 0.3) °C and RH (50 ± 1)%.

Material Conditioned at 23 °C ± (0.25 ÷ 0.3) °C and RH (50 ± 1)%	Sample Temperature 10 °C		Sample Temperature 25 °C		Sample Temperature 40 °C	
	$\lambda = f(Density)$	R^2	$\lambda = f(Density)$	R^2	$\lambda = f(Density)$	R^2
P1	$y = 0.0045x^2 - 0.4872x + 46.85$	0.97	$y = 0.0047x^2 - 0.5034x + 50.052$	0.97	$y = 0.0051x^2 - 0.5706x + 55.084$	0.98
P2	$y = 0.003x^2 - 0.3775x + 47.308$	0.92	$y = 0.005x^2 - 0.5988x + 55.32$	0.99	$y = 0.0041x^2 - 0.5128x + 56.485$	0.98
P3	$y = 0.489x + 50.809$	0.99	$y = 0.5024x + 54.4$	0.99	$y = 0.5481x + 56.682$	0.99
P4	$y = 0.0031x^2 - 0.3843x + 49.652$	0.95	$y = 0.0021x^2 - 0.2408x + 49.443$	0.93	$y = 0.0018x^2 - 0.2059x + 50.745$	0.99
P5	$y = 0.0039x^2 - 0.38x + 46.878$	0.99	$y = 0.0039x^2 - 0.4263x + 52.667$	0.95	$y = 0.0036x^2 - 0.4167x + 55.488$	0.96
P6	$y = 0.0042x^2 - 0.7433x + 78.387$	1.00	$y = 0.0032x^2 - 0.5817x + 76.898$	1.00	$y = 0.0036x^2 - 0.7015x + 87.968$	1.00
P7	$y = 0.0008x^2 - 0.1617x + 64.769$	1.00	$y = 0.0005x^2 - 0.0989x + 65.97$	1.00	$y = 0.0008x^2 - 0.1919x + 77.077$	1.00
P8	$y = 0.0022x^2 - 0.2147x + 46.37$	1.00	$y = 0.0025x^2 - 0.2451x + 51.515$	1.00	$y = 0.0037x^2 - 0.4469x + 62.038$	1.00
P9	$y = 0.006x^2 - 1.4232x + 131.05$	0.91	$y = 0.0082x^2 - 1.9444x + 165.66$	0.99	$y = 0.0094x^2 - 2.2358x + 184.11$	0.99
P10	$y = 0.002x^2 - 0.3411x + 45.721$	0.98	$y = 0.0021x^2 - 0.3837x + 50.267$	0.98	$y = 0.0022x^2 - 0.4147x + 55.211$	0.96

Table 5. Calculated density (kg/m^3) for which the thermal conductivity coefficient (W/mK) is minimum, for dry material, i.e., conditioned at 23 °C ± (0.25 ÷ 0.3) °C and RH (50 ± 1)%.

Material	Dry						Conditioned at 23 °C ± (0.25 ÷ 0.3) °C and RH (50 ± 1)%					
	10 °C		25 °C		40 °C		10 °C		25 °C		40 °C	
	kg/m³	W/mK	kg/m³	W/mK	kg/m³	W/mK	kg/m³	W/mK	kg/m³	W/mK	kg/m³	W/mK
P1	52.8	0.0330	54.1	0.0345	55.3	0.0370	54.1	0.0337	53.6	0.0366	55.9	0.0391
P2	58.3	0.0353	60.3	0.0367	61.9	0.0392	62.9	0.0354	59.9	0.0374	62.5	0.0405
P3 *	-	-	-	-	-	-	-	-	-	-	-	-
P4	49.7	0.0398	54.0	0.0418	59.0	0.0436	62.0	0.0374	57.3	0.0425	57.2	0.0449
P5	52.5	0.0372	52.7	0.0389	55.1	0.0411	48.7	0.0376	54.7	0.0410	57.9	0.0434
P6	86.1	0.0468	89.1	0.0492	91.9	0.0522	88.5	0.0455	90.9	0.0505	97.4	0.0538
P7 *	-	-	-	-	-	-	-	-	-	-	-	-
P8	45.9	0.0421	51.3	0.0622	54.0	0.0462	48.8	0.0411	49.0	0.0455	60.4	0.0485
P9	112.1	0.0440	112.4	0.0456	110.8	0.0501	118.6	0.0467	118.6	0.0504	118.9	0.0512
P10	88.0	0.0302	87.7	0.0320	88.7	0.0338	85.3	0.0312	91.4	0.0327	94.3	0.0357

* The calculation of corresponding values for materials P3 and P7 was not performed due to the monotonically increasing nature of the modelling functions. This means that the lowest thermal conductivity coefficient value is found at the material's natural density, specifically, the density under its own weight.

Table 6. Typical thermal conductivity values for insulation materials.

Material	Density (kg/m³)	Thermal Conductivity (W/mK)	Material	Density (kg/m³)	Thermal Conductivity (W/mK)
EPS foam	30	0.0375	P3	38.9	0.070
Rock wool	60	0.040	P4	35.1	0.040
Polyurethane rigid foam	30	0.032	P5	24.3	0.040
Cork slab	150	0.049	P6	55.8	0.049
Cellulose fibre	50	0.040	P7	125.5	0.055
Wood wool	180	0.070	P8	48.0	0.042
P1	21.2	0.038	P9	106.0	0.045
P2	31.7	0.038	P10	20.7	0.039

Note: For the materials analysed (P1–P10), the thermal conductivity coefficient, $\lambda_{10,ct}$, determined for the dry material, at natural density (without settling), determined at a sample average temperature of 10 °C, was given.

These findings align with similar results documented in the specialised literature [25,46,60,65–67].

3.2. Characterisation of Materials in Terms of Hygroscopicity Performance

The hygroscopicity of the thermal insulation materials evaluated was analysed using as a measurable indicator the variation of the specimen mass as a function of relative air humidity, $w(\%) = f(RH(\%))$. The average amount of adsorbed/desorbed water, expressed as a percentage of constant specimen mass, for each material type is shown in Figure 4; the sorption/desorption curves are shown in Figure 5.

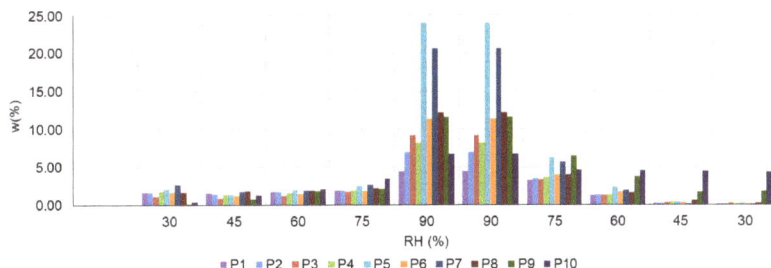

Figure 4. Average amount of adsorbed/desorbed water, expressed as a percentage of the constant mass of the specimen, recorded at RH of 30%, 45%, 60%, 75% and 90% during the sorption/desorption cycle.

Analysing the evolution of the indicators presented in Figure 4, it can be said that, in general, for all the 10 types of material analysed, there is a relative constancy of the sorption phenomenon, with a slightly increasing trend, up to and including RH 75%. When the specimens are exposed to RH 90%, the sorption phenomenon is greatly amplified for all materials, with the most evident intensification of sorption being recorded for P5 and P7 materials. Subsequently, going through the relative humidity decrease diagram induces, as expected, the desorption phenomenon in each material type. In this area, a hindrance of water loss is noticed at RH 75%. Specifically, it is shown that materials exposed to RH 75% in the desorption zone (gradual RH decrease) store more water than the same specimens in the situation where they have crossed the sorption diagram (gradual RH increase). Therefore, part of the water adsorbed at maximum RH (90%) is no longer delivered to the environment when RH is reduced to 75%. A similar phenomenon occurs for samples P9 and P10 in the RH 60–30% range. For all the other materials, in the desorption zone at RH 60%, 45% or 30%, there is a strong reduction of stored water, i.e., a very good capacity to release water as the relative humidity decreases.

Analysing the positioning of the sorption/desorption curves with respect to each other (Figure 5), for the case of P1–P8 materials, the existence of an intersection point is observed each time. Initially the sorption curve has a relatively linear evolution and is placed above the desorption curve. In the vicinity of relative air humidity RH = 60% (with small variations from one material to another) and up to RH = 90%, the position of the curves changes, the desorption curve being placed above. This behaviour indicates the strong influence that the relative humidity has on the ability of these materials to regulate the humidity of the indoor environment. On the other hand, the behaviour of P9 and P10 materials is different. In these cases, the intersection of the two curves, sorption and desorption, is no longer recorded, the sorption curve being placed below the desorption curve. This behaviour indicates the limited capacity of these materials to release the stored water in order to contribute to the regulation of this parameter under conditions of reduced relative humidity of the air in the interior space.

By analysing the behaviour of all 10 material types, it can be hypothesised that the type of material and its characteristics significantly influence its ability to contribute to air humidity regulation (indoor air quality parameter).

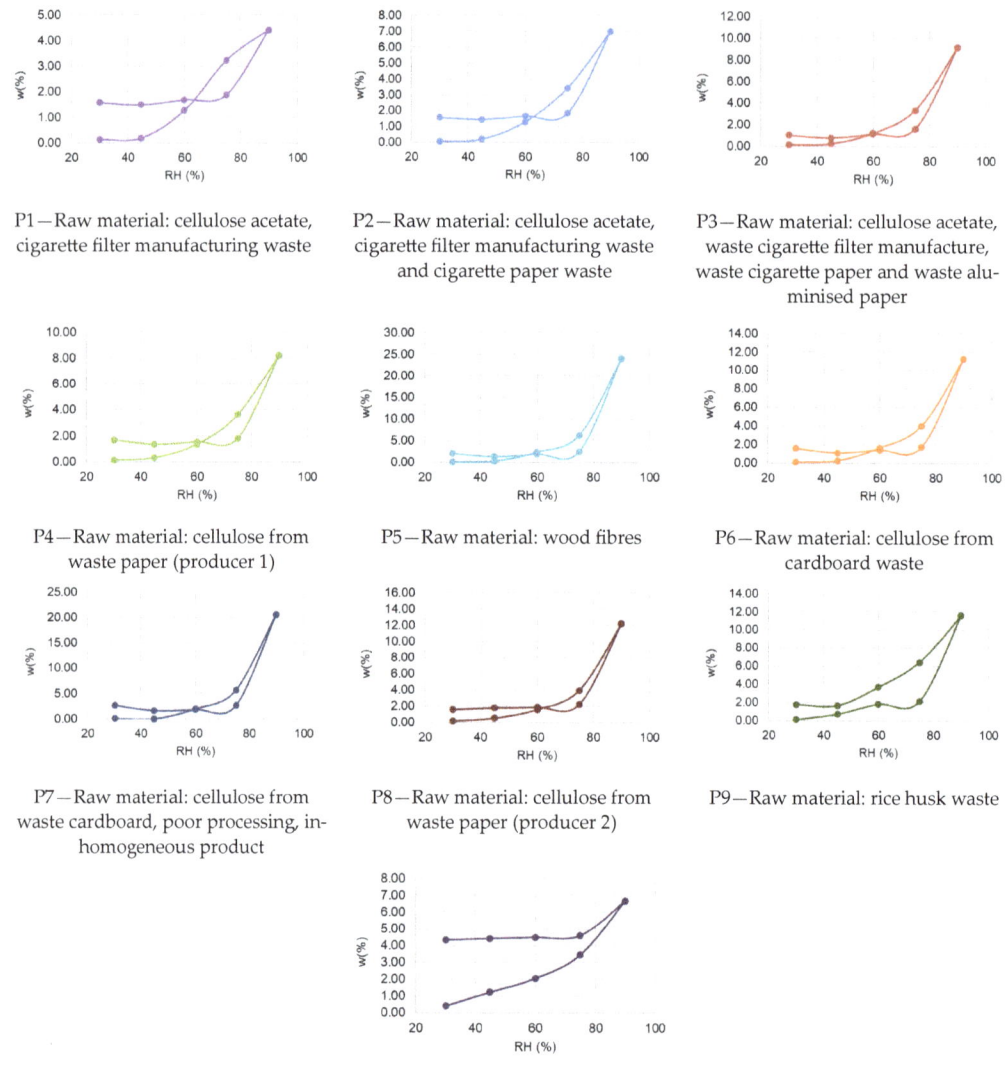

Figure 5. Sorption/desorption curves recorded for thermal insulation materials.

On the other hand, according to the literature, due to the limited capacity to release adsorbed moisture, there is a possibility of accumulation of this moisture in the insulation material. This accumulation of moisture not only does not help to regulate the humidity of the air in the interior space but may even lead to the creation of an environment favourable to the growth of mould, as shown in studies carried out mainly in northern climates [43–45,68–70]. Therefore, considering that residual humidity contributes to creating a suitable environment for the development of microorganisms, as presented in Section 3.2, it is considered that the analysed thermal insulation materials would be more suitable for use for the function of thermal insulation rather than for the function of regulating the humidity of the air in the interior space, by applying a vapour diffusion barrier to avoid mass transfer to the thermal insulation layer.

3.3. Characterisation of Materials in Terms of Resistance to the Action of Micro-Organisms

As the samples of materials analysed were taken from the footprint of the building materials, it is not possible in this case to analyse whether the existence of an antifungal treatment influences their behaviour. The existence of this antifungal treatment, as well as details of the treatment solution, is specific to each manufacturer and often confidential information, which provides elements that contribute to the degree of competitiveness among the manufacturers of these niche materials.

3.3.1. The Risk of Mould Growth If Thermal Insulation Materials Are Exposed to High Humidity Conditions

Visual and microscopic analysis of thermal insulation materials made by recycling selected agro-industrial wastes were carried out after 5 days, 10 days, 15 days, 20 days, 25 days and 28 days, respectively, of exposure under conditions of high humidity and constant temperature. The experimental results are shown in Figure 6 and summarised in Table 7 by indicating the rating class of fungal growth and performance category in terms of nutrient content conducive to the growth of micro-organisms, according to SR EN ISO 846 [55].

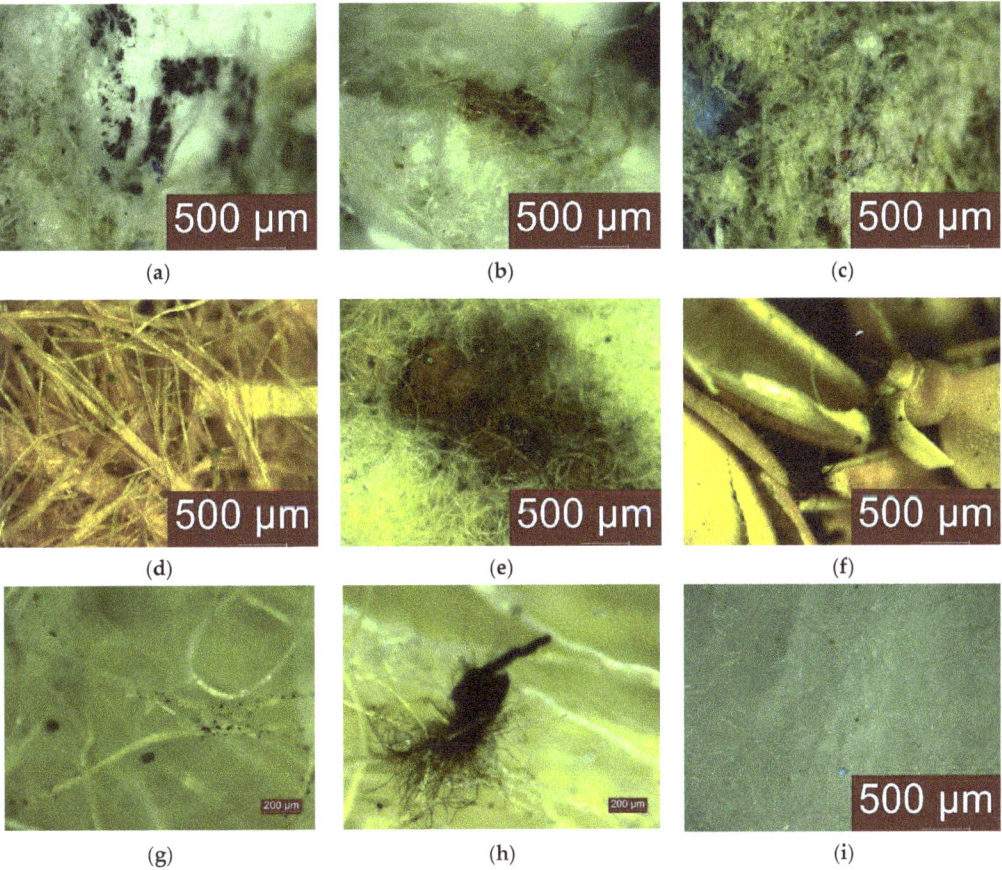

Figure 6. Example images of microscopic mould identification. (**a**) P8 (cellulose from wastepaper, producer 2)/10 days, (**b**) P3 (cellulose acetate, waste cigarette filter manufacture, waste cigarette paper and waste aluminised paper)/20 days, (**c**) P4 (cellulose from wastepaper, producer 1)/15 days,

(d) P5 (wood fibres)/28 days, (e) P2 (cellulose acetate, cigarette filter manufacturing waste and cigarette paper waste)/28 days, (f) P9 (rice husk waste)/28 days, (g) P10 (composite based on sheep wool, recycled PET fibres and cellulosic fibres for the textile industry)/15 days, (h) P10 (composite based on sheep wool, recycled PET fibres and cellulosic fibres for the textile industry)/28 days, (i) P1 (cellulose acetate, cigarette filter manufacturing waste)/28 days.

Table 7. Fungal growth rating class and performance category of the material in terms of nutrient content conducive to the growth of microorganisms.

Sample Code	P1	P2	P3	P4	P5	P6	P7	P8	P9	P10
	Fungal Growth Rating Class/Performance Category									
5 days exposure	0/0	0/0	1/1	1/1	0/0	0/0	0/0	0/0	0/0	0/0
10 days exposure	0/0	1/1	1/1	1+/1	0/0	1/1	0/0	1/1	1/1	0/0
15 days exposure	0/0	1+/1	1+/1	1+/1	1/1	1+/1	1/1	1+/1	1+/1	1/1
20 days exposure	0/0	1+/1	1+/1	1+/1	1+/1	1+/1	1+/1	1+/1	1+/1	1+/1
25 days exposure	0/0	1+/1	1+/1	1+/1	1+/1	1+/1	1+/1	1+/1	1+/1	2/2
30 days exposure	1/1	2/2	2/2	2/2	1+/1	1+/1	2/2	2/2	2/2	2+/2

Visual analysis, without optical magnification means, indicates a superficial degradation of the materials; they show slight changes in appearance, mostly in terms of colour, with a "slightly moist and damp" appearance.

Microscopic analysis of the materials showed signs of mould growth from the evaluation after exposure to damp conditions for 5 days in the case of samples P3 and P4, and from the evaluation after 10 days in the case of samples P2, P6, P8, P9. Evaluation after exposure for 15 days in a wet environment showed the presence of mould in all samples except P1. Moreover, in this sample P1, increased resistance was observed, with the first signs of mould being microscopically identified only after 28 days of exposure in a wet environment.

On the other hand, the microscopic analysis of the samples showed, in general, quantitative growth in mould from one evaluation to the next, with some of the samples being classified in the next class of fungal growth rating, i.e., class 2. Due to the shape of the material, most of which is loose, it is very difficult to assess the mould surface area proportionally, but for all these samples classified in fungal growth rating class 2, a material performance category of minimum 2 can be assumed, i.e., the material can be assessed as containing nutrients that allow mould growth.

Of the 10 materials analysed, none was identified as having a behaviour compatible with maintaining fungal growth rating class 0 (no sign of growth on microscopic examination) and performance category in terms of nutrient content conducive to the growth of microorganisms 0 (not a nutrient medium for microorganisms—inert or fungistatic). Therefore, it is considered that for all the cases analysed an antifungal treatment is required or, if this treatment existed in the production process, it is insufficient. Also, based on the results, it is considered necessary that, when these products are put into operation, the technology should provide for the use of specific protections to reduce the contact of the material with the humidity in the environment.

3.3.2. Resistance of Heat-Insulating Materials under Conditions Contaminated with Micro-Organisms

Both cultivation techniques showed an interesting behaviour of the tested insulation. The test for native microflora revealed the growth of fungi on all the insulations, regardless of the type (Table 8 and shown in Figure 7). Sample P1 showed class 2 growth of native fungi, with the development of small colonies around the insulation. The same class was established in inoculated conditions, both fungal species showing growth but with colonies under 1 cm in diameter. Sample P2 permitted the development of numerous colonies around the insulation, with the native microflora established as class 4, but in controlled inoculation, this insulation maintained the colonies of both species under 1 cm in diameter. The third type of insulation showed multiple small colonies (class 3), which developed as native microflora. In controlled inoculation, this insulation did not permit the growth of *Penicillium notatum*, unlike *Aspergillus niger* which proliferated on media. Insulation P4 showed abundant mycelium growth (native microflora), but blocked the development of *A. niger*. The same insulation restricted the development of *P. notatum* as a long and thin colony. The P5 insulation did not permit the development of either inoculated species but was sensitive to the native microflora test—where multiple confluent colonies were observed. The same results were obtained for insulations P6 and P7 in the native test, both showing high dimension colonies. The inoculation test for both insulations showed medium resistance to fungal contaminants, with small colonies developed—under 1 cm in diameter. Insulation P8 was considered as class 4 resistant to native microflora test, compared to insulations P9 and P10, which were considered as class 5. All three insulations showed well-developed colonies emerged from the insulation up to the complete coverage of fungal mycelium. In terms of controlled inoculation, all three insulations showed a different reaction. Insulation P8 was affected by both inoculated species, each having an equal share on culture media. Both P9 and P10 showed a resistance to *P. notatum*, but no resistance to *A. niger*.

Table 8. Fungal growth rating class and performance category of the material under growing medium conditions.

Sample		P1	P2	P3	P4	P5	P6	P7	P8	P9	P10
Native microflora		2	4	3	3/4	4	4	4	4	5	5
Inoculation test	*Penicillium notatum*	2	2	-	2	-	2	2	4/5	2	2
	Aspergillus niger	2	2	4	-	-	2	2	4/5	5	5

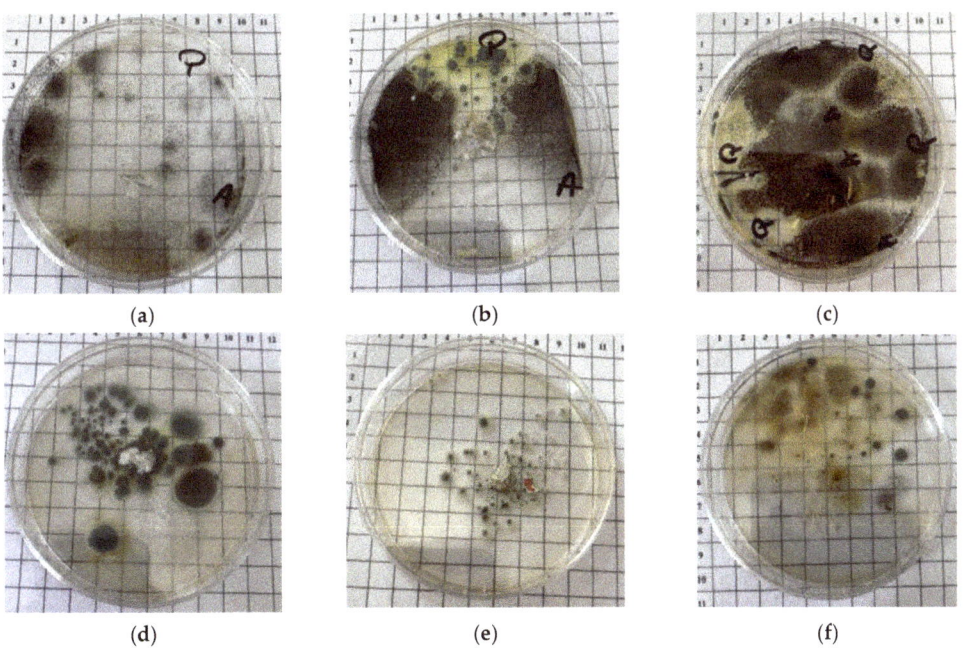

(a) (b) (c)

(d) (e) (f)

Figure 7. Cont.

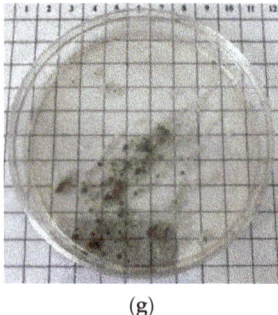
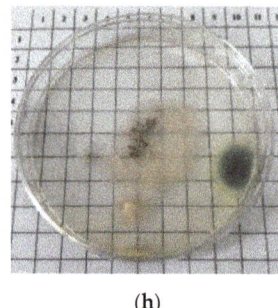
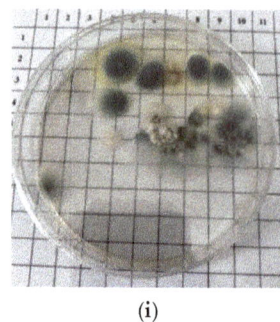

Figure 7. Development of inoculated and native microflora on culture media (*A—Aspergillus niger*; *P—Penicillium notatum*). (**a**) P3 (cellulose acetate, waste cigarette filter manufacture, waste cigarette paper and waste aluminised paper)—inoculated, (**b**) P8 (cellulose from waste paper, producer 2)—inoculated, (**c**) P10 (composite based on sheep wool, recycled PET fibres and cellulosic fibres for the textile industry)—inoculated, (**d**) P2 (cellulose acetate, cigarette filter manufacturing waste and cigarette paper waste)—native, (**e**) P3 (cellulose acetate, waste cigarette filter manufacture, waste cigarette paper and waste aluminised paper)—native, (**f**) P5 (wood fibres)—native, (**g**) P6 (cellulose from cardboard waste)—native, (**h**) P7 (cellulose from waste cardboard, poor processing, inhomogeneous product)—native, (**i**) P8 (cellulose from waste paper, producer 2)—native.

3.4. Benefits and Challenges of Loose-Fill Insulation in Construction

Loose-fill insulation has become popular due to its installation facility, cost-effectiveness, and suitability for insulating rather complex spaces. For example, insulating the voids between roof rafter structures is greatly accelerated when blowing loose-fill insulation, compared to the more labor-intensive installation of traditional insulation [57] that presents as rolls or boards. A remarkable use of loose-fill insulation is for house attics, a process proven to be quick and that minimises material waste [71]. The installation process of this thermal insulation method is simple, which helps reduce labour costs and makes it cost-effective overall. These features are appealing to construction companies and other relevant stakeholders. As a result of the specific behaviour and efficiency of eco-innovative thermal insulation materials, in the specialised literature, some reviews indicate the possibility of use, for example, for bedroom wall insulation (which could provide benefits of even 5–30% in terms of energy consumption) as indicated by [72,73]. Other studies [74,75] suggest that this insulation material would be suitable for insulating buildings with air conditioning and humidity control systems.

Additionally, incorporating loose-fill insulation aligns with sustainable and environmentally conscious construction practices. However, sourcing and producing these insulating materials locally is crucial to meet the eco-friendly goals. This approach significantly reduces both environmental and transportation costs associated with the product. By doing so, we can minimise transportation's carbon footprint while also strengthening local economies. It is important to note that produced loose-fill insulating materials may have unique properties due to variations in local resources [76,77] and manufacturing techniques [78,79]. One target of this research is to present some differences, where similar products or identical raw materials used in conjunction with different technologies yield notably different results. Consequently, conducting thorough testing of these products is vital to ensure their quality, performance, and safety. Thorough testing can identify the product's characteristics, such as thermal conductivity and density, allowing these materials to be used where they perform best.

4. Conclusions

The aim of this work was to analyse the performance and behaviour of 10 types of thermal insulation materials available on the building materials market in the "niche" area, given their origin from agro-industrial waste raw materials. The experimental requirements followed three directions:

1. Impact on the energy consumption of buildings and indoor air quality through their thermal insulation performance, quantified by the thermal conductivity coefficient determined on dry material, and on conditioned material at a temperature of 23 °C $\pm$ (0.25 $\div$ 0.3) °C and a relative air humidity (RH) of (50 $\pm$ 1)%.
2. Impact on indoor air quality through its ability to regulate humidity through the phenomenon of moisture sorption-desorption.
3. Impact on indoor air quality in terms of contamination by micro-organisms, through an analysis of the risk of mould growth and resistance to mould attack by *Aspergillus niger* and *Penicillium notatum*.

Based on the experimental results obtained, the following can be appreciated:

- In terms of thermal performance, the 10 types of materials analysed correspond to the criteria imposed by the intended field of use, with thermal conductivity coefficients comparable to those of common insulating materials;
- The coefficient of thermal conductivity is influenced by a variety of factors. These include, evidently, the type of raw material used in the production of the material, the process by which it is manufactured, its density, its moisture content, and the temperature at which the test is conducted. Typically, the presence of moisture has a negative impact on thermal performance. An increase in density, usually achieved by compressing the material, tends to reduce thermal conductivity for materials with uniform distributed air-gaps. Our experimental results show that product P3 (composite containing cellulose acetate, waste cigarette filter manufacture, waste cigarette paper and waste aluminised paper), as well as P8 (produced from waste paper by producer 2), perform best at their natural densities. For these materials, the most effective and practical thermal conductivity coefficient is achieved without altering their natural density;
- In terms of sorption-desorption capacity, a relative consistency of sorption was observed, with a slightly increasing trend, up to and including RH 75%, followed by an amplification of the phenomenon in the area of RH >75%. The most obvious intensification of sorption was recorded in the case of the thermal insulation material made from wood fibres (P5) and the one made from cellulose from waste cardboard, poor processing, inhomogeneous product (P7). The desorption curves show a lower capacity for loss than for accumulation of atmospheric moisture, i.e., following desorption, a residual amount of the adsorbed water remains in the material. Therefore, the materials analysed showed a certain capacity to contribute to the regulation of indoor air humidity, but the residual water indicated by the desorption curves not only does not contribute to the regulation of indoor air humidity, but may even lead to the creation of an environment favourable to mould growth;
- In terms of resistance to mould growth in a high humidity environment, the experimental results showed that all the materials analysed are at risk of mould growth after a longer or shorter period of exposure to a humid atmosphere, depending on the nature of the material;
- Each type of tested insulation carries a specific microflora, which can grow in optimum conditions from small-isolated colonies around the material up to complete cover of it. The artificial inoculation of insulation with two fungal species revealed a gradual resistance to colonisation from zero growth up to complete cover of one or both species;
- For none of the analysed materials, the preservation of the fungal growth rating class 0 and the performance category in terms of nutrient content conducive to the growth of microorganisms 0 was not identified; therefore a more effective antifungal treatment is required.

The scientific contribution of this study can be summarised as follows:

- It analyses a significant number of thermal insulation materials made by recycling agro-industrial waste, materials that are present in a construction materials market that is still underdeveloped;
- The experimental analysis is interdisciplinary, considering several aspects: thermal insulation performance both under dry conditions and under conditions of conditioning at a temperature of 23 °C ± (0.25 ÷ 0.3) °C and RH (50 ± 1)%; thermal insulation performance both under natural density conditions and for several degrees of compaction; the capacity for moisture absorption/desorption depending on the relative humidity of the air and resistance to mould action;
- It draws attention to weak points, particularly the need for antifungal treatments;
- Although there are some weak points, opportunities for improvement are identified and the possibility of obtaining materials that simultaneously respond to two pressing needs is highlighted: environmental protection through waste recycling and reducing energy consumption through the thermal insulation of spaces dedicated to human activities.

Author Contributions: Conceptualization, C.P., A.H. and V.S.; methodology, C.P., A.H. and V.S.; validation, C.P., A.H., V.S., C.S.D., A.A.C. and A.-V.L.; formal analysis, C.P., A.H., V.S. and A.A.C.; investigation, C.P., A.H., V.S. and C.F.; resources, C.P., V.S. and C.S.D.; data curation, C.P., A.H., V.S., A.A.C., A.-V.L. and C.F.; writing—original draft preparation, A.H. and A.-V.L.; writing—review and editing, C.P., V.S., C.S.D., A.A.C. and A.-V.L.; visualization, C.P., C.S.D., A.A.C., A.-V.L. and C.F.; supervision, A.H. and C.S.D.; funding acquisition, C.P., V.S. and C.S.D. All authors have read and agreed to the published version of the manuscript.

Funding: This research was funded by the Romanian Government Ministry of Research Innovation and Digitization, project No. PN 23 35 02 01 "Synergies of innovation and digitalization in the design of eco materials and multifunctional products for sustainable constructions, with an impact on the environment and the circular economy".

Institutional Review Board Statement: Not applicable.

Informed Consent Statement: Not applicable.

Data Availability Statement: The data presented in this study are available on request from the corresponding authors.

Conflicts of Interest: The authors declare no conflict of interest.

References

1. Haleem Khan, A.A.; Mohan Karuppayil, S. Fungal pollution of indoor environments and its management. *Saudi J. Biol. Sci.* **2012**, *19*, 405–426. [CrossRef]
2. Maskell, D.; da Silva, C.F.; Mower, K.; Rana, C.; Dengel, A.; Ball, R.J.; Ansell, M.P.; Walker, P.J.; Shea, A. Properties of bio-based insulation materials and their potential impact on indoor air quality. *Acad. J. Civ. Eng.* **2015**, *33*, 156–163.
3. Klinge, A.; Roswag-Klinge, E.; Ziegert, C.; Fontana, P.; Richter, M.; Hoppe, J. Naturally Ventilated Earth Timber Constructions. In *Expanding Boundaries: Systems Thinking for the Built Environment, Sustainable Built Environment (SBE) Regional Conference*; Habert, G., Schlueter, A., Eds.; VDF Hochschulverlag AG at the ETH Zurich: Zurich, Switzerland, 2016.
4. Ebbehoj, N.E.; Hansen, M.O.; Sigsgaard, T.; Larsen, L. Building-related symptoms and molds: A two-step intervention study. *Indoor Air* **2002**, *12*, 273–277. [CrossRef] [PubMed]
5. Zeliger, H.I. Toxic effects of chemical mixtures. *Arch. Environ. Health* **2003**, *58*, 23–29. [CrossRef] [PubMed]
6. Peuhkuri, R.; Viitanen, H.; Ojanen, T.; Vinha, J.; Lähdesmäki, K. Resistance against mould of building materials under constant conditions. In Proceedings of the Finnish Building Physics Symposium 2009: The Newest Research Results and Good Practical Solutions, Tampere, Finland, 27–29 October 2009.
7. Hope, J. A Review of the Mechanism of Injury and Treatment Approaches for Illness Resulting from Exposure to Water-Damaged Buildings, Mold, and Mycotoxins. *Sci. World J.* **2013**, *2013*, 767482. [CrossRef]
8. Zukiewicz-Sobczak, W.; Sobczak, P.; Krasowska, E.; Zwolinski, J.; Chmielewska-Badora, J.; Galinska, E.M. Allergenic potential of moulds isolated from buildings. *Ann. Agric. Environ. Med.* **2013**, *20*, 500–503. [PubMed]
9. Agag, B.I. Mycotoxins in foods and feeds. *Ass. Univ. Bull. Environ. Res.* **2004**, *7*, 3554731.
10. Rosen, E.; Heseltine, J. *WHO Guidelines for Indoor Air Quality: Dampness and Mould*; WHO Regional Office for Europe: Copenhagen, Denmark, 2009.

11. Fisk, W.J.; Lei-Gomez, Q.; Mendell, M.J. Meta-analyses of the associations of respiratory health effects with dampness and mold in homes. *Indoor Air* **2007**, *17*, 284–296. [CrossRef]
12. Mudarri, D.; Fisk, W.J. Public health and economic impact of dampness and mold. *Indoor Air* **2007**, *17*, 226–235. [CrossRef]
13. Jaakkola, J.J.K.; Hwang, B.F.; Jaakkola, N. Home dampness and molds, parental atopy, and asthma in childhood: A sixyear population-based cohort study. *Environ. Health Perspect.* **2005**, *113*, 357–361. [CrossRef]
14. Hope, J.H.; Hope, B.E. A review of the diagnosis and treatment of Ochratoxin A inhalational exposure associated with human illness and kidney disease including focal segmental glomerulosclerosis. *J. Environ. Res. Public Health* **2012**, *2012*, 835059. [CrossRef]
15. Karvala, K.; Toskala, E.; Luukkonen, R.; Lappalainen, S.; Uitti, J.; Nordman, H. New-onset adult asthma in relation to damp and moldy workplaces. *Int. Arch. Occup. Environ. Health* **2010**, *83*, 855–865. [CrossRef] [PubMed]
16. Doi, K.; Uetsuka, K. Mechanisms of mycotoxin-induced neurotoxicity through oxidative stressassociated pathways. *Int. J. Mol. Sci.* **2011**, *12*, 5213–5237. [CrossRef] [PubMed]
17. Haverinen-Shaughnessy, U. Prevalence of dampness and mold in European housing stock. *J. Expo. Sci. Environ. Epidemiol.* **2012**, *22*, 461–467. [CrossRef]
18. Andersen, B.; Frisvad, J.C.; Søndergaard, I.; Rasmussen, S.; Larsen, L.S. Associations between Fungal Species and Water-Damaged Building Materials. *Appl. Environ. Microbiol.* **2011**, *77*, 4180–4188. [CrossRef]
19. Baxter, D.M.; Perkins, J.L.; McGhee, C.R.; Seltzer, J.M. A Regional Comparison of Mold Spore Concentrations Outdoors and Inside "Clean" and "Mold Contaminated" Southern California Buildings. *J. Occup. Environ. Hyg.* **2005**, *2*, 8–18. [CrossRef]
20. Dieckmann, E.; Onsiong, R.; Nagy, B.; Sheldrick, L.; Cheesemann, C. Valorization of Waste Feathers in the Production of New Thermal Insulation Materials. *Waste Biomass. Valor.* **2021**, *12*, 1119–1131. [CrossRef]
21. Berardi, U.; Iannace, G. Acoustic characterization of natural fibers for sound absorption applications. *Build. Environ.* **2015**, *94*, 840–852. [CrossRef]
22. Danihelová, A.; Bubeníková, T.; Bednár, M.; Gergel, T. Acoustic properties of selected thermal insulation materials and their contribution to pollution of environment. *Akustika* **2018**, *30*, 35–41.
23. Hadded, A.; Benltoufa, S.; Fayala, F.J. Thermo physical characterisation of recycled textile materials used for building insulating. *Build. Eng.* **2016**, *5*, 34–40. [CrossRef]
24. Rabbat, C.; Awad, S.; Villot, A.; Rollet, D.; Andrès, Y. Sustainability of biomass-based insulation materials in buildings: Current status in France, end-of-life projections and energy recovery potentials. *Renew. Sust. Energ. Rev.* **2022**, *156*, 111962. [CrossRef]
25. Asdrubali, F.; D'Alessandro, F.; Schiavoni, S. A review of unconventional sustainable building insulation materials. *Sustain. Mater. Technol.* **2015**, *4*, 1–17. [CrossRef]
26. Illera, D.; Mesa, J.; Gomez, H.; Maury, H. Cellulose aerogels for thermal insulation in buildings: Trends and challenges. *Coatings* **2018**, *8*, 345. [CrossRef]
27. Danihelová, A.; Nemec, M.; Gergel, T.; Gejdoš, M.; Gordanová, J.; Scensný, P. Usage of Recycled Technical Textiles as Thermal Insulation and an Acoustic Absorber. *Sustainability* **2019**, *11*, 2968. [CrossRef]
28. Zeinab, S.; AbdeL-Rehim, Z.S.; Saad, M.M.; El-Shakankery, M.; Hanafy, I. Textile fabrics as thermal insulators. *Autex Res. J.* **2016**, *6*, 148–161.
29. Valverde, I.C.; Castilla, L.H.; Nunez, D.F.; Rodriguez-Senın, E.; de la Mano Ferreira, R. Development of New Insulation Panels Based on Textile Recycled Fibers. *Waste Biomass. Valor.* **2013**, *4*, 139–146. [CrossRef]
30. Patnaik, A.; Mvubu, M.; Muniyasamy, S.; Botha, A.; Anandjiwala, R.D. Thermal and sound insulation materials from waste wool and recycled polyester fibers and their biodegradation studies. *Energy Build.* **2015**, *92*, 161–169. [CrossRef]
31. Drochytka, R.; Dvorakova, M.; Hodna, J. Performance Evaluation and Research of Alternative Thermal Insulation Based on Waste Polyester Fibers. *Procedia Eng.* **2017**, *195*, 236–243. [CrossRef]
32. Hurtado, P.L.; Rouilly, A.; Raynaud, C.; Vandenbossche, V. The properties of cellulose insulation applied via the wet spray process. *Build. Environ.* **2016**, *107*, 43–51. [CrossRef]
33. Vejelis, S.; Gnipas, I.; Kersulis, V. Performance of loose-fill cellulose insulation. *Mater. Sci.* **2006**, *12*, 338–340.
34. Hurtado, P.L.; Rouilly, A.; Vandenbossche, V.; Raynaud, C. A review on the properties of cellulose fibre insulation. *Build. Environ.* **2016**, *96*, 170–177. [CrossRef]
35. Hubbe, M.A.; Venditti, R.A.; Rojas, O.J. What happens to cellulosic fibres during papermaking and recycling? A review. *Bioresources* **2007**, *2*, 739–788.
36. Nicolajsen, A. Thermal transmittance of a cellulose loose-fill insulation material. *Build. Environ.* **2005**, *40*, 907–914. [CrossRef]
37. Sprague, R.W.; Shen, K.K. The use of boron products in cellulose insulation. *J. Build. Phys.* **1979**, *2*, 161–174. [CrossRef]
38. Talukdar, P.; Osanyintola, O.F.; Olutimayin, S.O.; Simonson, C.J. An experimental data set for benchmarking 1-D, transient heat and moisture transfer models of hygroscopic building materials. Part II: Experimental, numerical and analytical data. *Int. J. Heat Mass Transf.* **2007**, *50*, 4915–4926. [CrossRef]
39. Sandberg, P.I. Determination of the effects of moisture content on the thermal transmisivity of cellulose fiber loose-fill insulation. In Proceedings of the Thermal Performance of the Exterior Envelopes of Building III, ASHRAE/DOE/BTECC/CIBSE Conference, Clearwater Beach, FL, USA, 2–5 December 1992; pp. 517–525.
40. Cerolini, S.; D'Orazio, M.; Di Perna, C.; Stazi, A. Moisture buffering capacity of highly absorbing materials. *Energy Build.* **2009**, *41*, 164–168. [CrossRef]

41. Day, M.; Suprunchuk, T.; Wiles, D.M. The fire properties of cellulose insulation. *J. Build. Phys.* **1981**, *4*, 157–170. [CrossRef]
42. Herrera, J. Assessment of fungal growth on sodium polyborate-treated cellulose insulation. *J. Occup. Environ. Hyg.* **2005**, *2*, 626–632. [CrossRef]
43. SR EN 12667; Thermal Performance of Building Materials and Products—Determination of Thermal Resistance by the Guarded Hot Plate Method and the Fluxmetric Method—Products with High and Medium Thermal Resistance. The National Organization for Standardization from Romania ASRO: Bucharest, Romania, 2002.
44. SR EN 10456; Building Materials and Products—Hygrothermal Properties—Tabulated Design Values and Procedures for Determining Declared and Design Thermal Values. The National Organization for Standardization from Romania ASRO: Bucharest, Romania, 2008.
45. Abdou, A.; Budaiwi, I. The Variation of Thermal Conductivity of Fibrous Insulation Materials under Different Levels of Moisture Content. *Constr. Build. Mater.* **2013**, *43*, 533–544. [CrossRef]
46. Brzyski, P.; Kosiński, P.; Skoratko, A.; Motacki, W. Thermal Properties of Cellulose Fiber as Insulation Material in a Loose State. *AIP Conf. Proc.* **2019**, *2133*, 020006.
47. Petcu, C.; Petran, H.A.; Vasile, V.; Toderasc, M.C. Materials from renewable sources as thermal insulation for nearly zero energy buildings (nZEB). In *Nearly Zero Energy Communities: Proceedings of the Conference for Sustainable Energy (CSE) 2017, Brasov, Romania, 19–21 October 2017*; Springer International Publishing: Cham, Switzerland, 2018; pp. 159–167.
48. Lambda-Messtechnik GmbH. *Instruction Manual for the Thermal Conductivity Test Tool λ-Meter EP500e*; Lambda-Messtechnik GmbH: Dresden, Germany, 2020.
49. Zhao, D.; Qian, X.; Gu, X.; Jajja, S.A.; Yang, R. Measurement Techniques for Thermal Conductivity and Interfacial Thermal Conductance of Bulk and Thin Film Materials. *J. Electron. Packag.* **2016**, *138*, 040802. [CrossRef]
50. SR EN 15101-1+A1; Thermal Insulation Products for Building Applications. Cellulose Formed-in-Place Thermal Insulation (LFCI). The National Organization for Standardization from Romania ASRO: Bucharest, Romania, 2019.
51. Wang, Y.; Zhang, S.; Wang, D.; Liu, Y. Experimental study on the influence of temperature and humidity on the thermal conductivity of building insulation materials. *Energy Built Environ.* **2023**, *4*, 386–398.
52. SR EN ISO 12571; Hygrothermal Performance of Building Materials and Products—Determination of Hygroscopic Sorption Properties. The National Organization for Standardization from Romania ASRO: Bucharest, Romania, 2021.
53. EAD 040138-01-1201; Factory-Made Thermal and/or Acoustic Insulation Products Made of Vegetable or Animal Fibres. EOTA: Brussels, Belgium, 2018. Available online: https://www.eota.eu/download?file=/2013/13-04-0005/ead%20for%20ojeu/ead-040005-00-1201-factory-made-thermal-acoustic-insulation-products-vegetable-animal-fibres-2015rev-2016rev_ojeu.pdf (accessed on 14 July 2023).
54. EAD 040005-00-1201; In-Situ Formed Loose Fill Thermal and/or Acoustic Insulation Products Made of Vegetable Fibres. EOTA: Brussels, Belgium, 2015. Available online: https://www.eota.eu/download?file=/2017/17-04-0880/for%20ojeu/ead%20040138-01-1201_ojeu2018.pdf (accessed on 14 July 2023).
55. SR EN ISO 846; Plastics—Evaluation of the Action of Micro-Organisms. The National Organization for Standardization from Romania ASRO: Bucharest, Romania, 2019.
56. Forsberg, C.H. *Heat Transfer Principles and Applications*; Academic Press: Cambridge, MA, USA, 2020; pp. 343–389.
57. Kivioja, H.; Vinha, J. Hot-Box Measurements to Investigate the Internal Convection of Highly Insulated Loose-Fill Insulation Roof Structures. *Energy Build.* **2020**, *216*, 109934. [CrossRef]
58. Vėjelis, S.; Skulskis, V.; Kremensas, A.; Vaitkus, S.; Kairytė, A. Performance of Thermal Insulation Material Produced from Lithuanian Sheep Wool. *J. Nat. Fibers* **2023**, *20*, 2. [CrossRef]
59. Carslaw, H.S.; Jaeger, J.C. *Conduction of Heat in Solids*; Oxford University Press: New York, NY, USA, 1959; p. 510.
60. Pásztory, Z. An Overview of Factors Influencing Thermal Conductivity of Building Insulation Materials. *J. Build. Eng.* **2021**, *44*, 102604.
61. Nield, D.A.; Bejan, A. *Convection in Porous Media*; Springer: New York, NY, USA, 2006; Volume 3, pp. 629–982.
62. Vėjelienė, J. Impact of Technological Factors on the Structure and Properties of Thermal Insulation Materials from Renewable Resources. Ph.D. Thesis, Vilnius Gediminas Technical University, Vilnius, Lithuania, 2012. Available online: http://dspace.vgtu.lt/bitstream/1/1526/1/2077_VEJELIENE%20Summary_WEB.pdf (accessed on 7 July 2023).
63. Tsai, P.P.; Yan, Y. The Influence of Fiber and Fabric Properties on Nonwoven Performance. In *Applications of Nonwovens in Technical Textiles*; Woodhead Publishing: Cambridge, UK, 2010; pp. 18–45.
64. Petcu, C.; Vasile, V. Traditional building materials for sustainable thermal insulating of building elements. *Rom. J. Mater./Rev. Romana Mater.* **2022**, *52*, 66–74.
65. Koh, C.H.; Gauvin, F.; Schollbach, K.; Brouwers, H.J.H. Investigation of Material Characteristics and Hygrothermal Performances of Different Bio-Based Insulation Composites. *Constr. Build. Mater.* **2022**, *346*, 128440. [CrossRef]
66. Zach, J.; Korjenic, A.; Petránek, V.; Hroudová, J.; Bednar, T. Performance evaluation and research of alternative thermal insulations based on sheep wool. *Energy Build.* **2012**, *49*, 246–253. [CrossRef]
67. Florea, I.; Manea, D.L. Analysis of thermal insulation building materials based on natural fibers. *Procedia Manuf.* **2019**, *32*, 230–235. [CrossRef]
68. Rode, C. Organic insulation materials, the effect on indoor humidity, and the necessity of a vapor barrier. *Proceed. Therm. Perform. Exter. Envel. Build.* **1998**, *VII*, 109–121.

69. Hagentoft, C.-E.; Harderup, E. Moisture conditions in a north facing wall with cellulose loose fill insulation: Constructions with and without vapor retarder and air leakage. *J. Build. Phys.* **1996**, *19*, 228–243. [CrossRef]
70. Vrana, T.; Gudmundsson, K. Comparison of fibrous insulations cellulose and stone wool in terms of moisture properties resulting from condensation and ice formation. *Constr. Build. Mater.* **2010**, *24*, 1151–1157. [CrossRef]
71. Ciucasu, C.; Gilles, J.; Ober, D.; Petrie, T.; Arquis, E. Convection phenomena in loose-fill attics—Tests and simulations. In *Research in Building Physics and Building Engineering*; CRC Press: Boca Raton, FL, USA, 2006; pp. 465–472.
72. Osanyintola, O.F.; Simonson, C.J. Moisture buffering capacity of hygroscopic building materials: Experimental facilities and energy impact. *Energy Build.* **2006**, *38*, 1270–1282. [CrossRef]
73. Simonson, C.J.; Salonvaara, M.; Ojanen, T. Moderating indoor conditions with hygroscopic building materials and outdoor ventilation/DISCUSSION. *Build. Eng.* **2004**, *110*, 804–819.
74. Kumar, D.; Alam, M.; Zou, P.X.; Sanjayan, J.G.; Memon, R.A. Comparative analysis of building insulation material properties and performance. *Renew. Sustain. Energy Rev.* **2020**, *131*, 110038. [CrossRef]
75. Palumbo, M.; Lacasta, A.M.; Giraldo, M.P.; Haurie, L.; Correal, E. Bio-based insulation materials and their hygrothermal performance in a building envelope system (ETICS). *Energy Build.* **2018**, *174*, 147–155. [CrossRef]
76. Malet-Damour, B.; Habas, J.P.; Bigot, D. Is Loose-Fill Plastic Waste an Opportunity for Thermal Insulation in Cold and Humid Tropical Climates? *Sustainability* **2023**, *15*, 9483. [CrossRef]
77. Augaitis, N.; Šeputytė-Jucikė, J.; Członka, S.; Kremensas, A.; Kairytė, A.; Vėjelis, S.; Balčiūnas, G.; Vaitkus, S. Performance Analysis of Loose-Fill Thermal Insulation from Wood Scobs Coated with Liquid Glass, Tung Oil, and Expandable Graphite Mixture. *Materials* **2023**, *16*, 3326. [CrossRef]
78. Cosentino, L.; Fernandes, J.; Mateus, R. A Review of Natural Bio-Based Insulation Materials. *Energies* **2023**, *16*, 4676. [CrossRef]
79. Sana, A.W.; Noerati, N.; Sugiyana, D.; Sukardan, M.D. Eco-Benign Thermal Insulator Made from Nonwoven Fabrics Based on Calotropis Gigantea Fiber. *AIP Conf. Proc.* **2023**, *2738*, 030018.

Disclaimer/Publisher's Note: The statements, opinions and data contained in all publications are solely those of the individual author(s) and contributor(s) and not of MDPI and/or the editor(s). MDPI and/or the editor(s) disclaim responsibility for any injury to people or property resulting from any ideas, methods, instructions or products referred to in the content.

Article

Electrochemical Study of Semiconductor Properties for Bismuth Silicate-Based Photocatalysts Obtained via Hydro-/Solvothermal Approach

Anastasiia V. Shabalina [1,*], Ekaterina Y. Gotovtseva [1], Yulia A. Belik [1], Sergey M. Kuzmin [2], Tamara S. Kharlamova [1], Sergei A. Kulinich [3], Valery A. Svetlichnyi [1,*] and Olga V. Vodyankina [1]

1. Tomsk State University, Lenin Av. 36, Tomsk 634050, Russia; kara4578@mail.ru (E.Y.G.); belik99q@gmail.com (Y.A.B.); kharlamova83@gmail.com (T.S.K.); vodyankina_o@mail.ru (O.V.V.)
2. G.A. Krestov Institute of Solution Chemistry of the Russian Academy of Science, Akademicheskaja 1, Ivanovo 153045, Russia; smk@isc-ras.ru
3. Research Institute of Science & Technology, Tokai University, Hiratsuka 259-1292, Kanagawa, Japan; skulinich@tokai-u.jp
* Correspondence: shabalinaav@gmail.com (A.V.S.); v_svetlichnyi@bk.ru (V.A.S.)

Abstract: Three bismuth silicate-based photocatalysts (composites of Bi_2SiO_5 and $Bi_{12}SiO_{20}$) prepared via the hydro-/solvothermal approach were studied using electrochemical methods. The characteristic parameters of semiconductors, such as flat band potential, donor density, and mobility of their charge carriers, were obtained and compared with the materials' photocatalytic activity. An attempt was made to study the effect of solution components on the semiconductor/liquid interface (SLI). In particular, the Mott–Schottky characterization was made in a common model electrolyte (Na_2SO_4) and with the addition of glycerol as a model organic compound for photocatalysis. Thus, a medium close to those in photocatalytic experiments was simulated, at least within the limits allowed by electrochemical measurements. Zeta-potential measurements and electrochemical impedance spectroscopy were used to reveal the processes taking place at the SLI. It was found that the medium in which measurements were carried out dramatically impacted the results. The flat band potential values (E_{fb}) obtained via the Mott–Schottky technique were shown to differ significantly depending on the solution used in the experiment, which is explained by different processes taking place at the SLI. A strong influence of specific adsorption of commonly used sulfate ions and neutral molecules on the measured values of E_{fb} was shown.

Keywords: bismuth silicates; photocatalyst; semiconductor in liquid; interface; electrochemical study; electric double layer; zeta-potential

1. Introduction

Currently, catalysts based on nanostructured semiconductor materials are widely used for a variety of processes, ranging from organic chemical synthesis [1,2] to solar-induced hydrogen production [3]. So far, many efficient semiconductor photocatalytic systems have been developed or are in progress. Among them are bismuth-based nanomaterials that keep attracting more and more attention from researchers [4,5]. Bismuth silicates (Bi_2SiO_5, $Bi_{12}SiO_{20}$, $Bi_4Si_3O_{12}$, etc.) and their composites seem to be promising photocatalysts [6–9], which is due to their activity in the visible range [10] and high polarization and internal electric field [11,12], as well as their ability to form effective junctions (S-scheme [13], type-II [14]), and so on. Previously, we obtained bismuth silicate (BSO)-based materials via mechanical activation [15] and laser fragmentation [16]. The latter material demonstrated catalytic activity in decomposing organic substances under visible light irradiation [16]. We also prepared photocatalytic BSO materials via a new hydro-/solvothermal approach and studied their activity for Rhodamine B (RhB) and phenol photo-decomposition [17].

The development of new photocatalysts would be unimaginable without studying carefully their morphology, chemical and phase composition, and drawing mechanisms of their interaction with target molecules under different conditions. The majority of photocatalytic reactions are carried out in liquids, e.g., degradation of water pollutants, oxygen or hydrogen evolution reactions, and so on. In this case, the catalytic process takes place at the semiconductor/liquid interface (SLI), which is formed when the semiconductor is immersed in a liquid. Several important processes are known to take place at the SLI, among which are: (1) charge transfer between semiconductor and components of the liquid; (2) formation of the electric double layer (EDL) consisting of a space charge (SC) layer in the semiconductor and the Helmholtz and Gouy layers located in the liquid; and (3) adsorption and formation of "surface states" [18]. Generally, when a semiconductor is placed into a liquid, the position of its conduction (CB) and the valence (VB) bands does not change, but its Fermi level (E_F) moves towards the equilibrium state, resulting in the so-called "band bending" [19]. An equilibrium in the semiconductor/liquid system is achieved when the E_F of the semiconductor and the E_{redox} level of the solution are equalized. A flat band potential (E_{fb}), at which the bands are flat, is a characteristic of such an equilibrium [20]. This is achieved using the SLI processes listed above. Therefore, investigating these processes is important for a better understanding of the effectiveness of the photocatalyst and the conversion mechanisms it is involved in.

Various electrochemical methods are commonly used to study SLI-related processes [21,22]. Voltammetry, photocurrent and/or open circuit potential measurements, electrochemical impedance spectroscopy (EIS), and other methods are applied to study the behavior of charge carriers and their participation in targeted photocatalytic processes. EIS occupies a special place in modern electrochemistry as one of its most informative and prospective analytic tools available. As a method, it reveals valuable information about electrochemical processes near the equilibrium state (and not only) of the studied system [23] (see Section S1, Supplementary Material).

Both EIS and calculations based on the Mott–Schottky (M–S) equation are actively used to reveal the semiconductor's properties and mechanisms of processes occurring at its SLI. These approaches permit obtaining the E_{fb} value and the concentration of charge carriers [19], both characteristics being necessary for further understanding of target photocatalytic processes, their effectiveness and mechanisms. Therefore, the above two techniques are widely applied to semiconductor-based systems, even though deviation from the M–S behavior is highly probable for a number of reasons, such as structural and compositional inhomogeneity of the material, appearance of surface states, and so on. [24,25]. The application of the EIS and M–S methods requires careful experimentation and extremely accurate handling of results. In this regard, clear and extensive recommendations can be found in the work of Hankin et al. [24].

Based on the above, this work aimed to obtain characteristics such as flat band potential, donor density, and mobility of charge carriers by applying electrochemical methods to three BSO-based photocatalysts prepared and described in our previous work [17]. The obtained results were then compared with the photocatalytic activity of the same samples, which allowed us to draw conclusions on the effect of solution components on the processes taking place at the SLI. Since the electrochemical parameters, such as E_{fb}, are governed by components dissolved in the liquid, the mechanisms of photocatalytic decomposition of organic dyes are concluded to be studied electrochemically in the presence of the same organic molecules and in the same solvent that is used in photocatalytic experiments.

2. Materials and Methods

2.1. Reagents

A sodium silicate lump ($Na_2O \times 3SiO_2$) was obtained from CheMondis GmbH (Köln, Germany), while the other reagents (TEOS, ethylene glycol, $Bi(NO_3)_3$, NaOH, Na_2SO_4, glycerol, polystyrene, 1,2-dichloroethane, HNO_3, KCl, $K_3Fe(CN)_6$, and $K_4Fe(CN)_6$) were purchased from Merck (Germany) and used as received.

2.2. Synthesis of BSO Materials

All three BSO materials investigated in this work were obtained via the hydro-/solvothermal approach. Their preparation procedure, previously described elsewhere [17], is schematically presented in Figure S1. Below, the materials are denoted as BSO/TEOS, BSO/OH, and BSO/NaSi, since TEOS, NaOH and $Na_2Si_3O_7$ were used as distinct reagents during their preparation, respectively (see Figure S1). In our previous work, these samples were marked as BSO_600, BSO_OH_600, and BSO_NaSi_OH_600, respectively [17].

2.3. Characterization of BSO Materials

Phase composition and size of coherent scattering regions (CSRs) were studied using an XRD 6000 diffractometer (Shimadzu, Japan) and the PDF-4 database. XPS analysis was performed with an ES-300 device (Kratos Analytical, Manchester, UK). The qualitative content of sample surfaces was studied by means of survey spectra, in accordance with the standard protocol, using the most intense spectral lines and taking into account the atomic sensitivity factors (ASF) for each element. The BET surface area was measured on the gas adsorption analyzer TriStar II 3020 (Micromeritics, Norcross, GA, USA). Before analysis, samples were degassed in vacuum (10^{-2} Torr) at 200 °C for 2 h. Scanning electron microscopy (SEM) images of electrode surfaces were obtained on a Vega 3H instrument (Tescan, Brno, Czech Republic) in back-scattered electron (BSE)–collecting mode.

2.4. Zeta-Potential Measurements

The zeta-potential of semiconductor particles was measured using an Omni S/N (Brookhaven, Holtsville, NY, USA) in ZetaPALS mode with a BI-ZTU autotitrator (Brookhaven, Holtsville, NY, USA). To measure zeta-potential values, samples were dispersed as powders (0.5 mg/mL) in distilled water or in a solution of interest: 1 mM aqueous Na_2SO_4, 1 mM aqueous glycerol, joint (1:1 molar ratio) aqueous solution of Na_2SO_4 (1 mM) and glycerol (1 mM). The pH values were adjusted by adding HNO_3 solution.

2.5. Electrochemical Studies

2.5.1. Electrochemical Measurements in Liquids

Electrochemical measurements in solutions were carried out using a CHI 660E electrochemical workstation (CH Instruments, Bee Cave, TX, USA). A three-electrode electrochemical cell was used, with a homemade graphite electrode (impregnated with polyethylene and paraffin) as a working electrode coated with semiconductor powder samples on its surface. A reference Ag/AgCl (1 M KCl) electrode and auxiliary Pt plate electrode were also applied. The electrodes were immersed in degassed (with Ar, 15 min) electrolyte solution that was placed in a sealed glass cell of 50 mL. Before proceeding to EIS and M–S measurements, linear sweep voltammograms in the region from −1 V to +1 V were recorded in each solution at a rate of 0.03 V/s.

The EIS measurements were carried out under a sinusoidal alternating voltage amplitude of 10 mV in the frequency range from 1 Hz to 100 kHz. The open circuit potential (OCP, E_{OC}) was chosen as an initial E value for each measurement. The EIS data were analyzed and simulated using the ZView software (Scribner Associates Inc., Southern Pines, NC, USA).

M–S plots were registered using the impedance-potential mode of the CHI Electrochemical Station in the potential window chosen from voltammetry data at the frequency of 1 kHz. The capacity was calculated using Z'' values [26] and was normalized to the geometric electrode area in accordance with standard protocols.

2.5.2. Solutions for Electrochemical Measurements

All solutions used were prepared with distilled water (pH of 6.0, specific conductivity of 1.3 µS/cm, ion content below 3.2, 5.2, 0.6, and 20 µM for NO_3^-, SO_4^{2-}, Cl^- and Ca^{2+}, respectively). To estimate the electroactive surface area of the working electrode (S_{EA}), 0.025 M equimolar aqueous solution of $K_3Fe(CN)_6$ and $K_4Fe(CN)_6$ with 0.1 M KCl was used as a redox probe for cyclic voltammetry measurements. For the EIS measurements, two liquids were used: 0.5 M aqueous Na_2SO_4 (pH of 4.8) and a mixed aqueous solution (10:1 molar ratio) of 0.5 M Na_2SO_4 and 0.05 M glycerol (pH of 5.8).

2.5.3. Preparation of Electrodes with BSO Material

The working graphite electrode was covered with a thick coating based on the studied material. For this, 18 mg of BSO powder was mixed with 2 mg of polystyrene and 100 µL of 1,2-dichloroethane. The obtained dispersion was sonicated for 3 min. Then, 5 µL of the obtained dispersion in polymer solution was applied to the graphite electrode's surface and dried at room temperature for 12 h to get rid of the solvent. Thus, a relatively thick non-porous dielectric polymer coating covered the electrode surface, completely preventing any direct contact of the graphite electrode with the tested liquid. Hence, the entire electric contact was through the coating consisting of semiconductor particles dispersed in a polymer binder. For every measurement, at least three freshly prepared electrodes were used, after which the results were averaged, and accuracy was estimated.

2.5.4. Electroactive Surface Area Measurements

To calculate the electroactive surface area (S_{EA}), cyclic voltammetry measurements were carried out in the solution with a redox probe ($K_3/_4Fe(CN)_6$), and voltammograms were recorded in the potential region from -1 V to $+1$ V at a scan rate of 0.03 V/s. The current of the anodic peak of the redox probe oxidation was measured, and S_{EA} was calculated using the Rendles–Shevchik equation [27]:

$$I = 2.69 \times 10^8 \, n^{3/2} \, S_{EA} \, D^{1/2} \, v^{1/2} \, C \tag{1}$$

where I is the peak current (in A); D is the diffusion coefficient (m^2/s); v is the scan rate (V/s); C is the concentration of the redox probe (M); n is the number of the electrons involved in the oxidation process; and S_{EA} expressed in m^2.

2.5.5. Measurements of Charge Carrier Mobility

A cell filled with a powder sample (so-called "solid electrochemical cell") was used for the calculation of the mobility of charge carriers via the Mott–Gurney law [28,29]. It consisted of two flat steel electrodes ($d = 5$ mm). A powder sample was placed between the electrodes and fixed there with spring clamps with a constant force to a thickness of 1 mm (the mass of each material depended on its bulk density). The I–V curves (from 0 to +10 V) and PEIS data (eight potential steps at 0; 0.2; 1; 2; 3; 4; 6; and 8 V with 100 mV amplitude in the frequency range of 1–700 kHz) were registered using an SP150 potentiostat (BioLogic, Seyssinet-Pariset, France). The ZView (Scribner, New York, NY, USA) and EC-Lab (BioLogic, Seyssinet-Pariset, France) software were used for data simulation and calculations.

3. Results and Discussion

3.1. Characterization of BSO Samples

The phase composition and structural features (the size of coherent scattering regions, CSRs) obtained from XRD data are presented in Table 1. Two main phases of bismuth silicates, Bi_2SiO_5 and $Bi_{12}SiO_{20}$, are seen to be present in all the samples. Only sample BSO/TEOS was found to contain two additional oxide phases, α-Bi_2O_3 and β-Bi_2O_3. (The presence of a small amount of amorphous SiO_2 at grain boundaries or between silicate layers was also possible in all the samples). Sample BSO/NaSi was found to have the highest content in the Bi_2SiO_5 phase. This sample also demonstrates the largest CSRs of 233 nm and, consequently, the lowest BET surface area among the three samples, i.e., 0.4 m^2/g. The BET surface area of samples was found to decrease as follows: BSO/TEOS > BSO/OH > BSO/NaSi. However, the electroactive surface area values for both samples BSO/TEOS and BSO/OH were very similar, and the largest one was demonstrated by sample BSO/NaSi. In addition, Table 1 also presents the values of band gap energies (E_g) obtained earlier for all three samples [17].

Table 1. Composition and structural characteristics of BSO samples.

Sample	Phase Composition (Content, %) [a]	Structural Features [b], nm	BET Surface Area [a], m^2/g	Electroactive Surface Area [c], cm^2	E_g [a], eV
BSO/TEOS	Bi$_2$SiO$_5$ (79) Bi$_{12}$SiO$_{20}$ (16) α-Bi$_2$O$_3$ (3) β-Bi$_2$O$_3$ (2)	150	12 ± 2	(3.2 ± 0.1) × 10^{-4}	3.24 2.81 – 2.10
BSO/OH	Bi$_2$SiO$_5$ (85) Bi$_{12}$SiO$_{20}$ (15)	227	2.0 ± 0.4	(3.24 ± 0.08) × 10^{-4}	3.41 3.02
BSO/NaSi	Bi$_2$SiO$_5$ (96) Bi$_{12}$SiO$_{20}$ (4)	233	0.40 ± 0.08	(13.0 ± 0.9) × 10^{-4}	3.27 2.88

[a] Data previously published elsewhere [17]. [b] CSR values obtained from XRD data were used to characterize the microstructure of the main phase in the materials (Bi$_2$SiO$_5$). [c] Geometric area of the working electrode was 0.28 cm^2.

The XPS results on the relative content of Bi, Si, O, and Na found on the samples' surfaces are presented in Table S1. The presence of trance amounts of Na found in two samples is explained by the use of NaOH during their preparation. Additionally, here, one should keep in mind that the surface may also contain a thin layer of SiO$_2$, which could affect the results. Despite this, it is seen that the element ratio closest to that in Bi$_2$SiO$_5$ was observed for sample BSO/NaSi. The composition of its counterpart BSO/OH is seen to slightly deviate from the ratio of 2:1:5. Finally, sample BSO/TEOS demonstrated its surface composition to be the farthest from that of bismuth silicates. So, the XPS results are well consistent with those obtained by XRD.

The observed values of S_{EA} are much lower than the geometric area of corresponding powders because of the working electrode's design. The polystyrene layer covering the graphite surface only allowed for electric contact through the powder dispersed in it. Therefore, it was only the particles contacting the liquid that determined the S_{EA}. This also explains the discrepancy between the BET and electroactive surface areas. The powder only partly contacted the solution being imbedded into the polystyrene binder, and the reduced contact area was a result of power pores filled with polymer.

The surfaces of the prepared electrodes were characterized by SEM. Figure 1 presents SEM images obtained by collecting back-scattered electrons (in the so-called "Z-contrast" mode). In this case, larger atoms (i.e., with higher Z) are displayed as brighter spots in the image. Because of the polymer binder used, the electrode surfaces were smooth and pore-free (the BSO powders used were not porous, based on their low S_{BET}). Thus, the bright spots seen in Figure 1 present BSO particles and their agglomerates, while the dark locations stand for polystyrene. Because of a relatively low binder fraction used, the dispersed BSO particles were in close contact with each other, providing good electric contact between the graphite electrode and tested liquid. It is seen in Figure 1 that all the working electrodes are characterized by a coral-like structure of their BSO material buried in polystyrene. The BSO/TEOS sample shows the most inhomogeneous structure, with denser and sparser areas (Figure 1a). Its BSO/OH counterpart in Figure 1b is seen to contain lamellar structures mixed with corals. The presence of denser and lamellar structures led to lower S_{EA} values in these two samples. Finally, sample BSO/NaSi exhibits the most homogeneous distribution of small coral-like particles dispersed in the polymer (Figure 1c), which is why its S_{EA} value was the largest.

Figure 1. SEM images of samples recorded in BSE mode (Z-contrast): (**a**) BSO/TEOS, (**b**) BSO/OH, and (**c**) BSO/NaSi.

3.2. Correlation between Electrochemically Determined Semiconducting Properties and Photocatalytic Activity of BSO Materials

To choose the region of potentials for the M–S measurements, corresponding I–V curves were obtained for the samples (Figure S2). The chosen regions were linear and located to the left of the turning point, as seen from the Tafel representation (as an example, see the case of sample BSO/TEOS in Figure 2a). This implies that the same mechanisms governed the processes at the SLI during M–S data collection. Additionally, the region without redox peaks was chosen for each sample in order to make sure that all the signals detected during measurements belonged to EDL and SC processes.

As mentioned above, the M–S technique is one of the most widely used methods for obtaining the E_{fb} value. For all the samples, impedance was measured at different potentials in the chosen region at a fixed value of frequency (1 kHz), and C_{sc} was determined for each potential. As a result, C_{sc}^{-2}–E curves were obtained (see an example in Figure 2b) with the slope direction characteristic of n-type semiconductors. The resulting E_{fb} values calculated from the obtained plots for the three BSO samples are presented in Table 2. Measurements conducted in widely used Na_2SO_4 solution resulted in either positive (for samples BSO/OH and BSO/NaSi) or slightly negative (sample BSO/TEOS) E_{fb} values.

Using the E_{fb} values obtained in this work and the E_g values provided in our previous work (Table 1), then taking that for n-type semiconductors $E_{CB} \approx E_{fb}$ [30] and assuming that the E_{fb} for the materials was determined by a dominant phase (Bi_2SiO_5 for all the samples, see Table 1), we constructed energy band diagrams for the materials. The diagrams are shown in Figure 2c, clearly indicating that generating ·OH radicals is possible under photoexcitation of all the three materials, while their CB bottom is located too low for active ·O_2^- radical formation.

Using the slope of the linear part of the M–S curve, donor density values were calculated following standard approach (Table 2). The mobility of charge carriers (μ) was also calculated, being within the range 30–45 cm^2/V·s for all the samples (Table 2). Of all the samples, the BSO/TEOS one is seen to exhibit the highest mobility of its charge carriers even though its donor density was the lowest and its CB bottom is seen in Figure 2c to be the closest to the potential of active ·O_2^- radical formation. Thus, sample BSO/TEOS appears to be the most promising material for photocatalytic applications.

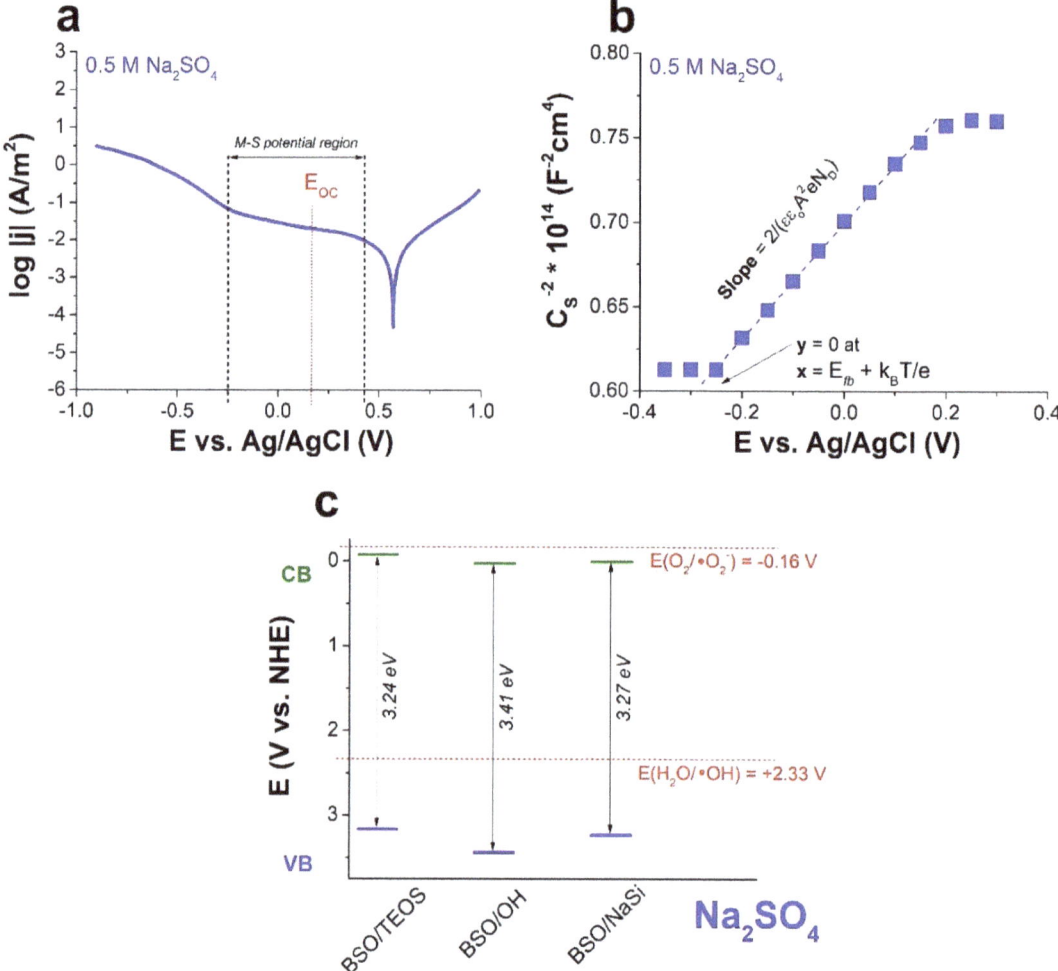

Figure 2. Results obtained by electrochemical methods for sample BSO/TEOS in 0.5 M aqueous Na$_2$SO$_4$: (**a**) Tafel representation of I–V curve; (**b**) Mott–Schottky plot; (**c**) Energy band diagram based on E_{fb} values, with potentials of active radical formation marked with red (for ·OH and ·O$_2^-$ species).

Table 2. Data from electrochemical measurements in aqueous Na$_2$SO$_4$ (E_{fb} and N_D) and for powder samples (μ).

Sample	E_{fb} [a], V vs. NHE	N_D [a], m^{-3}	μ, cm^2/V·s
BSO/TEOS	−0.076 ± 0.008	10^{22}	45
BSO/OH	+0.024 ± 0.009	10^{23}	43.9
BSO/NaSi	+0.004 ± 0.001	10^{23}	30.2

[a] From M–S results obtained in 0.5 M Na$_2$SO$_4$.

The results of photocatalytic tests previously reported for the three samples are briefly given in Table 3. It should be mentioned that the photocatalytic activity of the materials was tested for 5 cycles, during which no significant decrease in their performance was observed. Table 3 shows that indeed sample BSO/TEOS exhibited the highest activity for both RhB and phenol degradation. Sample BSO/OH demonstrated quite close results, while sample BSO/NaSi was relatively inactive. This is well consistent with the values of charge carrier mobility obtained for the same samples tested in the solid electrochemical cell. However, according to the M–S measurements, the activity of all the materials (at least that of samples BSO/TEOS and BSO/OH) should have been much lower as they should not generate $\cdot O_2^-$ species. Hence, the question arises: why was their observed catalytic activity better than expected?

Table 3. Photocatalytic activity of the BSO samples from [17].

Sample	Photocatalytic Conversion [a], %		
	Rhodamine B		Phenol
	Xe [b]	LEDs [c]	LEDs [c]
BSO/TEOS	88	100	28
BSO/OH	75	100	27
BSO/NaSi	52	66	7

[a] For a period of 4 h. [b] Full spectrum. [c] Wavelength of 378 nm.

To answer this question, we considered all possible factors that could influence the results observed during either the electrochemical measurements or the photocatalytic process itself. One of the most obvious conditions that differed in the two above processes was the medium. More specifically, while the electrochemical measurements were carried out in quite a concentrated (0.5 M) solution of a strong electrolyte, the photocatalytic experiments were run in diluted solutions of organics (10^{-5} M). Therefore, it was decided to repeat similar electrochemical measurements in presence of organic molecules to see how they may affect the results.

In principle, both RhB and phenol can exhibit electrochemical activity. Rh B is known to undergo electrochemical oxidation, in particular in Na_2SO_4 solutions [31,32]. In turn, phenol exhibits electro-transformations and can form different products starting from 0.08 V [33,34]. That is why, to avoid additional electrochemical signals, we chose glycerol as another model organic substance that can only be electro-oxidized either at higher potential or in presence of complex electro-catalysts [35]. Being a polyol, aqueous glycerol is also used as a model sacrificial agent and a hole scavenger for photocatalytic hydrogen evolution processes [36,37]. Additionally, it is a non-electrolyte, thus exhibiting different nature and behavior than aqueous Na_2SO_4. Hence, we duplicated all electrochemical measurements in a joint Na_2SO_4-glycerol solution.

The Tafel presentation of *I–V* curve and M–S plot obtained for sample BSO/TEOS in presence of glycerol are provided in Figure 3, while E_{fb} values obtained for all the samples in presence of glycerol are listed in Table 4. It is seen that adding 0.05 M glycerol to the electrolyte lowered the E_{fb} at least by 200 mV. Correcting the CB bottom location in Figure 2c using these values, one can find out that the formation of active oxygen radicals is possible for all the samples photoexcited in presence of glycerol (Figure 3c). Thus, E_{fb} values obtained during electrochemical measurements are seen to be very sensitive to the medium used. However, is it the effect of presence of organic molecules/non-electrolyte, or just an experimental inaccuracy? Do the values obtained in joint solution describe the SLI state during photocatalytic experiments better? These are the questions to be answered in the following sections.

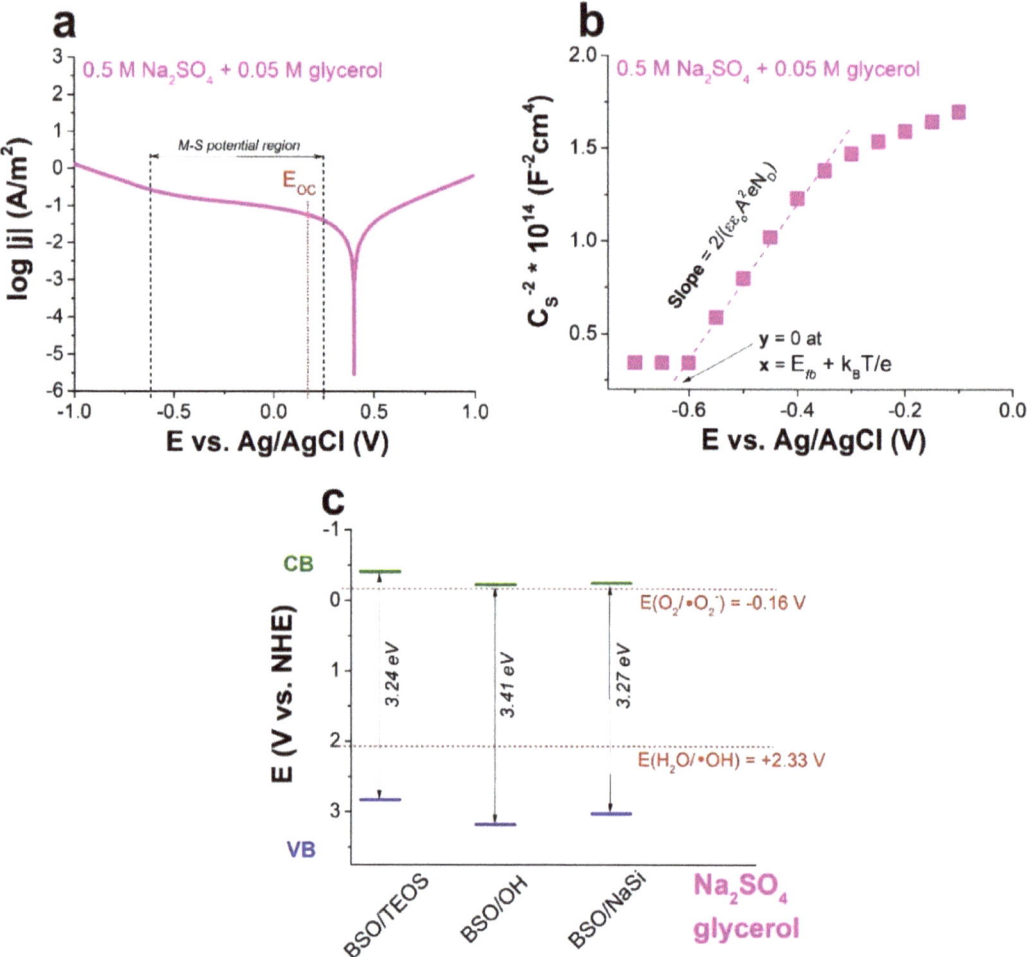

Figure 3. Results obtained by electrochemical methods for sample BSO/TEOS in 0.5 M aqueous Na$_2$SO$_4$ in presence of 0.05 M glycerol: (**a**) Tafel representation of *I–V* curve; (**b**) Mott–Schottky plot; (**c**) Energy band diagram based on E_{fb} values (the potentials of active radical formation are marked with red).

Table 4. Flat band potential values for BSO samples measured in presence of glycerol.

Liquid	E_{fb} (Mott–Shottky), V, vs. NHE		
	BSO/TEOS	BSO/OH	BSO/NaSi
Glycerol and Na$_2$SO$_4$ in H$_2$O	−0.409 ± 0.006	−0.225 ± 0.004	−0.242 ± 0.001

3.3. SLI State and Processes at SLI for BSO Materials in Two Media

The flat-band potential is one of the fundamental properties of any semiconductor–electrolyte system [38], and the Mott–Schottky technique is a common method for its determination. In fact, a lot of factors lead to deviations from ideal Mott–Schottky behavior, some of which were described, for example, by Cardon and Gomes [39]. Fynlason et al. measured flat-band potential values for three different forms of CdS (single crystal, thin film and powder) by three different methods, demonstrating some difference in obtained results [40]. Ge et al. also pointed out the high importance of the effect of surface states during E_{fb} studies [41].

Sharon and Sinha studied the effect of electrolytes on the flat-band potential of the n-BaTiO$_3$ semiconductor [42]. They focused on the influence of pH and redox potential of electrolyte on the measured results and revealed that the negative flat-band potential increases with the increase in the redox potential of the electrolyte used. Additionally, for electrolytes containing Fe(CN)$_6^{4-}$ and I$^-$ species, the specific adsorption of the anions was found to cause significant changes in Helmholtz potential, thus affecting flat-band potential values [42].

Typically, the majority of electrochemical measurements (including EIS and M–S measurements) are carried out in model electrolyte solutions. Sodium sulfate is one of such solutions [43–45], while KOH, H$_2$SO$_4$, and phosphate-buffered solutions are also used (see, for example, [9,46,47]). However, when a semiconductor, which was characterized electrochemically in a Na$_2$SO$_4$ solution, is immersed in another liquid, the processes on its surface might change dramatically. Consequently, in another liquid medium, the resulting target process (e.g., photocatalytic hydrogen production or dye decomposition) might change its mechanism, slow down, or even completely stop. This implies that it is probably incorrect to describe and explain processes taking place in pure water or in the presence of sacrificial agent(s) via parameters obtained for the same semiconductor characterized in aqueous Na$_2$SO$_4$ (or another model electrolyte).

Sulfate anions are known to be prone to strong, specific, and spatially inhomogeneous adsorption [48] on semiconducting surfaces, even at large potential bias [49]. This leads to changes in EDL capacitance and surface charge and even affects the kinetics of electrochemical reactions at surfaces covered with SO$_4^{2-}$ [50]. Therefore, we suggest that the SLI of a semiconducting material should be electrochemically characterized in the same solution where it is used as a catalyst. So far, there were few works dealing with the liquids used in the present study, such as water [51], glycerol [37], and tris-HCl [52]. To the best of our knowledge, no systematic experimental work was carried out to compare EIS and M–S data of semiconductors characterized in liquids that contained components of photocatalytic media, and only a few relevant reports could be found. For example, a decreased accuracy of E_{fb} determined in the presence of a hole scavenger (i.e., H$_2$O$_2$) was observed by Hankin and co-workers [24].

In the present work, we used the EIS method and zeta-potential measurements to study the difference in SLI state for the same BSO samples characterized in two solutions: model electrolyte (aqueous Na$_2$SO$_4$) and joint solution of Na$_2$SO$_4$ and model organic compound (glycerol) that is used in photocatalytic processes. For more information, the zeta-potential was also measured in water and glycerol solutions.

3.3.1. Electrochemical Impedance Spectroscopy Studies

A brief overview of some EIS studies on semiconductors characterized in different liquid media is given in Section S2 (Supplementary Material). Our results obtained for all samples were apparently quite similar. Before EIS measurements, LSV data were analyzed. Figure S2 demonstrates rectifying behavior of all the materials with classic current-potential curves characteristic of n-type semiconductor in the dark [53]. Then, EIS measurements at open circuit potentials listed in Table 5 were carried out. Typical EIS data, along with simulation results, obtained for sample BSO/TEOS in two liquids are presented in Figure 4.

Table 5. OCP values for electrodes with BSO samples in two media.

Liquid	E_{OC}, V, vs. Ag/AgCl		
	BSO/TEOS	BSO/OH	BSO/NaSi
Na$_2$SO$_4$ in H$_2$O	+0.17 ± 0.02	+0.170 ± 0.005	+0.16 ± 0.02
Glycerol and Na$_2$SO$_4$ in H$_2$O	+0.170 ± 0.005	+0.112 ± 0.002	+0.24 ± 0.03

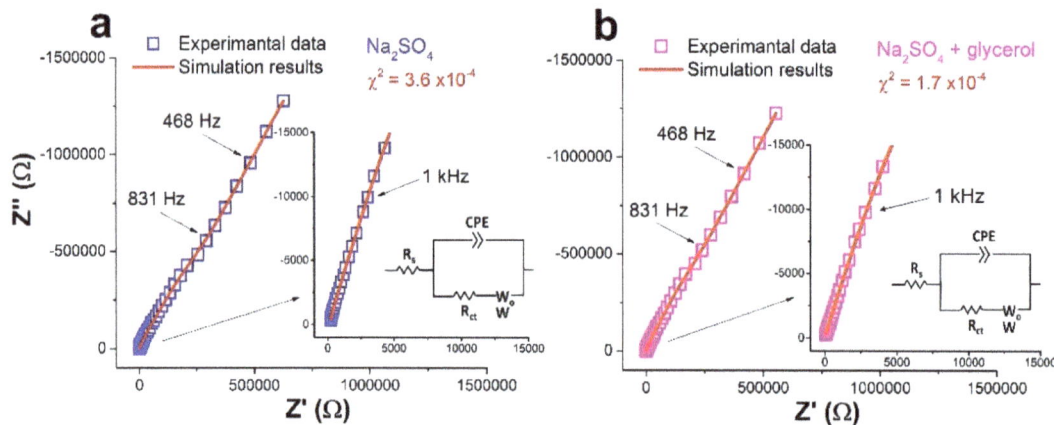

Figure 4. Typical Nyquist plots and simulated curves for sample BSO/TEOS in two liquids: (a) Na$_2$SO$_4$ solution; (b) joint Na$_2$SO$_4$ and glycerol solution. Corresponding equivalent circuits are given in insets.

As a next step, processes occurring at the SLI in two media were simulated for all the three samples, for which modified Randles equivalent circuits were constructed to fit the data observed both in aqueous Na$_2$SO$_4$ and Na$_2$SO$_4$ with glycerol [54,55]. The curves simulated using the ZView software showed good agreement with the experimental data, with χ^2 factors being on the order of 10^{-4}. Detailed parameters of all circuit elements are summarized in Table 6.

Table 6. Fitting results for EIS experimental data of BSO samples tested in two liquids.

Sample		BSO/TEOS		BSO/OH		BSO/NaSi	
	Liquid	Na$_2$SO$_4$	Na$_2$SO$_4$ Glycerol	Na$_2$SO$_4$	Na$_2$SO$_4$ Glycerol	Na$_2$SO$_4$	Na$_2$SO$_4$ Glycerol
	R_s, Ω	130	20	50	50	30	30
	CPE-T, $\Omega^{-1}s^\alpha$	4×10^{-8}	5×10^{-8}	9×10^{-8}	4×10^{-8}	2×10^{-8}	2×10^{-7}
	CPE-P (α)	0.85	0.84	0.86	0.85	0.90	0.81
	R_{ct}, Ω	40	30	20	30	40	20
Parameters	C_{dl}, F	5×10^{-5}	9×10^{-6}	4×10^{-5}	2×10^{-5}	4×10^{-6}	5×10^{-5}
	Wo-R, Ω	8×10^5	5×10^5	2×10^6	8×10^5	1×10^6	1×10^6
	Wo-T, s	9×10^{-3}	2×10^{-3}	5×10^{-2}	2×10^{-2}	4×10^{-3}	4×10^{-3}
	Wo-P	0.29	0.28	0.18	0.35	0.28	0.69
	χ^2	3.6×10^{-4}	1.7×10^{-4}	2.3×10^{-4}	3.0×10^{-4}	4.5×10^{-4}	9.6×10^{-4}

When the CPE-P value is between 0.8 and 1, the CPE-T characterizes a capacitance of EDL (C_{dl}) [56]. At the series connection of CPE and R_s elements, the value of C_{dl} can be defined by the following equation [57,58]:

$$C_{dl} = [A_{dl} \times R_s^{-(\alpha-1)}]^{1/\alpha} \qquad (2)$$

where C_{dl} is the double-layer capacitance of the electrode/electrolyte surface; A_{dl} is a parameter of the constant phase element (CPE-T); R_s is the electrolyte resistance; and α is the CPE-P parameter [59]. Thus, the C_{dl} values are also provided in Table 6.

The electrical double layer (EDL) is an important part of SLI. This section will discuss the EDL capacitance and composition, while results on two other components of the

EIS circuit (solution resistance and diffusion) are discussed in Supplementary Material (Sections S5 and S6).

The EDL is known to be characterized by the thickness of its dense and diffuse layers. The diffuse part is solely determined by the electrolyte itself, while the dense part is influenced by adsorption phenomena. Shimizu and co-authors studied ion adsorption on the hematite surface using EIS [60], revealing that the specific adsorption of ions modified the surface's structure and affected the EDL's thickness. It is known that SO_4^{2-} ions tend to specifically adsorb on semiconducting oxides, including bismuth oxide [61]. Some other ions possibly present in the water used in our experiments (NO_3^-, SO_4^{2-}, Cl^- and Ca^{2+}) might also be prone to adsorption. Commonly, specific adsorption decreases the EDL capacitance.

As Table 6 shows, the double-layer capacitance of BSO-covered electrodes was in the range of $10^{-5} \div 10^{-6}$ F and generally differed for aqueous Na_2SO_4 medium with and without glycerol added. Only sample BSO/OH showed similar values of C_{dl} for the two liquids tested. For sample BSO/TEOS, the EDL capacitance decreased about one order of magnitude in the presence of glycerol. On the contrary, for sample BSO/NaSi, its C_{dl} increased one order of magnitude when glycerol was added. To explain these diverse data obtained for different samples, we addressed the electro-kinetic properties obtained from zeta-potential measurements and analysis.

3.3.2. Results of Zeta-Potential Measurement

To reveal the specificity of the surface state of studied materials in the presence of solution components, we studied the zeta-potential and its change for all the samples in four different liquids. First of all, we used pure water to find out the surface state of the materials in the absence of other components. Then, we tested the most common electrolyte (Na_2SO_4), which is normally used for electrochemical studies, including those of E_{fb}. Then, tests were run in a joint solution of Na_2SO_4 and glycerol to reveal the difference. Finally, we also tested how glycerol alone influences the surface state of the materials.

The obtained zeta-potential values measured in the above-mentioned media are exhibited in Figure 5 and Table S2, and those obtained at different pH are presented in Figure S3. All the materials demonstrated quite similar results, which indicates their similar nature.

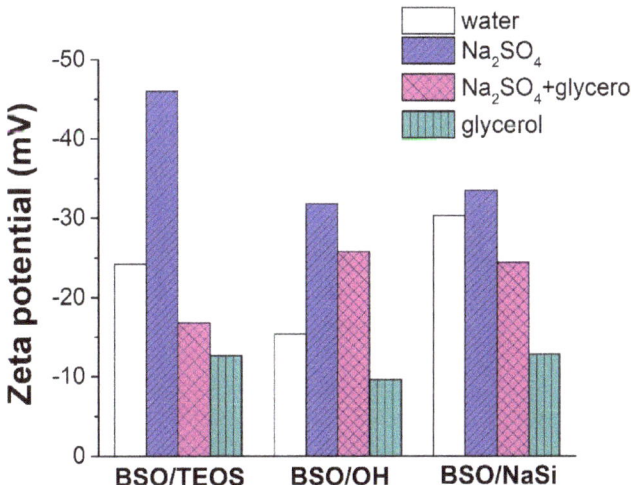

Figure 5. Results on zeta-potential measurement in four different liquids.

As clearly seen in Figure 5 and Table S2, the most negative and the largest by their absolute value zeta-potentials were observed in the aqueous Na_2SO_4 medium. For samples

BSO/TEOS and BSO/OH, their values were found to be almost twice larger than in water (compare blue and white bars in Figure 5). At the same time, the least negative and the lowest by absolute value zeta-potentials were detected in glycerol solution (see green bars in Figure 5). More details can be found in Section S3 (Supplementary Material).

Thus, it is obvious that the electro-kinetic properties of the studied BSO materials are affected by the surrounding liquid. Hence, depending on the medium chosen for the photocatalytic process, the catalyst may exhibit quite a different surface state. It is also possible that some components of the liquid medium can either increase or decrease material's activity, even though they do not take part in the process themselves (unlike, for example, hole scavengers which do participate in the photocatalytic process).

When the studied BSO materials are moved from one solution to another, obviously, the composition of the adsorbed surface layer changes. At different pH values, the ratio of species absorbed on the surface varies as a result of their electric and other properties. Therefore, changes in pH lead to a gradual replacement of one type of surface molecule with others. The extrema in the zeta-potential–pH curves seen in Figure S3 are believed to indicate structural rearrangements of the adsorption layer. For convenience, Figure S4 schematically presents the surface composition of different liquids.

The BSO surface is positively charged (n-type semiconductor) when in contact with the aqueous solution. This positive charge is known to be located at the Bi^{3+}, $[Bi_2O_2]^{2+}$ or other surface centers and interacts with ions from the solution. While more details and reactions expected on the surface are provided in Section S4 (Supplementary Materials), below, we give a short account of our findings.

The EDL formed in water should consist of relatively small single-charged ions (see Figure S4). In the glycerol environment, the zeta-potential is expected to be more positive. In this case, the polyol can be adsorbed through the interaction with positively charged surface centers [62,63]. As a result, polyol functional groups become potential-determining species. Then, because glycerol molecules do not dissociate (as a non-electrolyte) and are quite large, the negative surface charge of the BSO particles in glycerol solution decreases in comparison with the same BSO particles immersed in pure water (Figure 5). The zeta-potential decreased when the samples were immersed in Na_2SO_4 solution, and the low IEP values might be connected with the surface's recharging, which must be caused by SO_4^{2-} anions adsorbed on positively charged surface sites (Figure S4).

It is thus seen that the presence of large double-charged ions on the electrode's surface increases the C_{dl} values for both solutions of Na_2SO_4 and Na_2SO_4 with glycerol. Hence, the results obtained for two-component solutions point to competitive adsorption of Na_2SO_4 and glycerol with sulfate anions in a leading position (Figure S4).

Thus, in the case of water-glycerol dispersions, zeta-potential could probably change due to the adsorption of polyol molecules and ions from water. In contrast, the results obtained in aqueous Na_2SO_4 and in two-component solutions can be explained by the semiconductor surface that only interacts with electrolytes, which normally increases C_{dl} values.

So, the electrochemical measurements in solutions containing Na_2SO_4 characterize not the semiconductor's surface itself but the SLI containing specifically adsorbed sulfate anions that affect EDL's composition and capacity. That is why, when photocatalytic experiments are carried out, the SLI is different.

3.3.3. Discussion and Findings

The SLI can be presented by two capacitors—C_H and C_{sc}—connected in a series [64]. The C_H component stands for the capacitance of the Helmholtz layer in the electrolyte, and the C_{SC} is the semiconductor's space charge capacitance. In the case we investigated, $C_H \gg C_{SC}$ was ensured by the adsorption of anions/negatively charged fragments of molecules in the solution and on the surface of the studied sample (see Section S4). Thus, it is assumed that $C_{dl} \approx C_{SC}$, and generally, the M–S approach can be used to study the semiconductor's properties in the systems under investigation.

Specific adsorption plays an important role in processes occurring at the SLI since it affects the thickness of the dense EDL part. Large neutral molecules of glycerol provide a thicker layer than smaller SO_4^{2-} ions. So, we suggest that the difference in the E_{fb} and N_D values measured in different liquids could be caused by the specific adsorption of glycerol and/or sulfate anions, which immediately leads to changes in the SLI.

Getting back to the zeta-potential values (see Figure 5 and Table S2), it is seen that the lower the surface charging (i.e., the closer the surface to its neutral state), the larger potential should be applied to achieve flat-band conditions (the case of two-component Na_2SO_4-glycerol solution). Additionally, vice versa, when the surface charge is maximal (due to SO_4^{2-} anion adsorption), a smaller negative potential shift is required to stop net charge transfer. Thus, specific adsorption of ionic or neutral species is seen to dramatically affect the measured E_{fb} values.

So, obviously, the presence in the solution of a strong electrolyte, non-electrolyte, or both affected the E_{fb} values obtained during electrochemical measurements. All of the processes taking place at the SLI contributed to the semiconductor's surface state, also influencing the characteristics of the material during its analysis. Consequently, it is believed that during photocatalytic tests, the results were affected not only by the absence of strong electrolyte but also by changes in the semiconductor's surface charge (in the presence or absence of ions or large molecules of organics, dyes, etc.) and adsorption processes (that influence the SLI composition). Thus, to better understand the mechanism of any photocatalytic process, the semiconductor–liquid characteristics should be studied not only in a model solution but also in the presence of other components involved (such as target contaminants, sacrificial agents, etc.).

3.4. Difference in EDL for Three BSO Samples

According to Figure 5 and Table S2, the BSO/NaSi sample exhibited the highest zeta-potential in water and the smallest increase in its negative potential when Na_2SO_4 was added (an increment of 10% vs. 90–100% for the other samples). In addition, it showed the highest potential increment when glycerol was added to water, with about a 60% decrease in its negative potential. Therefore, sample BSO/NaSi was more strongly affected by glycerol's presence than by Na_2SO_4's presence. This could be a reason why the E_{fb} value obtained in aqueous Na_2SO_4 described its photocatalytic activity quite well. For the same reason, sample BSO/NaSi exhibited the lowest photocatalytic activity: together with low charge carriers' mobility, its SLI state is much affected by the presence of organic molecules.

To understand the reasons for unexpected behavior observed for sample BSO/NaSi, one should investigate better the nature of this sample. Its phase composition is quite close to that of sample BSO/OH, but the content of phase Bi_2SiO_5 in sample BSO/NaSi is the highest among the three materials, while the content of phase $Bi_{12}SiO_{20}$ is minimal. This combination could explain the different behavior demonstrated by the electrode coated with sample BSO/NaSi in glycerol-containing media.

Moreover, according to the SEM images, the BSO/NaSi powder was homogeneously distributed in polystyrene and, consequently, on the electrode's surface. Even though it demonstrated the lowest BET surface area (measured for its powder), it exhibited the highest electroactive surface area (measured for its electrode). This could explain why it adsorbed glycerol quantitatively better, which affected its zeta-potential, EDL, as well as processes occurring on its electrode surface.

4. Conclusions

Three nanomaterials based on bismuth silicates obtained via a hydro-/solvothermal approach were studied using electrochemical methods (voltammetry and electrochemical impedance spectroscopy) both in liquid-containing and solid electrochemical cells. Such characteristic parameters as flat-band potential, donor density and mobility of charge carriers were obtained for all the materials. The obtained results were compared with materials' photocatalytic activity, revealing certain inconsistencies. In particular, according to the E_{fb} values obtained in aqueous Na_2SO_4, the observed photoactivity of the materials was better than it was expected.

Thereafter, we attempted to simulate the medium used in photocatalytic experiments, at least within the limits that are acceptable for electrochemical measurements. For that, we added a model organic substance (glycerol) to the model electrolyte and obtained dramatically different values of E_{fb}. Zeta-potential and EIS measurements were also carried out in the presence and in the absence of glycerol. Based on the obtained results, we suggest that two aspects should be taken into account when semiconductor photocatalysts are studied by electrochemical methods. On one hand, correct data can only be obtained in a liquid with a strong electrolyte from the experimental setup point of view since high ionic conductivity is required for high accuracy. On the other hand, however, the specific adsorption of ions (SO_4^{2-} and/or others) and other species (organic molecules, surfactants, etc.) present in electrolytes may affect the processes occurring at the SLI. Consequently, conclusions on the semiconductor's behavior drawn from systems' "electrode-liquid" with model solutions will not necessarily be applicable to systems with "real" liquids used for photocatalytic processes. The presence of components of interest (the target contaminants, sacrificial agents, etc.) during the photocatalytic process and their influence on the SLI should be taken into account.

Supplementary Materials: The following supporting information can be downloaded at: https://www.mdpi.com/article/10.3390/ma15124099/s1, Section S1: EIS theory concise basics; Figure S1: Scheme of synthesis for the BSO nanomaterials used; Table S1: Relative content of elements on sample surfaces according to XPS analysis; Figure S2: LSVs for the electrodes with the BSO samples in two liquids; Section S2: An overview of some results of EIS study of semiconductors in different solutions; Section S3: More details on the results of electro-kinetic properties study; Figure S3: Change in zeta-potential values as the colloids' pH increased; Table S2: Zeta-potential values of the particles of BSO samples in different media; Section S4: Surface composition discussion; Section S5: Solution resistance study; Section S6: Diffusion process at the SLI.

Author Contributions: Methodology, A.V.S. and S.M.K.; Conceptualization, A.V.S.; writing—original draft, A.V.S., Y.A.B., E.Y.G. and T.S.K.; Investigation, A.V.S., T.S.K., Y.A.B. and E.Y.G.; Data curation, V.A.S.; Formal analysis: S.M.K., S.A.K. and O.V.V.; writing—review and editing, A.V.S., S.M.K., S.A.K., Y.A.B., E.Y.G., V.A.S. and O.V.V.; Supervision, V.A.S. and O.V.V.; Project administration, V.A.S. and O.V.V. All authors have read and agreed to the published version of the manuscript.

Funding: This work was supported by the Russian Science Foundation (project no. 19-73-30026).

Acknowledgments: The authors thank S.V. Koscheev, Boreskov Institute of Catalysis (SB RAS, Novosibirsk), for help with XPS analyses.

Conflicts of Interest: The authors declare no conflict of interest.

Abbreviations and Designations

BSO	bithmuth silicon oxides (bismuth silicates)
RhB	Rhodamine B
SLI	semiconductor/liquid interface
EDL	electric double layer
SC	space charge layer
CB	conduction band
VB	valance band
E_F	Fermi level
E_{redox}	redox potential of the electrolyte
E_{fb}	flat band potential
EIS	electrochemical impedance spectroscopy
M–S	Mott–Schottky
CSRs	coherent scattering regions

SEM	scanning electron microscopy
BSE	back-scattered electrons
OCP, E_{OC}	open circuit potential
S_{EA}	electroactive surface area
TEOS	tetraethoxysilane
XRD	X-ray diffraction
E_g	band gap energy
S_{BET}	specific surface area according to BET
NHE	Normal Hydrogen Electrode
N_D	donor density
μ	mobility of charge carriers
LEDs	light-emitting diodes
C_{dl}	EDL capacitance
C_H	Helmholtz layer capacitance
C_{SC}	space charge layer capacitance
CPE-T, Rs, and other EIS simulation parameters.	are described in Section S1 (Supplementary Material).
IEP	isoelectric point

References

1. Friedmann, D.; Hakki, A.; Kim, H.; Choic, W.; Bahnemannd, D. Heterogeneous photocatalytic organic synthesis: State-of-the-art and future perspectives. *Curr. Green Chem.* **2016**, *18*, 5391–5411. [CrossRef]
2. Kobielusz, M.; Mikrut, P.; Macyk, W. Photocatalytic synthesis of chemicals. *Adv. Inorg. Chem.* **2018**, *72*, 93–144. [CrossRef]
3. Zhang, C.; Liu, B.; Cheng, X.; Guo, Z.; Zhuang, T.; Lv, Z. A CdS@NiS reinforced concrete structure derived from nickel foam for efficient visible-light H2 production. *Chem. Eng. J.* **2020**, *393*, 124774. [CrossRef]
4. Yang, X.; Zhang, X.; Wu, T.; Gao, P.; Zhu, G.; Fan, J. Novel approach for preparation of three-dimensional BiOBr/BiOI hybrid nanocomposites and their removal performance of antibiotics in water. *Colloid. Surf. A* **2020**, *605*, 125344. [CrossRef]
5. Yu, W.; Ji, N.; Tian, N.; Bai, L.; Ou, H.; Huang, H. BiOI/Bi$_2$O$_2$[BO$_2$(OH)] heterojunction with boosted photocatalytic degradation performance for diverse pollutants under visible light irradiation. *Colloids Surf. A Physicochem. Eng. Asp.* **2020**, *603*, 125–184. [CrossRef]
6. Batool, S.S.; Hassan, S.; Imran, Z.; Rasool, K.; Ahmad, M.; Rafiq, M.A. Comparison of different phases of bismuth silicate nanofibers for photodegradation of organic dyes. *Int. J. Environ. Sci. Technol.* **2016**, *13*, 1497–1504. [CrossRef]
7. Shabalina, A.V.; Fakhrutdinova, E.D.; Golubovskaya, A.G.; Kuzmin, S.M.; Koscheev, S.V.; Kulinich, S.A.; Svetlichnyi, V.A.; Vodyankina, O.V. Laser-assisted preparation of highly-efficient photocatalytic nanomaterial based on bismuth silicate. *Appl. Surf. Sci.* **2022**, *575*, 151722. [CrossRef]
8. Zaitsev, A.V.; Kirichenko, E.A.; Kaminsky, O.I.; Makarevich, K.S. Investigation into the efficiency of photocatalytic oxidation of aqueous solutions of organic toxins in a unit with an automatically cleaning bismuthsilicate photocatalyst. *J. Water Process Eng.* **2020**, *37*, 101468. [CrossRef]
9. Hwang, S.W.; Seo, D.H.; Kim, J.U.; Lee, D.K.; Choi, K S.; Jeon, C.; Yu, H.K.; Cho, I.S. Bismuth vanadate photoanode synthesized by electron-beam evaporation of a single precursor source for enhanced solar water-splitting. *Appl. Surf. Sci.* **2020**, *528*, 146906. [CrossRef]
10. Gu, W.; Teng, F.; Liu, Z.; Liu, Z.; Yang, J.; Teng, Y. Synthesis and photocatalytic properties of Bi$_2$SiO$_5$ and Bi$_{12}$SiO$_{20}$. *J. Photochem. Photobiol. A Chem.* **2018**, *353*, 395–400. [CrossRef]
11. Al-Keisy, A.; Ren, L.; Zheng, T.; Xu, X.; Higgins, M.; Hao, W.; Du, Y. Enhancement of charge separation in ferroelectric heterogeneous photocatalyst Bi$_4$(SiO$_4$)$_3$/Bi$_2$SiO$_5$ nanostructures. *Dalton Trans.* **2017**, *46*, 15582–15588. [CrossRef] [PubMed]
12. Ding, S.; Xiong, X.; Liu, X.; Shi, Y.; Jiang, Q.; Hu, J. Synthesis and characterization of single-crystalline Bi$_2$O$_2$SiO$_3$ nanosheets with exposed {001} facets. *Catal. Sci. Technol.* **2017**, *7*, 3791–3801. [CrossRef]
13. Dou, L.; Jin, X.; Chen, J.; Zhong, J.; Li, J.; Zeng, Y.; Duan, R. One-pot solvothermal fabrication of S-scheme OVs-Bi$_2$O$_3$/Bi$_2$SiO$_5$ microsphere heterojunctions with enhanced photocatalytic performance toward decontamination of organic pollutants. *Appl. Surf. Sci.* **2020**, *527*, 146775. [CrossRef]
14. Chai, B.; Yan, J.; Fan, G.; Song, G.; Wang, C. In-situ construction of Bi$_2$SiO$_5$/BiOBr heterojunction with significantly improved photocatalytic activity under visible light. *J. Alloys Compd.* **2019**, *802*, 301–309. [CrossRef]
15. Belik, Y.; Kharlamova, T.; Vodyankin, A.; Svetlichnyi, V.; Vodyankina, O. Mechanical activation for soft synthesis of bismuth silicates. *Ceram. Int.* **2020**, *46*, 10797–10806. [CrossRef]
16. Svetlichnyi, V.; Belik, Y.; Vodyankin, A.; Fakhrutdinova, E.; Vodyankina, O. Laser fragmentation of photocatalyst particles based on bismuth silicates. *Proc. SPIE* **2019**, *11322*, 113221D. [CrossRef]
17. Belik, Y.A.; Vodyankin, A.A.; Fakhrutdinova, E.D.; Svetlichnyi, V.A.; Vodyankina, O.V. Photoactive bismuth silicate catalysts: Role of preparation method. *J. Photochem. Photobiol. A Chem.* **2022**, *425*, 113670. [CrossRef]

18. Pleskov, Y.V. *Electric Double Layer on the Semiconductor-Electrolyte Interface*; Springer Science + Business Media: New York, NY, USA, 1980; pp. 57–93.
19. Bott, A.W. Electrochemistry of semiconductors. *Curr. Sep.* **1998**, *17*, 87–91.
20. Zhang, Z.; Yates, J.T., Jr. Band bending in semiconductors: Chemical and physical consequences at surfaces and interfaces. *Chem. Rev.* **2012**, *112*, 5520–5551. [CrossRef]
21. Memming, R. *Semiconductor Electrochemistry*; WILEY-VCH Verlag GmbH & Co. KGaA: Weinheim, Germany, 2015.
22. Sato, N. *Electrochemistry at Metal and Semiconductor Electrodes*; Elsevier Sience: Amsterdam, The Netherlands, 1998.
23. Barsoukov, E.; Macdonald, J.R. *Impedance Spectroscopy Theory, Experiment, and Applications*; John Wiley & Sons, Inc.: Hoboken, NJ, USA, 2005.
24. Hankin, A.; Bedoya-Lora, F.E.; Alexander, J.C.; Regoutz, A.; Kelsall, G.H. Flat band potential determination: Avoiding the pitfalls. *J. Mater. Chem. A* **2019**, *7*, 26162–26176. [CrossRef]
25. La Mantia, F.; Habazaki, H.; Santamaria, M.; Di Quarto, F. A critical assessment of the Mott–Schottky analysis for the characterisation of passive film-electrolyte junctions. *Russ. J. Electrochem.* **2010**, *46*, 1306–1322. [CrossRef]
26. Beranek, R. (Photo)electrochemical methods for the determination of the band edge positions of TiO_2-based nanomaterials. *Adv. Phys. Chem.* **2011**, *2011*, 786759. [CrossRef]
27. Kogikoski, S., Jr.; Sousa, C.P.; Liberato, M.S.; Andrade-Filho, T.; Prieto, T.; Ferreira, F.F.; Rocha, A.R.; Guha, S.; Alves, W.A. Multifunctional biosensors based on peptide-polyelectrolyte conjugates. *Phys. Chem. Chem. Phys.* **2016**, *18*, 3223–3233. [CrossRef] [PubMed]
28. Clymer, D.; Matin, M. Application of Mott-Gurney law to model the current voltage relationship of PPV/CN-PPV with a thin-metal anode buffer. *Proc. SPIE* **2005**, *5907*, 59070S. [CrossRef]
29. Wang, Z.B.; Helander, M.G.; Greiner, M.T.; Qiu, J.; Lu, Z.H. Carrier mobility of organic semiconductors based on current-voltage characteristics. *J. Appl. Phys.* **2010**, *107*, 034506. [CrossRef]
30. Ling, Y.; Dai, Y. Direct Z-scheme hierarchical WO_3/BiOBr with enhanced photocatalytic degradation performance under visible light. *Appl. Surf. Sci.* **2020**, *509*, 145201. [CrossRef]
31. Austin, J.M.; Harrison, I.R.; Quickenden, T. Electrochemical and photoelectrochemical properties of rhodamine B. *J. Phys. Chem.* **1986**, *90*, 1839–1843. [CrossRef]
32. Dai, Q.; Jiang, L.; Luo, X. Electrochemical oxidation of rhodamine B: Optimization and degradation mechanism. *Int. J. Electrochem. Sci.* **2017**, *12*, 4265–4276. [CrossRef]
33. Comninellis, C.; Pulgarin, C. Electrochemical oxidation of phenol for wastewater treatment using SnO_2 anodes. *J. Appl. Electrochem.* **1993**, *23*, 108–112. [CrossRef]
34. Enache, T.A.; Oliveira-Brett, A.M. Phenol and para-substituted phenols electrochemical oxidation pathways. *J. Electroanal. Chem.* **2011**, *655*, 9–16. [CrossRef]
35. Coutanceau, C.; Baranton, S.; Bitty Kouame, R.S. Selective electrooxidation of glycerol into value-added chemicals: A short overview. *Front. Chem.* **2019**, *7*, 100. [CrossRef] [PubMed]
36. Tahir, M.; Siraj, M.; Tahir, B.; Umer, M.; Alias, H.; Othman, N. Au-NPs embedded Z–scheme WO_3/TiO_2 nanocomposite for plasmon-assisted photocatalytic glycerol-water reforming towards enhanced H_2 evolution. *Appl. Surf. Sci.* **2020**, *503*, 144344. [CrossRef]
37. Sadanandam, G.; Zhang, L.; Scurrell, M.S. Enhanced photocatalytic hydrogen formation over Fe-loaded TiO_2 and g-C_3N_4 composites from mixed glycerol and water by solar irradiation. *J. Renew. Sustain. Energy* **2018**, *10*, 034703. [CrossRef]
38. Gelderman, K.; Lee, L.; Donne, S.W. Flat-band potential of a semiconductor: Using the Mott–Schottky equation. *J. Chem. Educ.* **2007**, *84*, 685–688. [CrossRef]
39. Cardon, F.; Gomes, W.P. On the determination of the flat-band potential of a semiconductor in contact with a metal or an electrolyte from the Mott-Schottky plot. *J. Phys. D Appl. Phys.* **1978**, *11*, L63–L67. [CrossRef]
40. Finlayson, M.F.; Wheeler, B.L.; Kakuta, N.; Park, K.H.; Bard, A.J.; Campion, A.; Fox, M.A.; Webber, S.E.; White, J.M. Determination of flat-band position of CdS crystals, films, and powders by photocurrent and impedance techniques, photoredox reaction mediated by intragap states. *J. Phys. Chem.* **1985**, *89*, 5676–5681. [CrossRef]
41. Ge, H.; Tian, H.; Zhou, Y.; Wu, S.; Liu, D.; Fu, X.; Song, X.M.; Shi, X.; Wang, X.; Li, N. Influence of surface states on the evaluation of the flat band potential of TiO_2. *ACS Appl. Mater. Interfaces* **2014**, *6*, 2401–2406. [CrossRef]
42. Sharon, M.; Sinha, A. Effect of electrolytes on flat-band potential of n-$BaTiO_3$ semiconductor. *Electrochim. Acta* **1983**, *28*, 1063–1066. [CrossRef]
43. Huang, L.; Yang, L.; Li, Y.; Wang, C.; Xu, Y.; Huang, L.; Song, Y. p-n BiOI/Bi_3O_4Cl hybrid junction with enhanced photocatalytic performance in removing methyl orange, bisphenol A, tetracycline and Escherichia coli. *Appl. Surf. Sci.* **2020**, *527*, 146748. [CrossRef]
44. Ma, Q.; Zhang, H.; Guo, R.; Li, B.; Zhang, X.; Cheng, X.; Xie, M.; Cheng, Q. Construction of CuS/TiO_2 nano-tube arrays photoelectrode and its enhanced visible light photoelectrocatalytic decomposition and mechanism of penicillin G. *Electrochim. Acta* **2018**, *283*, 1154–1162. [CrossRef]
45. Haider, Z.; Zheng, J.Y.; Kang, Y.S. Surfactant free fabrication and improved charge carrier separation induced enhanced photocatalytic activity of {001} facet exposed unique octagonal BiOCl nanosheets. *Phys. Chem. Chem. Phys.* **2016**, *18*, 19595–19604. [CrossRef] [PubMed]

46. Bera, B.; Chakraborty, A.; Kar, T.; Leuaa, P.; Neergat, M. Density of states, carrier concentration, and flat band potential derived from electrochemical impedance measurements of N-doped carbon and their influence on electrocatalysis of oxygen reduction reaction. *J. Phys. Chem. C* **2017**, *121*, 20850–20856. [CrossRef]
47. Liu, Q.; Wang, F.; Lin, H.; Xie, Y.; Tong, N.; Lin, J.J.; Zhang, X.; Zhang, Z.; Wang, X. Surface oxygen vacancy and defect engineering of WO_3 for improved visible light photocatalytic performance. *Catal. Sci. Technol.* **2018**, *8*, 4399–4406. [CrossRef]
48. Cuesta, A.; Kleinert, M.; Kolb, D. The adsorption of sulfate and phosphate on $Au_{(111)}$ and $Au_{(100)}$ electrodes: An in situ STM study. *Phys. Chem. Chem. Phys.* **2000**, *2*, 5684–5690. [CrossRef]
49. Kazarinov, V.; Andreev, V.; Mayorov, A. Investigation of the adsorption properties of the TiO_2 electrode by the radioactive tracer method. *J. Electroanal. Chem.* **1981**, *130*, 277–285. [CrossRef]
50. Ehrenburg, M.; Rybin, A.; Danilov, A. Effect of sulfate ion adsorption on the rate of copper electrocrystallization on $Ag_{(111)}$ face. *Russ. J. Electrochem.* **2010**, *46*, 714–720. [CrossRef]
51. Dadwal, U.; Ali, D.; Singh, R. Silicon-silver dendritic nanostructures for the enhanced photoelectrochemical splitting of natural water. *Int. J. Hydrogen Energy* **2018**, *43*, 22815–22826. [CrossRef]
52. Zhang, L.; Zhu, Y.C.; Liang, Y.Y.; Zhao, W.W.; Xu, J.J.; Chen, H.Y. Semiconducting CuO nanotubes: Synthesis, characterization, and bifunctional photocathodic enzymatic bioanalysis. *Anal. Chem.* **2018**, *90*, 5439–5444. [CrossRef]
53. Bard, A.; Stratmann, M.; Gileadi, E.; Urbakh, M.; Calvo, E. *Encyclopedia of Electrochemistry*; Wiley-VCH: New York, NY, USA, 2007.
54. Randles, J.E.B. Kinetics of rapid electrode reactions. *Faraday Discuss.* **1947**, *1*, 11–19. [CrossRef]
55. Khan, S.; Bazaz, S.A.; Aslam, D.M.; Chan, H.Y. Electrochemical interface characterization of MEMS based brain implantable all-diamond microelectrodes probe, Saudi Int. Electron. In Proceedings of the Communications and Photonics Conference, Riyadh, Saudi Arabia, 27–30 April 2013; pp. 1–4.
56. Xiao, P.; Garcia, B.B.; Guo, G.; Liu, D.; Cao, G. TiO_2 nanotube arrays fabricated by anodization in different electrolytes for biosensing. *Electrochem. Commun.* **2003**, *9*, 2441–2447. [CrossRef]
57. Jovic, V.D.; Jovic, B.M. EIS and differential capacitance measurements onto single crystal faces in different solutions Part I: $Ag_{(111)}$ in 0.01 M NaCl. *J. Electroanal. Chem.* **2003**, *541*, 1–11. [CrossRef]
58. Brug, G.J.; Van Den Eeden, A.L.G.; Sluysters-Renbach, M.; Sluysters, L.H. The analysis of electrode impedances complicated by the presence of a constant phase element. *J. Electroanal. Chem.* **1984**, *176*, 275–295. [CrossRef]
59. Huang, J. Diffusion impedance of electroactive materials, electrolytic solutions and porous electrodes: Warburg impedance and beyond. *Electrochim. Acta* **2018**, *281*, 170–188. [CrossRef]
60. Shimizu, K.; Nystro, J.; Geladi, P.; Lindholm-Sethsona, B.; Boilya, J.-F. Electrolyte ion adsorption and charge blocking effect at the hematite/aqueous solution interface: An electrochemical impedance study using multivariate data analysis. *Phys. Chem. Chem. Phys.* **2015**, *17*, 11560–11568. [CrossRef] [PubMed]
61. Horanyi, G. Radiotracer study of the adsorption of sulfate ions at a Bi_2O_3 powder/electrolyte solution interface. *J. Solid State Electrochem.* **2003**, *7*, 309–312. [CrossRef]
62. Feng, S.; Shixiang, Y.; Miura, H.; Nakatani, N.; Hada, M.; Shishido, T. Experimental and theoretical investigation of the role of bismuth in promoting the selective oxidation of glycerol over supported Pt-Bi catalyst under mild conditions. *ACS Catal.* **2020**, *10*, 6071–6083. [CrossRef]
63. Garcia, A.C.; Birdja, Y.Y.; Tremiliosi-Filho, G.; Koper, M.T.M. Glycerol electro-oxidation on bismuth-modified platinum single crystals. *J. Catal.* **2017**, *346*, 117–124. [CrossRef]
64. Rajeshwar, K. *Fundamentals of Semiconductor Electrochemistry and Photoelectrochemistry*; Wiley-VCH Verlag GmbH & Co.: Houston, TX, USA, 2007.

Article

Modifier Effect in Silica-Supported FePO$_4$ and Fe-Mo-O Catalysts for Propylene Glycol Oxidation

Darya Y. Savenko [1], Mikhail A. Salaev [1], Valerii V. Dutov [1], Sergei A. Kulinich [2,3] and Olga V. Vodyankina [1,*]

[1] Laboratory of Catalytic Research, Tomsk State University, 36 Lenin Avenue, 634050 Tomsk, Russia; darya_ebert@mail.ru (D.Y.S.); mihan555@yandex.ru (M.A.S.); dutov_valeriy@mail.ru (V.V.D.)
[2] Research Institute of Science & Technology, Tokai University, Hiratsuka 259-1292, Kanagawa, Japan; skulinich@tokai-u.jp
[3] School of Natural Sciences, Far Eastern Federal University, 690091 Vladivostok, Russia
* Correspondence: vodyankina_o@mail.ru

Abstract: Currently, catalytic processing of biorenewable raw materials into valuable products attracts more and more attention. In the present work, silica-supported FePO$_4$ and Fe-Mo-O catalysts are prepared, their phase composition, and catalytic properties are studied in the process of selective oxidation of propylene glycol into valuable mono- and bicarbonyl compounds, namely, hydroxyacetone and methylglyoxal. A comparative analysis of the main routes of propylene glycol adsorption with its subsequent oxidative conversion into carbonyl products is carried out. The DFT calculations show that in the presence of adsorbed oxygen atom, the introduction of the phosphate moiety to the Fe-containing site strengthens the alcohol adsorption on the catalyst surface with the formation of the 1,2-propanedioxy (–OCH(CH$_3$)CH$_2$O–) intermediate at the active site. The introduction of the molybdenum moiety to the Fe-containing site in the presence of the adsorbed oxygen atom is also energetically favorable, however, the interaction energy is found by 100 kJ/mol higher compared to the case with phosphate moiety that leads to an increase in the propylene glycol conversion while maintaining high selectivity towards C$_3$ products. The catalytic properties of the synthesized iron-containing catalysts are experimentally compared with those of Ag/SiO$_2$ sample. The synthesized FePO$_4$/SiO$_2$ and Fe-Mo-O/SiO$_2$ catalysts are not inferior to the silver-containing catalyst and provide ~70% selectivity towards C$_3$ products, while the main part of propylene glycol is converted into methylglyoxal in contrast to the Ag/SiO$_2$ catalyst featuring the selective transformation of only the secondary C-OH group in the substrate molecule under the studied conditions with the formation of hydroxyacetone. Thus, supported Fe-Mo-O/SiO$_2$ catalysts are promising for the selective oxidation of polyatomic alcohols under low-temperature conditions.

Keywords: bio-regenerable sources; alcohol selective oxidation; iron molybdate catalysts; iron phosphate; bicarbonyl compounds; DFT

Citation: Savenko, D.Y.; Salaev, M.A.; Dutov, V.V.; Kulinich, S.A.; Vodyankina, O.V. Modifier Effect in Silica-Supported FePO$_4$ and Fe-Mo-O Catalysts for Propylene Glycol Oxidation. *Materials* **2022**, *15*, 1906. https://doi.org/10.3390/ma15051906

Academic Editor: Barbara Pawelec

Received: 22 January 2022
Accepted: 28 February 2022
Published: 4 March 2022

Publisher's Note: MDPI stays neutral with regard to jurisdictional claims in published maps and institutional affiliations.

Copyright: © 2022 by the authors. Licensee MDPI, Basel, Switzerland. This article is an open access article distributed under the terms and conditions of the Creative Commons Attribution (CC BY) license (https://creativecommons.org/licenses/by/4.0/).

1. Introduction

Currently, the problem of the depletion of fossil natural resources with the accompanying complication of the global environmental situation is closely related to the increase in the demand for energy sources, fuels and chemicals. Therefore, the opportunity to use renewable raw materials, in particular biomass and biodiesel, to solve the challenges of modern energy as well as the development of new methods for biomass conversion into valuable organic intermediates used in various industries are among the trending research areas. Biomass is a renewable source of raw materials to produce fuels and a number of important organic compounds for chemical industries. Thus, polyatomic alcohols, in particular propylene glycol (PG) and glycerol, are the products of biomass conversion [1–3]. It is noteworthy that PG can also be obtained by the glycerol hydrogenolysis [4].

Experimental studies of the PG selective oxidation into carbonyl/carboxylic compounds (methylglyoxal, hydroxyacetone, etc.) are known in the literature for catalysts with supported metal nanoparticles, including Ag [5–7], Pt [8] and Pd [9–12]. The mechanism of PG conversion was theoretically studied over the Ag_4 cluster [13]. As in the case of ethylene glycol oxidation [9,14,15], the PG adsorption on the surface of single crystal Ag (110) and Pd (111) coated with oxygen led to the breaking of O–H bonds with the formation of the intermediate 1,2-propanedioxy (–OCH(CH_3)CH_2O–, PDO) [5,6,9]. As the temperature increased, the PDO transformations yielded a number of products, including hydroxyacetone or acetol (CH_3COCH_2OH), lactaldehyde ($CH_3CH(OH)CHO$), and methylglyoxal or pyruvaldehyde (CH_3COCHO). Moreover, methylglyoxal and hydroxyacetone are of particular interest as the key intermediates with unique chemical properties due to the presence of two oxygen-containing groups. In Refs. [5,6], it was shown that, first of all, the C–H bond breaking on Ag (110) occurred at the secondary carbon atom of the PDO intermediate as evidenced by the low temperature of hydroxyacetone desorption. In Ref. [9], using the adsorption of ethylene glycol and PG on Pd (111) as an example it was shown that the intermediate compounds were strongly bound to the surface with Pd (111) in a planar orientation; as a result, glyoxal and methylglyoxal were desorbed in small amounts at temperatures above 250 K, however, they were primarily decarbonylated to form CO, H_2 and methane. In Ref. [16], it was suggested that in the case of using the metal/bimetallic catalysts in the oxidation of polyatomic alcohols (e.g., PG and glycerol), the key role of the modifying metal was connected with controlling the reagent bond strength with the catalyst surface without a sharp change in the preferred geometry of the adsorbate, and such changes of binding energies can have a strong impact on the catalytic properties. Transition metals, such as Bi for Pd-Bi supported catalysts [12], are often used as a modifying metal added to those of the Pt or Au subgroups.

Catalysts based on transition metal compounds, such as Mo, Fe, V, are catalytically active in various oxidation processes making it possible to consider them as a promising alternative to those based on noble metals for selective conversion of alcohols and methane. In Ref. [17], the ethylene glycol reactions on the surface of the Mo (110) single crystal were considered in details by a complex of physical-chemical methods. It was shown that during the ethylene glycol adsorption on the surface of Mo (110), the intermediate compounds with the Mo-O bond were formed both in mono- (–OCH_2CH_2OH) and bidentate (–OCH_2CH_2O–) configurations. The authors found that bidentate species were more reactive due to their stability and greater propensity to form ethylene as compared to the potential transformations of the monodentate intermediates. High strength of the O–Mo bond caused the ethylene glycol deoxygenation to yield ethylene.

In Ref. [18], $FePO_4$ catalysts were studied in direct oxidation of methane to methanol. Using Mössbauer spectroscopy coupled with XRD, the authors studied in details the phase transformations of the $FePO_4$ catalysts in atmospheres of various oxidizers O_2, H_2O, and N_2O. It was shown that the $Fe_2P_2O_7$ phase dominated in the reduced catalyst. The use of H_2O and N_2O as oxidants in the selective conversion of methane to methanol promoted the formation of an active and selective mixed phase α-$Fe_3(P_2O_7)_2$ (Fe^{3+}, Fe^{2+}), while the formation of the less active phase β-$Fe_3(P_2O_7)_2$ was observed to a lesser extent, and the amount of the $Fe_2P_2O_7$ phase decreased accordingly. It is worth noting that the Fe valence states can strongly affect the performance of Fe-based catalysts [19,20].

Due to the important industrial applications, the study of the surfaces of oxide iron-molybdenum catalysts for selective methanol oxidation into formaldehyde continues to attract the research attention [21,22]. In Ref. [22], the influence of the preparation method on the stability of the synthesized and industrial iron–molybdenum catalysts were studied. Thus, the catalysts prepared by hydrothermal method contained an excess of molybdenum represented by the metastable h-MoO_3 phase, which, upon calcination, transformed into the thermodynamically stable α-MoO_3 phase. The authors found that the crystal structure, crystal size and/or morphology of MoO_3, taken in excess with respect to the iron molybdate, had a significant effect on the stability of the iron-molybdenum catalysts.

An approach involving the silver incorporation into the complex framework zirconium phosphates to prepare the catalysts active in the selective oxidation/dehydrogenation of ethanol to acetaldehyde was proposed in Ref. [23]. It was shown that the silver addition to complex zirconium phosphate catalysts led to a selectivity increase in the ethanol oxidation to acetaldehyde up to 74% at 330 °C at 93% ethanol conversion, significantly reducing the conversion along the dehydration route. The combination of redox and acid-base sites on the catalyst surface made it possible to control the main directions of transformations of polyatomic alcohols.

Thus, the studies aimed at finding the approaches and methods to assess the reactivity of different catalysts based on transition metal compounds, in particular Fe, using quantum-chemical calculations, are currently relevant to predict the surface properties of multicomponent catalysts. In previous works, when studying a series of supported $FePO_4/SiO_2$ [24] and $Fe-Mo-O/SiO_2$ catalysts [25,26], it was shown that such catalysts were active and selective in the reaction of vapor-phase PG oxidation to methylglyoxal.

The present work is devoted to a comparison of the catalytic properties of $FePO_4/SiO_2$ and $Fe-Mo-O/SiO_2$ catalysts using theoretical and experimental approaches. It has been shown that the theoretical estimation of the substrate–catalyst binding energy in comparison with the catalytic data can be a key to estimate the reactivity of intermediates depending on the chemical surrounding of the Fe sites.

2. Experimental Section

2.1. Catalyst Preparation

Silica gel of the KSKG brand (LLC "Salavat Catalyst Plant", Salavat, Russia, S_{BET} = 300 m^2/g), preliminarily dried in air at 110 °C for 15 h, was used as a support for the prepared catalysts. The amount of Fe in the $FePO_4/SiO_2$ and $Fe-Mo-O/SiO_2$ samples calculated per iron (III) oxide was constant and amounted to 2.5 wt.% (Table 1).

Table 1. Physical-chemical characteristics of prepared catalysts.

Sample	Fe_2O_3, wt.%	[P_2O_5 or MoO_3], wt.%	[P or Mo]/Fe, mol.	S_{BET}, m^2/g
$FePO_4/SiO_2$	2.4	5.1	2.4	289
$Fe-Mo-O/SiO_2$	2.5	8.7	1.9	240
SiO_2 support	-	-	-	300

Synthesis of the $FePO_4/SiO_2$ catalyst included two stages: impregnation of the silica gel fraction (0.25–0.5 mm) with a 0.2 M $Fe(NO_3)_3$ solution in terms of moisture capacity and subsequent impregnation with a $(NH_4)_3PO_4$ solution of the appropriate concentration. The amount of the introduced P corresponded to the molar ratio P/Fe = 2.4 (Table 1). The prepared catalyst was dried at 110 °C and calcined in air at 600 °C for 8 h.

The $Fe-Mo-O/SiO_2$ iron-molybdenum catalyst was prepared by co-impregnation with a citric acid solution containing both components, ammonium heptamolybdate (($NH_4)_3Mo_7O_{24} \cdot 4H_2O$) and iron (III) nitrate in the appropriate concentrations based on Mo/Fe = 2 (mol.). The synthesized catalyst was dried at 120 °C and calcined at 550 °C in an air stream.

The Ag-containing sample with a silver content of 5 wt.% was used as a reference sample. The Ag/SiO_2 catalyst was synthesized by impregnating the moisture capacity with an $AgNO_3$ solution of the appropriate concentration [27]. Then, the sample was dried at 70 °C for 12 h, then the deposited Ag precursor was subjected to high-temperature treatment in an air flow at 500 °C followed by reduction at 200 °C in the H_2/Ar flow. The specific surface area of the synthesized Ag/SiO_2 sample was 170 m^2/g.

Individual Fe_2O_3/SiO_2 oxide catalyst was prepared as a reference sample through the wetness impregnation. The amount of supported component in the model sample was 10 wt.%. The resulting sample was dried at 110 °C and calcined at 500 °C. The specific surface area of the synthesized Fe_2O_3/SiO_2 sample was 268 m^2/g.

2.2. Catalyst Characterization

The phase composition of the samples was studied on the Miniflex 600 diffractometer (Rigaku, Japan) with a Cu anode in the range of 2θ = 10°–80° at a scanning rate of 2°/min. The phase composition was identified using the PDF-2 database and the full profile analysis program POWDER CELL 2.4. Elemental analysis of the catalysts was carried out by X-ray fluorescence analysis (XRF) using the X-ray fluorescence-wave dispersive spectrometer (XRF-1800, Shimadzu, Kyoto, Japan). The source was the X-ray tube with an Rh anode, a voltage was 40 kV, a current was 95 mA, and a diaphragm was 10 mm.

The textural characteristics of the synthesized samples were studied using the method of low-temperature nitrogen adsorption at 196 °C on the TriStar II 3020 analyzer (Micromeritics, Norcross, GA, USA). All samples were subjected to degassing (10^{-2} Torr) at 200 °C for 2 h prior to experiments. The specific surface was determined by the BET method. The pore size distribution was calculated from the desorption branch of the adsorption-desorption isotherm by the Barrett-Joyner-Halend (BJH) method.

2.3. Computational Details

The theoretical interpretation of the vapor-phase PG oxidation to methylglyoxal on $FePO_4/SiO_2$ and $Fe-Mo-O/SiO_2$ catalysts was carried out by density functional theory (DFT) method using the Gaussian'09 software package installed on the SKIF "Cyberia" supercomputer of Tomsk State University [28]. The DFT calculations were carried out using the B3LYP functional. Modeling of the interactions of reagents (PG, O_2), intermediates and products obtained during the PG conversion was carried out using the DGDZVP basis set (Geometries of substrates are represented in the Supplementary Materials (Figure S1)). To provide the effectiveness of time and computer costs, $FePO_4H$ and $Fe_2(MoO_4)_3$ were taken as models of active sites for iron-phosphate and iron-molybdate catalysts, respectively. A hydrogen atom was added to $FePO_4$ moiety to ensure conversion. The geometries of the main substrates considered, as well as DFT-optimized cartesian coordinates for the main $FePO_4$- and $Fe_2(MoO_4)_3$-based models, are represented in the Supplementary Materials (Figures S2 and S3).

The interactions of the main process components were modeled in accordance with the procedure described in Ref. [13] as well as on the basis of the reactions presented in Refs. [5,6] that occurred during the PG oxidation to methylglyoxal. The geometries of all structures obtained were fully optimized. In all cases, the stationary point nature was verified by calculating the vibrational frequencies. Most of the optimized structures were in global energy minimum and featured only real frequencies. The absence of imaginary vibrational frequencies confirmed the stationary nature of the structures. When there were imaginary frequencies in the structure, the IRC calculations were carried out to determine the transition state. The calculated thermodynamic parameters of the molecules were corrected for zero-point vibrational energy (ZPVE) and brought to normal conditions (298.15 K, 1 atm) using the thermal corrections to enthalpy and free energy. The interaction energy of the gas-phase molecule/intermediate with the active site of the catalyst ($FePO_4$, $Fe_2(MoO_4)_3$) was determined as the difference between the total energy of the substrate/active site system and the sum of the energies of the isolated substrate and the active site of the catalyst. In the case of oxygen-containing systems, the interaction energies were determined as the difference between the total energy of the oxygen-containing active site/substrate system and the sum of the energies of the isolated oxygen-containing active site and substrate.

2.4. Testing of Catalytic Activity

The catalytic properties of the synthesized catalysts were studied in the vapour-phase PG oxidation to methylglyoxal at a temperature of 350 °C (composition of the incoming mixture: 3 vol.% $C_3H_6(OH)_2$, 3.7 vol.% O_2, 63.3 vol.% N_2, 30 vol.% H_2O). A sample weighing of 0.15 g was loaded into a quartz reactor with a fraction of 0.25–0.5 mm; the reaction mixture was fed at a rate of 1.5 L/min. The reaction products were analyzed by gas chromatography (GC) and high-performance liquid chromatography (HPLC). The PG conversion for iron-molybdenum catalysts was calculated according to formula (1):

$$X = \frac{\sum_i a n_i}{3n(PG)_{inlet}} \times 100\% \quad (1)$$

where n_i was the number of moles of the ith product (i—methylglyoxal (MeGO), hydroxyacetone (HA), formaldehyde (FD), CO and CO_2); a was the number of carbon atoms in the $i$thproduct. The selectivities towards the reaction products was calculated in accordance with the presented formulas (2)–(5):

$$S_{MeGO} = \frac{n_{MeGO}}{\sum_i a n_i} \times 100\%; \quad (2)$$

$$S_{HA} = \frac{n_{HA}}{\sum_i a n_i} \times 100\%; \quad (3)$$

$$S_{FD} = \frac{n_{FD}}{3\sum_i a n_i} \times 100\%; \quad (4)$$

$$S_{CO_x} = \frac{n_{CO_x}}{3\sum_i a n_i} \times 100\%. \quad (5)$$

3. Results and Discussion

Table 1 lists the designations of the synthesized catalysts, the amounts of deposited components according to the XRF results and the specific surface area. Both iron-containing catalysts are characterized by similar Fe content (per iron (III) oxide) as well as a close molar ratio [P or Mo]/Fe.

Figure 1 shows the XRD patterns for the supported catalysts and reference samples. The appearance of a wide halo in the range of 2Θ angles up to 30° associated with the amorphous state of SiO_2 is observed in the XRD patterns for all deposited samples. Figure 1 shows that no crystalline phases are observed by XRD in the Fe-Mo-O/SiO_2 sample. The XRD pattern of the $FePO_4$/SiO_2 sample features a reflection that corresponds to the $FePO_4$ iron orthophosphate phase (No. 01-077-0094). The low-intensity reflection in the range of 38° 2Θ in the XRD pattern for the Ag/SiO_2 reference sample corresponds to the metallic Ag(111) phase.

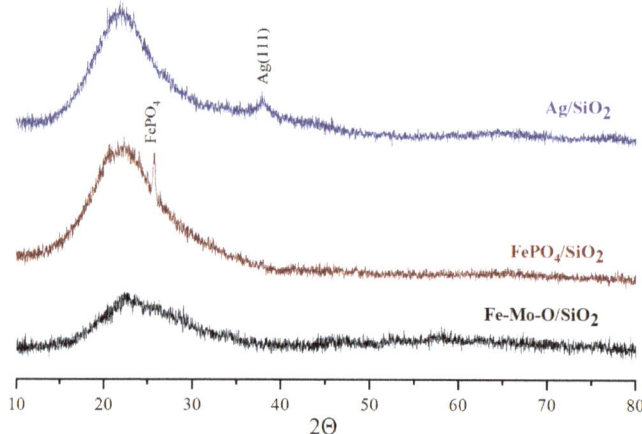

Figure 1. XRD patterns of studied catalysts.

The catalytic properties of the prepared catalysts $FePO_4$/SiO_2, Fe-Mo-O/SiO_2 and Ag/SiO_2 (reference sample) were studied in the reaction of vapor-phase PG oxidation to methylglyoxal under the same conditions for all catalysts: contact time was 0.011 s (m_{cat} = 0.15 g) and temperature was 350 °C. For the Ag/SiO$_2$ sample, the additional catalytic experiments were carried out at a decreased contact time of 0.007 s (m_{cat} = 0.1 g). Figure 2 represents the results obtained.

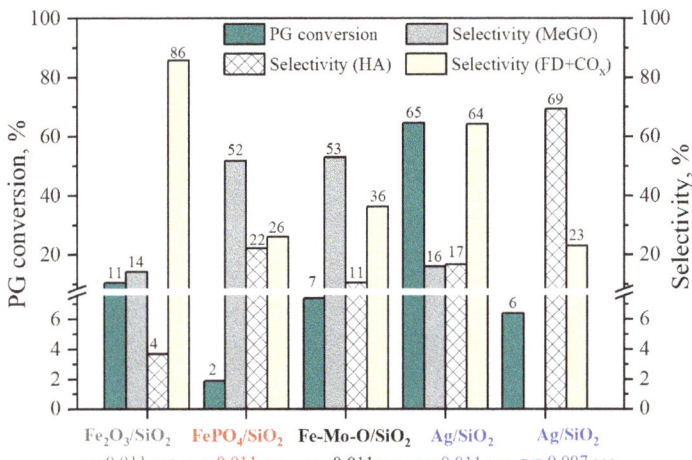

Figure 2. Catalytic properties of prepared catalysts in selective PG oxidation.

Thus, in the case of iron-containing catalysts, at the PG conversion of 2 to 7%, the total selectivity towards C$_3$ products is ~70%, including ~50% towards methylglyoxal for both FePO$_4$/SiO$_2$ and Fe-Mo-O/SiO$_2$ catalysts. The contribution of the side processes with the cleavage of the C–C bond yielding C$_1$ products (formaldehyde, CO$_x$) on these catalysts is 26% and 36%, respectively. A Fe$_2$O$_3$/SiO$_2$ reference sample shows higher activity in PG oxidation when compared with samples FePO$_4$/SiO$_2$ and Fe-Mo-O/SiO$_2$. However, in the case of pristine Fe$_2$O$_3$/SiO$_2$ catalyst, the C–C bond cleavage route is the dominant one, and the total selectivity of formation of the C–C bond cleavage products reaches 86% (Figure 2), while the selectivity towards the desired methylglyoxal does not exceed 20%. The use of FePO$_4$/SiO$_2$ and Fe-Mo-O/SiO$_2$ catalysts permits to achieve high selectivity towards methylglyoxal while keeping the PG conversion in the case of Mo introduction in the composition of Fe-containing catalyst.

For the Ag-containing catalyst at a contact time of 0.011 s, an increase in the PG conversion to up to 65% is accompanied by a decrease in the total selectivity towards C$_3$ products up to 33% (selectivities towards MeGO and HA are 16% and 17%). For correct comparison of the catalytic properties of the Fe-containing catalysts with Ag/SiO$_2$, the additional experiments are carried out for the Ag-containing catalyst with a contact time reduced to 0.007 s. At low PG conversion (6%), hydroxyacetone is the main reaction product with a selectivity of ~70%, and methylglyoxal is not formed at this contact time (Figure 2). The formation of products of C–C bond cleavage under these conditions is reduced to up to 23% that is comparable to iron-containing catalysts.

The iron phosphate catalyst exhibits the lowest activity in PG selective oxidation, while the selectivity towards the target product methylglyoxal reaches 52%, and the selectivity towards by-products does not exceed 30%. For the iron-molybdenum catalyst, a noticeable increase in the PG conversion (up to 7%) is observed, while maintaining high selectivity towards methylglyoxal (Figure 2). Such an effect can be associated with the energies of interaction of the reagents with the surface-active sites of the supported FePO$_4$ and Fe$_2$(MoO$_4$)$_3$ catalysts and the structure of the formed intermediate compounds.

Maintaining high selectivity towards C$_3$ products (hydroxyacetone) for the silver-containing supported catalyst is possible only if the contact time is reduced by almost a factor of 2 (from 0.011 to 0.007 s). At low contact time and 350 °C, the PG adsorption on the surface of the Ag/SiO_2 catalyst occurs predominantly in the monodentate configuration with the participation of OH groups on the silica surface [15].

Thus, at low PG conversion (<10%), the iron-containing catalysts are characterized by high selectivity towards C_3 products, methylglyoxal (50–53%) and hydroxyacetone (up to 22%), while high selectivity is observed on the Ag-containing catalyst only towards hydroxyacetone (~70%), methylglyoxal is not detected in the product composition. Thus, in the case of iron-containing catalysts, PG is adsorbed on the surface mainly in the bidentate configuration that contributes to the selective conversion of both hydroxyl groups with the formation of methylglyoxal. In the case of the Ag-containing catalyst, at low degree of PG conversion (up to 10%), only monodentate intermediates are adsorbed on the surface as evidenced by the high selectivity towards hydroxyacetone. As the contact time increases, PG is adsorbed on the surface of the Ag/SiO_2 catalyst both in the mono- and bidentate configurations as evidenced by the methylglyoxal appearance in the products. However, an increase in the contact time also leads to the readsorption of products on the surface of the Ag/SiO_2 catalyst, which is accompanied by high probability of the C–C bond cleavage; with an increase in the PG conversion, the selectivity towards C_1 products sharply increases.

To substantiate the obtained experimental results, the quantum-chemical calculations of the main transformations of PG and oxygen into C_3 products were carried out accounting for the binding energies with the Fe-containing sites. The configuration of the active site for each catalyst was chosen based on the obtained phase analysis results and Raman spectroscopy results in Refs. [13,25]. $FePO_4$ and $Fe_2(MoO_4)_3$ structures were used as models of active sites in iron phosphate and iron molybdenum catalysts, respectively.

Results of theoretical calculations show that the first stage of reagent conversion comprises the adsorption and dissociation of molecular oxygen on the active sites of the catalysts (see Supplementary Materials for the profile of interactions of the key reagents, intermediates (Figure S4), and products with the active sites along the reaction coordinate (Tables S1 and S2)). In both cases, the formation of atomic oxygen species adsorbed on Fe sites is observed. The PG is then adsorbed to form adsorbed PDO intermediate through the O–H bond breaking. Moreover, at the iron-phosphate site, a stronger interaction (−715 kJ/mol) occurs between the diol molecule and the Lewis Fe^{3+} center contrary to the iron-molybdenum system (Table 2). In this case, one should expect a decrease in the activity of the iron-phosphate catalyst due to an increase in the residence time of a strongly bound adsorption complex. In the case of the iron–molybdenum catalysts, the PG binding activates the O–H bond both on the Fe-containing sites and on the Fe-O-Mo moiety. The increase of the C–O bond length to up to 1.46 Å in the alcohol molecule (compared to 1.42 Å for the isolated molecule) also occurs at the Fe-containing sites.

Table 2. Calculated binding energies (kJ/mol) of key intermediates with model active sites of studied catalysts.

Intermediates	$FePO_4 + O_{ads}$	$FePO_4$	$Fe_2(MoO_4)_3 + O_{ads}$	$Fe_2(MoO_4)_3$
PDO	−715	−586	−576	−549
OPO	−360	−498	−284	−422
PPA	−439	−389	−480	−408

The subsequent PDO conversion both at the $FePO_4$ and $Fe_2(MoO_4)_3$ sites occurs via the formation of intermediates, namely, propan-1-al (PPA) and oxopropoxy (OPO). In turn, the PPA and OPO are the intermediates species in the formation of methylglyoxal, hydroxyacetone (and formaldehyde), respectively. The binding energy of the PPA structure (Table 2) is higher for the $Fe_2(MoO_4)_3$ site due to the assistance of adsorbed atomic oxygen species. The binding energies with the $Fe_2(MoO_4)_3$ site for the OPO and PPA intermediates are practically similar. However, the presence of the adsorbed oxygen atom strengthens the PPA binding energy with the catalyst surface, while for the OPO structure it results in its decrease. It is noteworthy that in the case of the $Fe_2(MoO_4)_3$ site, the strengthening of the binding energy with the catalyst surface occurs due to the breaking of the C_2–H bond followed by the methylglyoxal desorption and the formation of the Fe–O_{ads}H bond (O_{ads}—oxygen atom adsorbed on the Fe site).

As in the case of PDO intermediate formation, the strength of the OPO binding to the active site is higher for the $FePO_4$ site as compared to the $Fe_2(MoO_4)_3$ site (Table 2). It is noteworthy that the OPO interaction with both $FePO_4$ and $Fe_2(MoO_4)_3$ sites is more favorable in the absence of adsorbed atomic oxygen species. Further transformation of the OPO at the $Fe_2(MoO_4)_3$ site in the presence of atomic oxygen is accompanied by the elongation of the C_1–C_2 bond to up to 1.67 Å (1.59 Å in the isolated species) followed by its cleavage and desorption of the C_1 products due to the weak interaction with the active site. In this case, the parallel transformations of the OPO intermediate into hydroxyacetone and C_1 by-products occur depending on the presence of the adsorbed oxygen atom. Moreover, at the $FePO_4$ sites, both transformation routes of the OPO intermediate are implemented to similar extent that is also evidenced by the experimental data (Figure 2).

The methylglyoxal formation is more favorable during the PPA dehydrogenation, while hydroxyacetone is formed as a result of the OPO transformations through the oxidative route. Since the reduction of the surface sites in the H_2-TPR mode for the $Fe_2(MoO_4)_3$ catalyst [25] proceeds at higher temperatures (by 100 °C higher as compared to $FePO_4$ [24]) indicating higher strength of oxygen binding to the surface, the dehydrogenation route is more favorable. In the presence of the adsorbed oxygen atom, a competing process for the OPO selective oxidation to HA is the C–C bond cleavage in the OPO species yielding formaldehyde. At the same time, it is noteworthy that the oxygen presence in the reaction medium favorably affects the course of the oxidative processes as was shown in Ref. [29] exemplified by the oxidation of organic compounds (acrolein, formaldehyde, ethanol, etc.) on the oxide V–Ti catalysts.

Thus, the studied systems are characterized by a relatively similar position of the main intermediates formed during the PG oxidation. However, significant differences are found in the interaction energies of the key PDO intermediate with the active sites of the $FePO_4$ and $Fe_2(MoO_4)_3$. The PDO binding energy on the iron-phosphate site is much lower and amounts to -586 kJ/mol (-715.5 kJ/mol in the presence of atomic oxygen). In this case, the strong PDO interaction with the $FePO_4$ site leads to the increase of the residence time of the adsorbed intermediates on the active sites of the catalyst and, accordingly, to a decrease in the degree of PG conversion, which is consistent with the catalytic experiments. According to the results of theoretical calculations, on the iron-molybdenum site, an intermediate value of the PDO binding energy was obtained as compared to those over $FePO_4$. Thus, it can be assumed that the use of iron-molybdenum catalysts leads to an increase in the turnover frequency for the Fe-containing sites upon interaction with the intermediate compounds and, at the same time, facilitates further selective oxidation of PG without breaking the C–C bond.

4. Conclusions

In the present work, the experimental and theoretical studies of the catalytic properties of $FePO_4/SiO_2$ and Fe-Mo-O/SiO_2 catalysts in the selective oxidation of propylene glycol to methylglyoxal were carried out. The experimental comparison of the catalytic properties of the supported iron-containing catalysts with the silver-containing one was also carried out using the Ag/SiO_2 reference sample. It was shown that the iron-containing catalysts $FePO_4/SiO_2$ and Fe-Mo-O/SiO_2 were highly selective in the propylene glycol oxidation to methylglyoxal. While the Ag-containing sample was highly active, however, the composition of the reaction products was dominated by the contribution of the by-products of the C–C bond cleavage. Moreover, with a comparable propylene glycol conversion over all catalysts (up to 10%), for $FePO_4$ and Fe-Mo-O species the formation of the bidentate-bound intermediate with the methylglyoxal release was favorable. The results of the catalytic experiments were consistent with the theoretical calculations carried out for the main reactions of the propylene glycol oxidation to methylglyoxal at the $FePO_4$ and $Fe_2(MoO_4)_3$

sites. In accordance with the results obtained, it can be concluded that the supported iron-containing catalysts can be an alternative to conventional oxidation catalysts based on noble metals.

Supplementary Materials: The following supporting information is available online at https://www.mdpi.com/article/10.3390/ma15051906/s1, Figure S1: Geometries of substrates; Figure S2: Calculated structure of iron phosphate (FePO4H); Table S1: DFT-optimized cartesian coordinates for the main $FePO_4$-based models; Figure S3: Calculated structure of iron molybdate ($Fe_2(MoO_4)_3$); Table S2: DFT-optimized cartesian coordinates for the main $Fe_2(MoO_4)_3$-based models; Figure S4: Interactions of key reagents, intermediates and products with active sites along the reaction coordinate (B3LYP/DGDZVP level of theory).

Author Contributions: Methodology, D.Y.S. and M.A.S.; Conceptualization, D.Y.S. and O.V.V.; Writing—original draft, D.Y.S., M.A.S. and V.V.D.; Investigation, D.Y.S., M.A.S. and V.V.D.; Data curation, M.A.S. and V.V.D.; Formal analysis: M.A.S., S.A.K. and O.V.V.; Writing—review and editing, S.A.K. and O.V.V.; Supervision, O.V.V.; Project administration, O.V.V. All authors have read and agreed to the published version of the manuscript.

Funding: This study was supported by the Tomsk State University Development Programme ("Priority-2030").

Data Availability Statement: The data presented in this study are available on request from the corresponding author.

Conflicts of Interest: The authors declare no conflict of interest.

References

1. Jia, Y.; Suna, Q.; Liu, H. Selective hydrogenolysis of biomass-derived sorbitol to propylene glycol and ethylene glycol on in-situ formed PdZn alloy catalysts. *Appl. Catal. A* **2020**, *603*, 117770. [CrossRef]
2. Liu, C.; Shang, Y.; Wang, S.; Liu, X.; Wang, X.; Gui, J.; Zhang, C.; Zhu, Y.; Li, Y. Boron oxide modified bifunctional Cu/Al$_2$O$_3$ catalysts for the selective hydrogenolysis of glucose to 1,2-propanediol. *Mol. Catal.* **2020**, *485*, 110514. [CrossRef]
3. Wang, X.; Beine, A.K.; Palkovits, R. 1,2-Propylene glycol and ethylene glycol production from lignocellulosic biomass. *Stud. Surf. Sci. Catal.* **2019**, *178*, 173–193. [CrossRef]
4. Jiménez, R.X.; Young, A.F.; Fernandes, H.L.S. Propylene glycol from glycerol: Process evaluation and break-even price determination. *Renew. Energ.* **2020**, *158*, 181–191. [CrossRef]
5. Ayre, C.R.; Madix, R.J. C-C and C-H bond activation of 1,2-propanedioxy by atomic oxygen on Ag(110): Effects of co-adsorbed oxygen on reaction mechanism. *Surf. Sci.* **1994**, *303*, 297–311. [CrossRef]
6. Ayre, C.R.; Madix, R.J. The adsorption and reaction of 1,2-propanediol on Ag(110) under oxygen lean conditions. *Surf. Sci.* **1994**, *303*, 279–296. [CrossRef]
7. Feng, Y.; Yin, H.; Wang, A.; Xue, W. Selectively catalytic oxidation of 1,2-propanediol to lactic, formic, and acetic acids over Ag nanoparticles under mild reaction conditions. *J. Catal.* **2015**, *326*, 26–37. [CrossRef]
8. Pinxt, H.H.C.M.; Kuster, B.F.M.; Marin, G.B. Promoter effects in the Pt-catalyzed oxidation of propylene glycol. *Appl. Catal. A* **2000**, *191*, 45–54. [CrossRef]
9. Timofeev, K.L.; Vodyankina, O.V. Selective oxidation of bio-based platform molecules and their conversion products over metal nanoparticle catalysts: A review. *React. Chem. Eng.* **2021**, *6*, 418–440. [CrossRef]
10. Griffin, M.B.; Jorgensen, E.L.; Medlin, J.W. The adsorption and reaction of ethylene glycol and 1,2-propanediol on Pd(111): A TPD and HREELS study. *Surf. Sci.* **2010**, *604*, 1558–1564. [CrossRef]
11. Sugiyama, S.; Tanaka, H.; Bando, T.; Nakagawa, K.; Sotowa, K.-I.; Katou, Y.; Mori, T.; Yasukawa, T.; Ninomiya, W. Liquid-phase oxidation of propylene glycol using heavy-metal-free Pd/C under pressurized oxygen. *Catal. Today* **2013**, *203*, 116–121. [CrossRef]
12. Ten, S.; Torbina, V.V.; Zaikovskii, V.I.; Kulinich, S.A.; Vodyankina, O.V. Bimetallic AgPd/UiO-66 hybrid catalysts for propylene glycol oxidation into lactic acid. *Materials* **2020**, *13*, 5471. [CrossRef] [PubMed]
13. Salaev, M.A.; Poleshchuk, O.K.; Vodyankina, O.V. Propylene glycol oxidation over silver catalysts: A theoretical study. *J. Mol. Catal. A Chem.* **2016**, *417*, 36–42. [CrossRef]
14. Capote, A.J.; Madix, R.J. Oxygen-hydrogen and carbon-hydrogen bond activation in ethylene glycol by atomic oxygen on silver(110): Heterometallacycle formation and selective dehydrogenation to glyoxal. *J. Am. Chem. Soc.* **1989**, *111*, 3570–3577. [CrossRef]
15. Mamontov, G.V.; Knyazev, A.S.; Paukshtis, E.A.; Vodyankina, O.V. Adsorption and conversion of ethylene glycol on the surface of Ag-containing catalyst modified with phosphate. *Kinet. Catal.* **2013**, *54*, 735–743. [CrossRef]
16. Medlin, J.W. Understanding and controlling reactivity of unsaturated oxygenates and polyols on metal catalysts. *ACS Catal.* **2011**, *1*, 1284–1297. [CrossRef]

17. Queeney, K.T.; Arumainayagam, C.R.; Weldon, M.K.; Friend, C.M.; Blumberg, M.Q. Differential reactivity and structure of mono- and dialkoxides: The reactions of ethylene glycol on Mo(110). *J. Am. Chem. Soc.* **1996**, *118*, 3896–3904. [CrossRef]
18. Dasireddy, V.D.B.C.; Khan, F.B.; Hanzel, D.; Bharuth-Ram, K.; Likozar, B. Application of Mössbauer spectroscopy in industrial heterogeneous catalysis: Effect of oxidant on $FePO_4$ material phase transformations in direct methanol synthesis from methane. *Hyperfine Interact.* **2017**, *238*, 29. [CrossRef]
19. Feng, G.; Lu, Z.-H.; Yang, D.; Kong, D.; Liu, J. A first principle study on Fe incorporated MTW-type zeolite. *Microporous Mesoporous Mater.* **2014**, *199*, 83–92. [CrossRef]
20. Yu, X.; Zhang, X.; Jin, L.; Feng, G. CO adsorption, oxidation and carbonate formation mechanisms on Fe_3O_4 surfaces. *Phys. Chem. Chem. Phys.* **2017**, *19*, 17287–17299. [CrossRef]
21. Yeo, B.R.; Pudge, G.J.F.; Bugler, K.G.; Rushby, A.V.; Kondrat, S.; Bartley, J.; Golunski, S.; Taylor, S.H.; Gibson, E.; Wells, P.P.; et al. The surface of iron molybdate catalysts used for the selective oxidation of methanol. *Surf. Sci.* **2016**, *648*, 163–169. [CrossRef]
22. Raun, K.V.; Lundegaard, L.F.; Beato, P.; Appel, C.C.; Nielsen, K.; Thorhauge, M.; Schumann, M.; Jensen, A.D.; Grunwaldt, J.-D.; Høj, M. Stability of iron molybdate catalysts for selective oxidation of methanol to formaldehyde: Influence of preparation method. *Catal. Lett.* **2020**, *150*, 1434–1444. [CrossRef]
23. Dorofeeva, N.V.; Vodyankina, O.V.; Sobolev, V.I.; Koltunov, K.Y.; Zaykovskii, V.I. Main routes of ethanol conversion under aerobic/anaerobic conditions over Ag-containing zirconium phosphate catalyst. *Curr. Org. Synth.* **2017**, *14*, 389–393. [CrossRef]
24. Ebert, D.Y.; Savel'eva, A.S.; Dorofeeva, N.V.; Vodyankina, O.V. $FePO_4/SiO_2$ catalysts for propylene glycol oxidation. *Kinet. Catal.* **2017**, *58*, 720–725. [CrossRef]
25. Ebert, D.Y.; Dorofeeva, N.V.; Savel'eva, A.S.; Kharlamova, T.S.; Salaev, M.A.; Svetlichnyi, V.A.; Magaev, O.V.; Vodyankina, O.V. Silica-supported Fe-Mo-O catalysts for selective oxidation of propylene glycol. *Catal. Today* **2019**, *333*, 133–139. [CrossRef]
26. Savenko, D.Y.; Velieva, N.Y.; Svetlichnyi, V.A.; Vodyankina, O.V. The influence of the preparation method on catalytic properties of $Mo–Fe–O/SiO_2$ catalysts in selective oxidation of 1,2-propanediol. *Catal. Today* **2020**, *357*, 399–408. [CrossRef]
27. Dutov, V.V.; Mamontov, G.V.; Zaikovskii, V.I.; Vodyankina, O.V. The effect of support pretreatment on activity of Ag/SiO_2 catalysts in low-temperature CO oxidation. *Catal. Today* **2016**, *278*, 150–156. [CrossRef]
28. Frisch, M.J.; Trucks, G.W.; Schlegel, H.B.; Scuseria, G.E.; Robb, M.A.; Cheeseman, J.R.; Scalmani, G.; Barone, V.; Mennucci, B.; Petersson, G.A.; et al. *Gaussian 09, Revision C.01*; Gaussian, Inc.: Wallingford, CT, USA, 2010.
29. Andrushkevich, T.V.; Chesalov, Y.A. Mechanism of heterogeneous catalytic oxidation of organic compounds to carboxylic acids. *Russ. Chem. Rev.* **2018**, *87*, 586–603. [CrossRef]

MDPI AG
Grosspeteranlage 5
4052 Basel
Switzerland
Tel.: +41 61 683 77 34

Materials Editorial Office
E-mail: materials@mdpi.com
www.mdpi.com/journal/materials

Disclaimer/Publisher's Note: The statements, opinions and data contained in all publications are solely those of the individual author(s) and contributor(s) and not of MDPI and/or the editor(s). MDPI and/or the editor(s) disclaim responsibility for any injury to people or property resulting from any ideas, methods, instructions or products referred to in the content.

www.ingramcontent.com/pod-product-compliance
Lightning Source LLC
LaVergne TN
LVHW070507100526
838202LV00014B/1806